JEPPESEN®

AVIONICS
FUNDAMENTALS

Jeppesen

Published in the United States of America
Jeppesen
55 Inverness Drive East, Englewood, CO 80112-5498
www.jeppesen.com

ISBN-13: 978-0-88487-432-4
ISBN-10: 0-88487-432-X

Jeppesen
55 Inverness Drive East
Englewood, CO 80112-5498
Web site: www.jeppesen.com
Email: Captain@jeppesen.com
Published 1974, 2006

JS312661—003

TABLE OF CONTENTS

COURSE DESCRIPTION

Avionics I is part one of a two-part course. Part One includes an introduction to electronic concepts and components such as AC & DC bridge circuits, AC phase angle, logic, data busses, magnetic vectors, Synchros, tach rate generators and Gyros. Description and operation of many aircraft systems will be covered in detail such as compass systems, ADF, VOR, ILS, ATC, DME, MB LRRA. Reading and the use of the Primary Flight display (ADI) and Navigation Display (HSI) will be discussed and concepts such as attitude heading, course and bearing will be learned. In addition Part I explores many components and systems used with aircraft autopilot and navigation systems. Many of the learned components and systems will be used in discussion of autopilot function in the Avionics II portion of this course.

Avionics Fundamentals II is the second part of a two-part course and continues building on the material learned in part one. Topics include Air Data Systems, Theory of Flight and Flight Control Surfaces, Ground Proximity Warning System, Autopilot Rudder Channel (Yaw Damp), Autopilot / Flight Director Pitch Channel and Roll Channel functions, and Autoflight navigation using Area Nav (RNAV) and the Flight Management Computer Systems (FMCS).

COURSE OBJECTIVES

1. The student will assess "Human Factors" issues to the extent that what experiences and understanding one has of the relationship between the current month's "dirty dozen" theme and the safety precautions pertinent to the subjects being presented are collectively discussed as evaluated by the instructor, student and peers.

2. The student will recognize known safety hazards associated with the operation and maintenance of Aircraft Electrical Systems to the extent that these known hazards and precautions are collectively discussed as evaluated by the instructor, student and peers.

3. The student will recognize proper and safe operation of Aircraft autoflight and navigation systems to the extent that the potential for aircraft damage or bodily harm is minimized trough discussion and verbal questioning as evaluated by the instructor, student and peers.

4. The student will understand and discuss operation of basic avionics components including Transistors, Logic, Data Busses, Bridge circuits, Modulator/De-modulators, Synchros, Gyros, Servo Motors and Tach Rate Generators to the extent that a level of understanding is obtained that will enable the student to achieve a minimum passing score of 80% on post test "A" within 1.5 hours as evaluated by the instructor, student and peers.

5. The student will understand and discuss use and operation of aircraft avionics instruments and systems such as HSI, ADI, VOR, ILS, ADF, Radio Altimeter, Enhanced GPWS and ATC to the extent that system knowledge will aid in diagnoses of system faults. as evaluated by the instructor, student and peers.

6. The student will understand the function and interaction of flight deck systems such as Autopilot / Flight Director, Yaw Damper, Autopilot Pitch & Roll channel function, the Flight Management System and related navigation systems to the extent that the systems interaction can explained verbally or by using exercises as evaluated by the instructor, student and peers.

7. The student will discuss common faults and troubleshooting techniques for Autopilot / Flight director and related navigation systems to the extent that the student will be able to isolate the most probable component or system fault in written or verbal exercises; as evaluated by the instructor, student and peers.

8. The student will demonstrate their knowledge of the course material received and discussed; to the extent that a minimum score of 80% is obtained within 1.5 hours as evaluated by the instructor, student and peers.

PREFACE

In 1964, a Line Maintenance Training organization was formed by United Airlines in San Francisco, California. The new training group's task was to introduce the new Boeing 727. The instructors soon recognized a significant lack of knowledge regarding aviation electronics (AVIONICS), such as Air Data instrumentation, Flight Directors, and Autopilot functions and operation. Jim Sowden an instructor with United at the time was given the task to transform electronics technicians into avionics technicians. After looking for a suitable textbook to teach the required information, it was discovered that nothing adequate was available. Jim, along with the help several of assistants, set out to create a text that would teach the required background information needed to transform an electronics technician into an avionics technician. Because Jim was the primary contributor to Avionics Fundamentals, he should be acknowledged for his technical knowledge and contribution to Avionics Fundamentals.

As such, this book presents basic material which is prerequisite to Category II and Category III qualification training courses specific to particular aircraft types.

Many of the explanations of elementary avionics principles found in this text are not easily discovered elsewhere. The information may be useful to anyone interested in learning the fundamentals without the involved design details of: autopilots, flight directors, air data computers, radio altimeters, VORs, localizers, glideslopes, compasses, gyros, and indicators such as HSIs and RMIs.

These subjects are covered in a generalized way to bring the student up to the level at which the subjects are treated in the particular systems of advanced avionics maintenance courses.

Because the material presented is basic and generalized in nature, it is not intended for application in specific equipment or systems. Specific information on detailed operation procedures should be obtained from the manufacturer through his appropriate maintenance manuals, and followed in detail. The authors and publishers of Avionics Fundamentals are in no way to be held liable for the use or misuse in application of information contained herein.

This Study Guide represents the first major revision since the publication of the first Jeppesen Avionics Fundamentals book in the early1980's.

For inquiries or manual correction requests please contact Mike Restivo, United Airlines SFOED.

CHAPTER 1 - SOLID STATE DEVICES

Two Dimensional Crystal Lattice

This section briefly explains the behavior of semiconductor materials and the simple semiconductor devices symbolized in circuits drawn for line maintenance work.

Silicon and germanium are the two principal elements used in crystalline form as the basic semiconductor materials. These are alike in that an atom of each has four electrons in its outer shell and that a total of eight is required to complete the outer shell.

Figure 1-1 shows the silicon atom with two completed inner shells and the third shell with four electrons and room for four more. Figure 1-2 shows the germanium atom with four electrons in the fourth shell and room for four more.

When crystals are grown from pure silicon or pure germanium they will be combined as represented (in two dimensions) in Figure 1-3. Since any piece of a formed crystal has many millions of atoms, practically all of them would exist as shown in the center of this drawing, with each atom having a completed outer shell borrowing very stable non-conducting material.

Figure 1-1

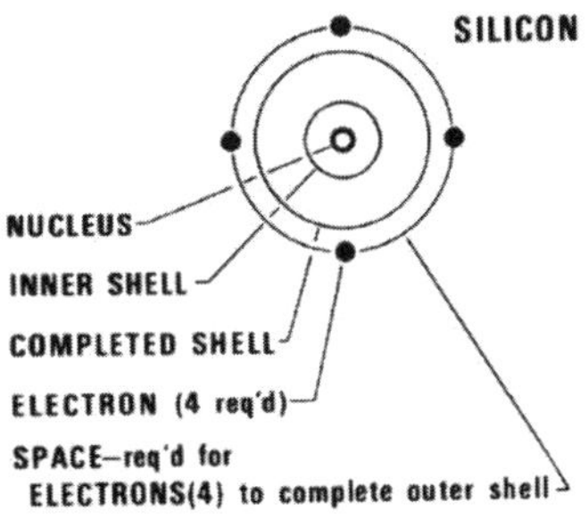

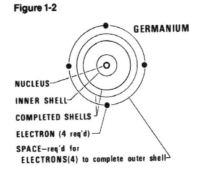

Figure 1-2

Figure 1-3

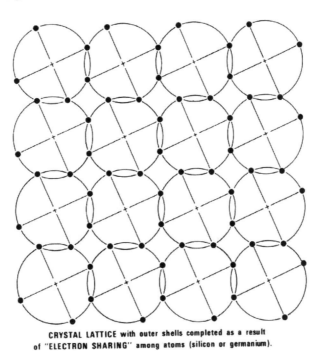

CRYSTAL LATTICE with outer shells completed as a result
of "ELECTRON SHARING" among atoms (silicon or germanium).

CRYSTAL LATTICE

"N" And "P" Type Material

In order for one of these materials to become a semiconductor it must be "doped" with an impurity. This impurity is added either at the time the crystal is grown or, for very thin sections, by highly complicated and delicate procedures afterward. Regardless of how it is achieved, the impurity atoms become an integral part of the crystal structure.

To produce an N type material (one with an excess of electrons), doping is done with an element having five electrons in its outer shell, such as antimony or arsenic (Figure 1-4). Since the crystal structure can use only four, the fifth electron becomes a so-called "free" electron. It has no place in particular to go and typically wanders about the crystal structure.

When an electron leaves an impurity atom, called a donor atom, that atom becomes a positive ion. In Figure 1-4 two donor atoms are shown in the crystal structure, one of which still has a fifth electron. That atom, therefore, is not ionized. The other atom is shown with its fifth electron gone; that atom has become a positive ion.

The ratio of impurity atoms in the crystal structure is on the order of one in ten million.

The N type material just described is so called because it has an excess of electrons.

To make a P type material, one with a deficiency of electrons (Figure 1-5), an impurity with only three electrons in the outer shell is used. This could be a substance such as gallium, indium or boron. When incorporated into the basic crystal structure, these atoms are called acceptor atoms. The P type crystal therefore exists with a deficiency of electrons.

Figure 1-5 shows one acceptor atom with its electron hole. This atom is not charged. But most atoms will become charged ions because electrons are rather fickle and don't particularly care where they reside, either permanently or temporarily. They are very likely to fill in a hole in an acceptor atom, leaving a hole in a basic crystal atom. In that case, the acceptor atom becomes negatively charged as a negative ion, and somewhere else in the structure is a corresponding electron hole.

These N and P type materials are called semiconductors because they may or may not conduct electricity, depending upon particular conditions.

Figure 1-4

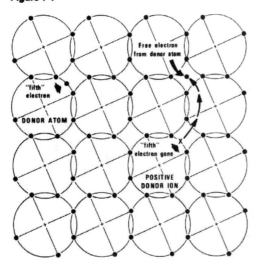

"DONOR" ATOM with FIVE ELECTRONS in outer shell becomes POSITIVELY CHARGED ion if "fifth" electron leaves. The departing electron becomes a "FREE" electron.

"N" TYPE MATERIAL

Figure 1-5

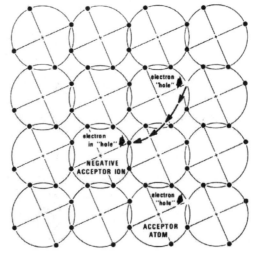

"ACCEPTOR" ATOM with THREE ELECTRONS in outer shell becomes NEGATIVELY CHARGED ion when electron fills "HOLE". Electron which fills acceptor "hole" leaves a NEW "HOLE" where it came from.

"P" TYPE MATERIAL

Voltage Across "N" Type Material

Figure 1-6 shows a section of N type material between the terminals of a DC voltage source. Since it is a characteristic of N type material that the free electrons prefer to wander away from home, most of the donor atoms become donor ions, as indicated by the many solid black circles contrasted with only one shaded circle.

For each free electron there will be one donor ion in a neutral charge crystal. These free electrons are capable of carrying current to and from the voltage source, as indicated by electron flow arrows.

Figure 1-6

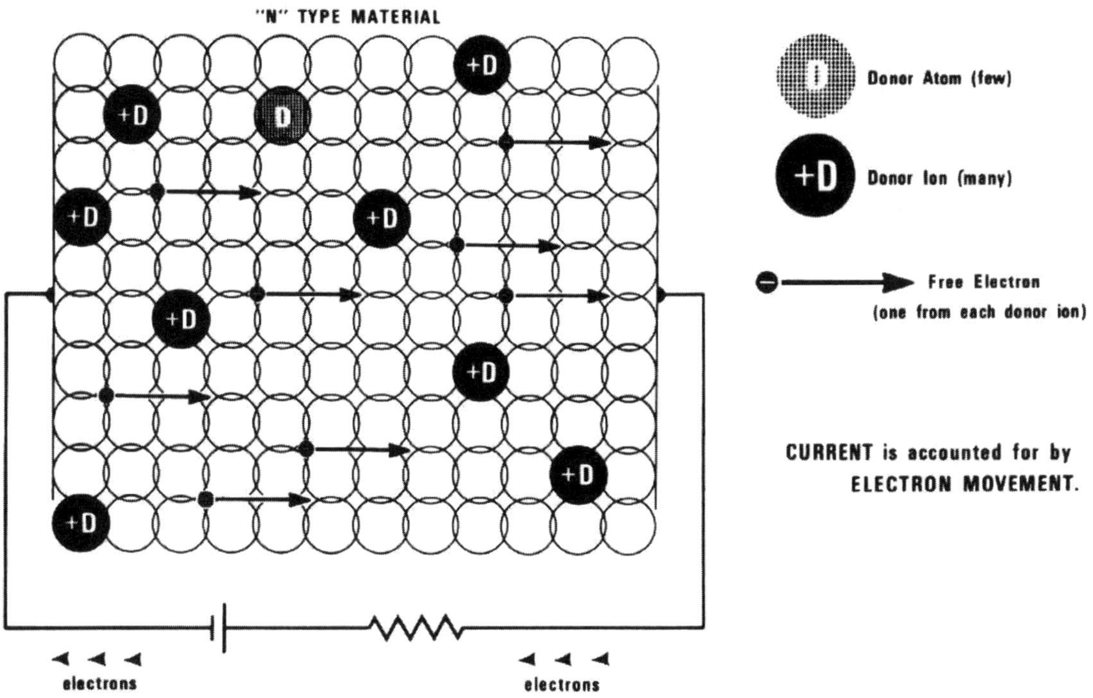

Voltage Across "P" Type Material

In a P type material (Figure 1-7), most acceptor atoms will have received an electron from somewhere else in the crystal. This means that most acceptor atoms have become acceptor ions, and that there are a corresponding number of holes elsewhere in the structure. The basic crystal electrons wander about as much as they do in the N type material, the difference being that in their wandering they leave holes at one place and fill a hole at another place. Consequently, we have wandering holes.

These holes are represented in Figure 1-7 by small black dots with a plus sign in the middle. A DC voltage is connected across P type material in this figure. Since electrons are moving from left to right through the material, the holes are moving from right to left.

Figure 1-8 schematically represents the linear progress of holes from right to left, while electrons are moving from left to right. At time-1, two electrons are about to move into two holes. At time-2, these electrons have moved, and the holes have been displaced to the left by one step. Now two other electrons are about to move into two holes. At time-3, these electrons have moved into the holes, and the holes have moved one more step to the left.

Figure 1-7

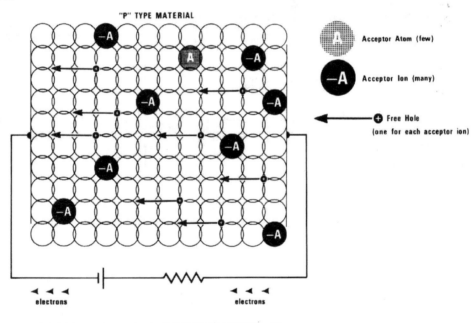

CURRENT is accounted for by ELECTRON MOVEMENT. This results in HOLE movement in the OPPOSITE DIRECTION.

Figure 1-8

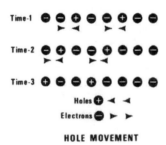

HOLE MOVEMENT

Diode Junction

Figure 1-9 represents a section of N material joined to a section of P material in such a manner that the crystal structure is intact throughout the junction. There are a variety of ways of accomplishing this, one of which is to change the melt from which the crystal is grown. Another is a fusion process similar to welding; for thin sections, varieties of diffusion processes are used. Regardless of how it is accomplished, the basic crystal structure is intact throughout.

N and P type materials joined in this way constitute a diode. With no external voltage applied, a depletion region exists on each side of the junction.

In the N material near the junction, free electrons are repelled by acceptor ions in the P material. In the P material, holes are repelled by the donor ions in the N material. Therefore, through a narrow space immediately adjacent to the junction (the depletion region), there are no current carrying elements; and for a short distance beyond that, there is a scarcity of current carrying elements.

Figure 1-9

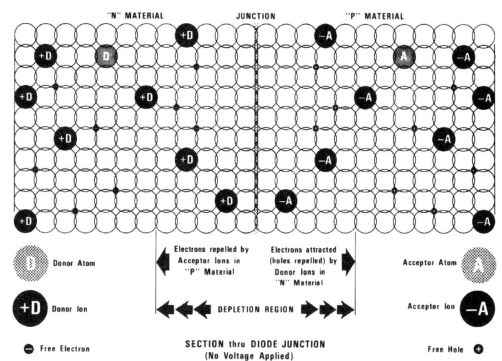

SECTION thru DIODE JUNCTION
(No Voltage Applied)

Forward Voltage Across A Diode

Figure 1-10 shows the diode connected across the terminals of a DC voltage source in the "forward" direction. In this case, forward direction means the direction which results in current flow.

Current flows because even a small voltage applied can cause free electrons in the N material to surge across the depletion region and across the junction where they will find holes to fill.

The continued movement of electrons across the P material from left to right results in hole movement in the P material from right to left. The amplitude of current flow is a function of applied voltage, external resistance (always present), and internal diode resistance.

Figure 1-10

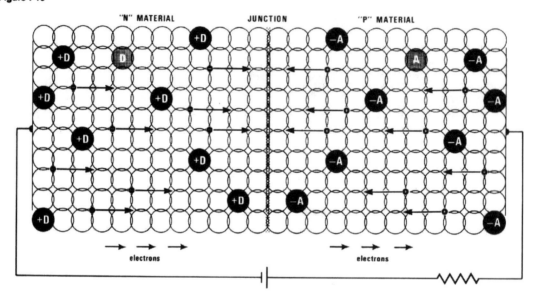

Changing Bias Across A Diode

Figure 1-11 schematically represents the depletion region of a diode with no bias applied.

Figure 1-12 shows why a diode will not conduct with the negative terminal of the voltage source connected to the P material of the diode (reverse biased). With a reverse bias applied, the depletion region is enlarged, increasing the size of the zone where no current carrying elements exist. It is enlarged because the negative voltage in the P material repels free electrons in the N material, while the positive voltage in the N material repels the free holes in the P material.

Figure 1-13 shows a larger reverse bias applied, and therefore a large depletion region. If a sufficiently large reverse bias is applied, electrons can be forced across the depletion region.

Some diodes would be destroyed by this action, others are especially constructed to break down at a particular voltage and subsequently heal them-selves if the current has not been excessive.

Figure 1-14 illustrates a diode constructed in this manner, called a zener diode. It is not destroyed if the reverse current is within design limits. It reverts to normal when current flow ceases. Within its normal operating limits it always breaks down at the same potential. It can be used for such functions as voltage regulation.

Figure 1-15 illustrates a diode and external resistor connected to a small voltage source in the forward direction. A small current flow results.

Figure 1-16 illustrates the same diode and resistor connected to a larger voltage flow.

Even though we have covered electron flow to this point, all schematics, drawings and figures from this point on will show current flow. Current flows from positive to negative.

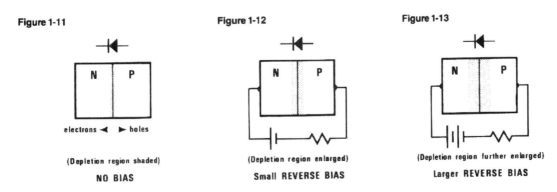

Figure 1-11 — (Depletion region shaded) — NO BIAS

Figure 1-12 — (Depletion region enlarged) — Small REVERSE BIAS

Figure 1-13 — (Depletion region further enlarged) — Larger REVERSE BIAS

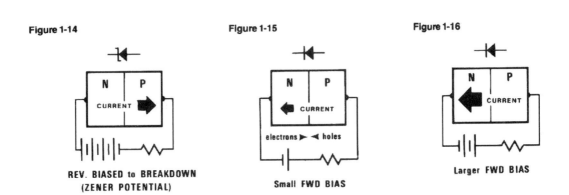

Figure 1-14 — REV. BIASED to BREAKDOWN (ZENER POTENTIAL)

Figure 1-15 — Small FWD BIAS

Figure 1-16 — Larger FWD BIAS

Transistor Conduction

Figure 1-17 illustrates schematically the construction of an NPN transistor. Above it is a schematic symbol. The arrow head represents the direction of current flow if the transistor is conducting. Electron flow, of course, is in the opposite direction.

In this figure the two depletion regions are equal because no external voltage is applied.

With an external voltage applied as illustrated in Figure 1-18, the size of the depletion regions changes because one of them is forward biased and the other is reverse biased. Current does not flow because the reverse biased depletion region is enlarged.

Figure 1-19 gives the terms commonly used to identify the three sections. The Emitter emits the current carriers, the Base controls the amount of current passed and the Collector collects the current. This is the same circuit as Figure 1-18, with a bias voltage added and supply voltage increased. The bias voltage is a small forward bias allowing current to flow from the Emitter to the Base, as explained in the following.

The P section is made very thin, too thin to illustrate realistically. The bias voltage is felt all through the long, thin P section; but the area of opportunity for bias current flow is limited since the connection is at one end.

Because of the bias voltage, electrons flow all across the junction of the Emitter and Base. Most of these electrons come from the supply, however, and are pushed on through to the Collector.

Continued on next page ...

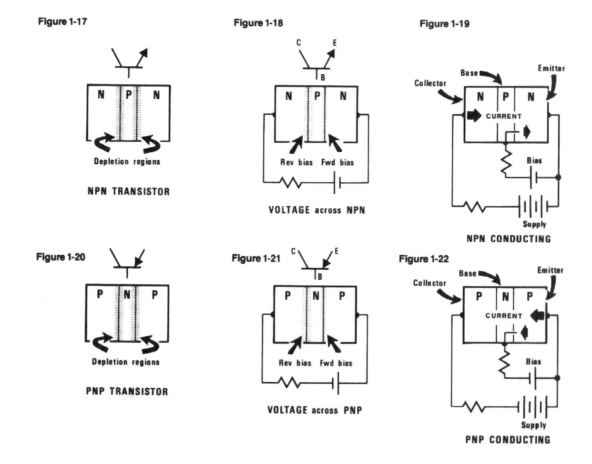

Figure 1-17
NPN TRANSISTOR
Depletion regions

Figure 1-18
VOLTAGE across NPN
Rev bias   Fwd bias

Figure 1-19
NPN CONDUCTING
Collector   Base   Emitter   CURRENT   Bias   Supply

Figure 1-20
PNP TRANSISTOR
Depletion regions

Figure 1-21
VOLTAGE across PNP
Rev bias   Fwd bias

Figure 1-22
PNP CONDUCTING
Collector   Base   Emitter   CURRENT   Bias   Supply

Transistor Conduction (cont'd.)

This is partly because the supply voltage is greater than the bias voltage. It is also partly due to bias electron flow being inhibited by the repelling negative acceptor ions, which remain fixed in the crystal structure. These effects and the thinness of the P section, from which the bias electrons have limited egress, result in most of the current coming from the supply source.

Only one-tenth to one-twentieth or less of the electrons in the Base pass to the bias source. Our transistor is, therefore, a current amplifier. By varying the bias current with a signal source we get a related but amplified current through the supply load.

A similar action takes place in a PNP transistor, illustrated without external voltage in Figure 1-20. The schematic symbol above represents, with its arrow head, the direction of current flow.

Figure 1-21 shows external voltage applied all across the transistor with its effect on the depletion regions. Current flow does not take place because of the enlarged depletion area on the left where reverse bias is felt.

Connected as shown in Figure 1-22, transistor conduction occurs for reasons similar to those that apply in the NPN.

Since the bias voltage is felt all across the Emitter to Base junction, electrons flow all across that face. Most of them come from the supply voltage, partly because bias electrons are repelled by sup-ply electrons, tending to limit bias electron penetration into the relatively great length of N material; partly also because the supply voltage is greater than the bias voltage.

These effects, combined with the thinness of the Base section to which the bias electrons have limited access, result in most of the current coming from the supply source.

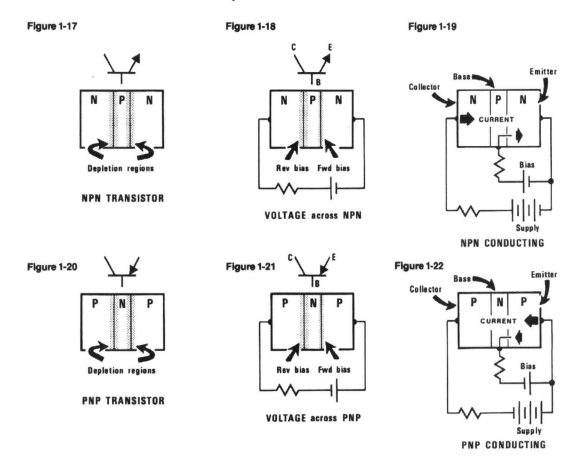

Figure 1-17 — NPN TRANSISTOR
Figure 1-18 — VOLTAGE across NPN
Figure 1-19 — NPN CONDUCTING
Figure 1-20 — PNP TRANSISTOR
Figure 1-21 — VOLTAGE across PNP
Figure 1-22 — PNP CONDUCTING

Changing Transistor Bias

Actual transistor circuits are never as simple as those which we have just illustrated, but in flight line applications we are seldom concerned with the actual black box circuitry. Therefore it is only occasionally that we will use a transistor to represent an amplifier. When we do, it will be simply symbolic and strictly a matter of convenience. Most of the time transistors will represent switches.

Figure 1-23 shows, on the left, an NPN transistor as an open switch. The next illustration shows a small bias causing a small current. In the third illustration, a large bias results in a large current. The one on the right shows a maximum bias and therefore represents a closed switch.

Figure 1-24 represents a PNP transistor used as a switch in a similar manner to Figure 1-23. Most of the time it will be more convenient to represent switches with NPNs since most switching functions are positive.

Figure 1-23

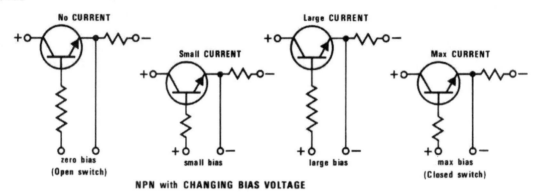

NPN with CHANGING BIAS VOLTAGE

Figure 1-24

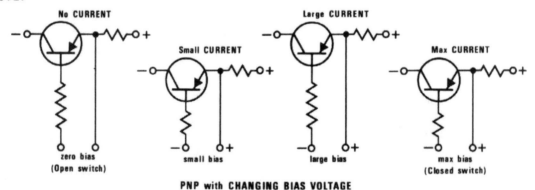

PNP with CHANGING BIAS VOLTAGE

Transistor Switches For + 28V DC

Figures 1-25 through 1-30 illustrate one transistor switch symbol which you may find in schematic diagrams. Although much of the time a 28 volt DC will be shown turning an NPN on, any positive voltage to the base of an NPN will turn the switch fully on. Any voltage less than + 28 volts DC to the base of a PNP will turn that switch fully on.

For convenience in simplifying circuits we will sometimes use these representations, even though in fact an actual transistor may require different voltages for operation.

Figure 1-25 shows a transistor switch not conducting 28 volts DC power, and Figure 1-26 shows it conducting.

Figure 1-27 shows a variation of usage; 28 volts DC appears at the output of the circuit only if the control voltage is not present; 28 volts DC does not appear at the output if the control voltage is present. This performs as a logic NOT circuit (discussed later).

Figure 1-28 shows a PNP not conducting because 28 volts is present at the base. Figure 1-29 shows the PNP conducting because less than 28 volts is present at the base.

Figure 1-30 shows a lock-up arrangement whereby a positive pulse to the base of the NPN triggers the PNP. It then continues to conduct because it provides a holding voltage to the base of the NPN.

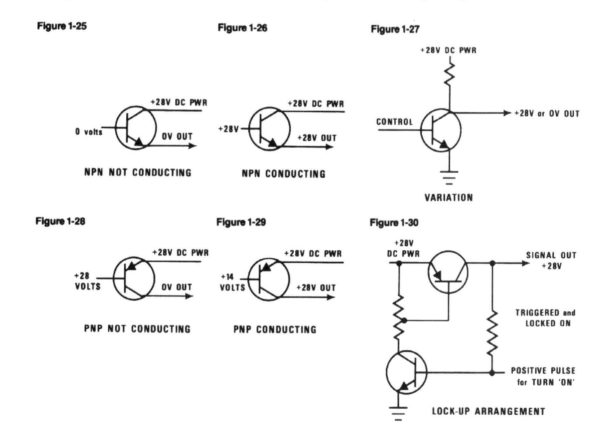

Transistor Switches In Schematics

Figure 1-31 shows transistor switch symbols used in a simplified manner which ignores actual transistor characteristics. For simplicity in connecting inputs, outputs, and base controlling volt-ages, voltage modifications which could be necessary for actual operation are omitted.

Figure 1-32 shows a more common way in which a transistor switch will be schematically represented. Usually it will represent an NPN and a voltage to the base, or the existence of a "named condition" turns the switch on.

The two illustrations on the right show a typical usage of this symbol. For example, a very high frequency omni-range (VOR) or localizer (LOC) deviation signal present on the left will not be seen at the output unless VOR capture or localizer capture occurs. Similarly, an altitude rate signal present on the left of the switch will not be seen at the output unless an altitude hold or vertical speed hold condition exists.

A further simplification exists in that even though it might be an AC signal which is switched, the signal is passed intact. You may find this done in early flight director or autopilot schematics.

Figure 1-31

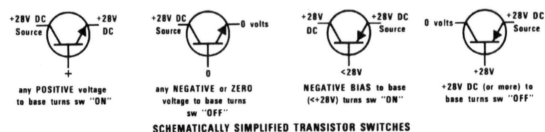

SCHEMATICALLY SIMPLIFIED TRANSISTOR SWITCHES

Figure 1-32

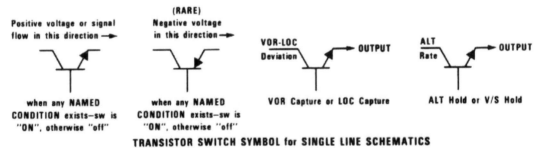

TRANSISTOR SWITCH SYMBOL for SINGLE LINE SCHEMATICS

Transistor Switches In Schematics

Present practice in switch symbology is shown in Figure 1-33. The switch contact will always be against the circle unless the named condition exists. In that case, the switch contact moves to the triangle.

Solid state logic switches: The solid state switch is a transistor which will change switch positions as a result of logic "1" input. See symbol in Figure 1-34.

Figure 1-33

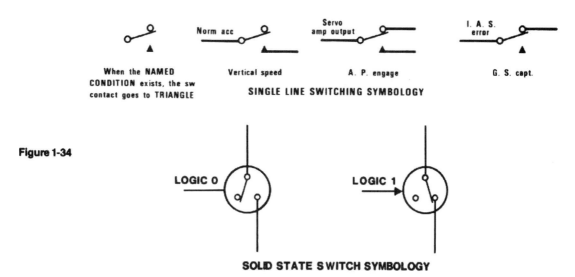

When the NAMED CONDITION exists, the sw contact goes to TRIANGLE

SINGLE LINE SWITCHING SYMBOLOGY

Figure 1-34

LOGIC 0   LOGIC 1

SOLID STATE SWITCH SYMBOLOGY

Silicon Controlled Rectifier

The silicon controlled rectifier becomes a very handy schematic symbol because we often want to represent the fact that a circuit can be turned on by a pulse and subsequently remain on even though the pulse does not last.

Figure 1-35 shows the SCR symbol. There will be no conduction through the symbol from positive to negative until the gate has seen a positive pulse.

Conduction will continue until the source voltage is removed either above or below the SCR symbol. Another gating voltage is then required for it to conduct again.

Figure 1-36 is a block schematic illustrating the construction of a typical SCR. There are four layers of P and N type material, with the gate at the lower P type layer.

Figure 1-35

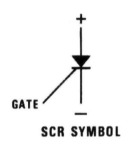

SCR SYMBOL

Figure 1-36

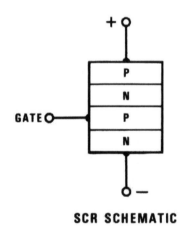

SCR SCHEMATIC

"AND" And "OR" Equivalents

Figure 1-37 shows an enlarged schematic AND symbol containing relays. These relays illustrate its logic function. Unless all three relays are energized there is no output voltage. Any logic AND circuit must have all of its inputs present for there to be an output.

Figure 1-38 shows the same AND circuit with transistors inside. For there to be an output from the AND circuit, all three input signals must be present.

Figure 1-39 shows the AND logic symbol the way you will find it in schematics. When you see this symbol you know there will be no output from the AND circuit unless all of the inputs are present.

Figure 1-40 shows an enlarged OR circuit containing relays. There is an output from this OR circuit whenever any one of the three inputs are present (any one of the three relays energized).

Figure 1-41 shows an enlarged OR circuit also has an output if any one of the inputs is present.

Figure 1-42 shows a diode OR circuit illustrating the same fact.

Figure 1-43 is the OR logic symbol as used in schematic diagrams. This symbol indicates that there is an output if one or more inputs are present.

Even though Figure 1-43 does not so indicate, in our logic circuits, all logic symbols are assumed to have their own internal power supply.

Figure 1-37

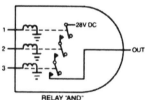

RELAY "AND"

Figure 1-38

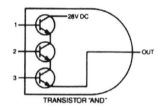

TRANSISTOR "AND"

Figure 1-39

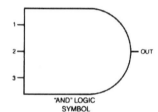

"AND" LOGIC SYMBOL

Figure 1-40

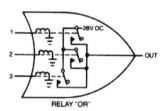

RELAY "OR"

Figure 1-41

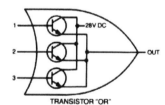

TRANSISTOR "OR"

Figure 1-42

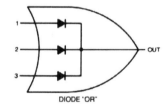

DIODE "OR"

Figure 1-43

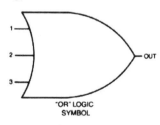

"OR" LOGIC SYMBOL

"AND" Circuits And "NOT" Circuits

Figure 1-44 shows examples of the logic NOT circuit. The circuit on the left is a relay NOT circuit. If there is no input to the relay, there is an output from the circuit. If there is a voltage applied to the relays, there is no output from the circuit.

The solid state NOT performs the same function. With no input signal to the base of the transistor, there is an output. If there is an input to the transistor, there is not output.

The schematic symbol for a NOT circuit is a simple open circle. This symbol is used only in conjunction with other logic symbols. When a separate NOT function is needed, it is attached to a triangle so as not to confuse it with a junction point.

"NOT" circuits, in common with our other logic symbols, are assumed to have their own internal power supply.

Figure 1-45 is an example of NOT functions used with an AND circuit. To obtain an output from the relay circuit on the left, the top relay must not have an input; the next one down must have an input; the third relay must not have an input; and the bottom one must have input.

The solid state circuit requires the same inputs in order to have an output.

The logic symbols on the right are the equivalent of either the relay or solid state circuits.

Figure 1-44

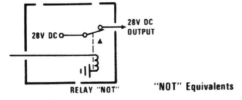
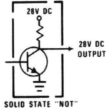

Figure 1-45

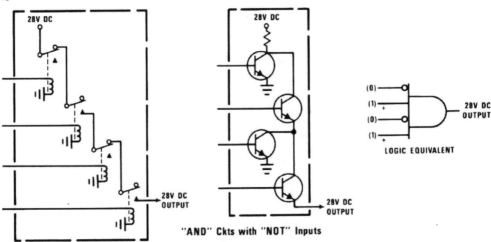

"OR" Circuits And "NOT" Circuits

In Figure 1-46, the two circuits on the left are the equivalent of the logic circuit on the right.

In the relay circuit, if the top relay is not energized, there is an output; or if the second relay is energized there is an output; or if the third relay down is not energized there is an output; or if the bottom relay is energized there is an output.

In the solid state equivalent, if the top transistor has no input there is an output; or if there is an input to the upper diode there is an output; or if there is no input to the lower transistor there is an output; or if there is an input to the lower diode there is an output. Any one or all of these input conditions result in an output.

Figure 1-46

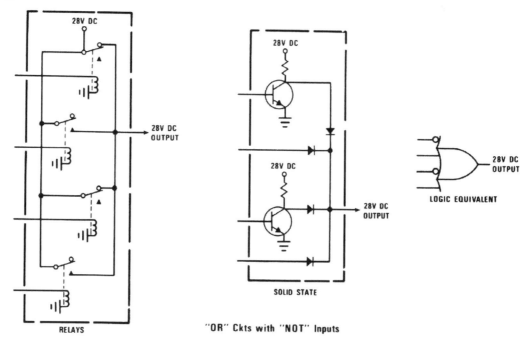

RELAYS   "OR" Ckts with "NOT" Inputs   LOGIC EQUIVALENT

Exclusive "OR"

Figure 1-47 shows an Exclusive "OR" gate as a device which requires the inputs to be different to obtain an output.

In Figure 1-48, the circuit shown illustrates the function of an Exclusive "OR" gate. Only when there is an input at A and not B or vice versa, can there be an OUTPUT at X. For example, if there is an input at A and not B, relay A is energized; and the lower switch above relay A is open; and the upper switch above A closes. Because there is no input at B, relay B is not energized. The lower switch above B is open and the upper switch is closed.

Therefore, with the upper switch above relay A closed, and the upper switch above B closed, the circuit is completed to get an output at X. Note only when both switches on the upper circuit are closed, or when both switches on the lower circuit are closed, is there an output at X.

Figure 1-49: The various combinations of inputs and the resulting outputs for the function circuit are shown.

Figure 1-47

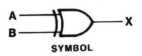

SYMBOL

Figure 1-48

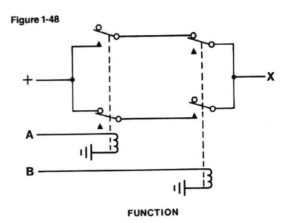

FUNCTION

Figure 1-49

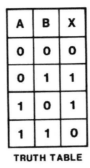

A	B	X
0	0	0
0	1	1
1	0	1
1	1	0

TRUTH TABLE

Logic Truth Table

Figure 1-50

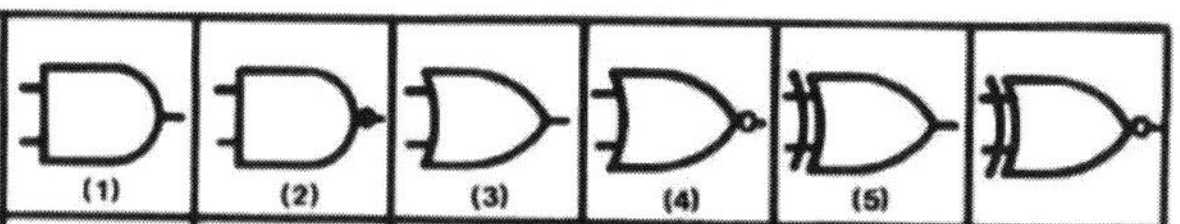

| INPUTS | | A | | | | | |
| | | B | | | | | |
		(1)	(2)	(3)	(4)	(5)	
A	B	AND	NAND	OR	NOR	Exclusive OR	Exclusive NOR
O	O	O	1	O	1	O	1
O	1	O	1	1	O	1	O
1	O	O	1	1	O	1	O
1	1	1	O	1	O	O	1

Logic Symbol Exercise

Figure 1-51 shows an AND circuit with NOTs on the three inputs. This circuit functions the same as the OR circuit with the NOT output. Any input to either circuit removes the output.

Figure 1-52 shows two other logic equivalents. The OR circuit on the left has an output if any one of the three NOT circuits does not have an input. Similarly, the NOT circuit on the end of the AND circuit will have an output if the AND circuit is not satisfied; and the AND is not satisfied if any one of the three inputs is not present.

Figure 1-53 is a sample logic circuit with which you can practice the use of these symbols.

Put random inputs on the left, and see what the result will be as you go from left to right to the last AND circuit.

Another way is to assume that the right AND circuit has an output and see what inputs are required, working your way to the left, to satisfy the right AND circuit.

These three basic logic symbols are very useful in our circuits for condensing information.

Although many other logic symbols are used elsewhere, these three are all that we have need for.

Figure 1-51   **Figure 1-52**

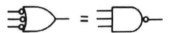

LOGIC EQUIVALENTS (in our LOGIC SYSTEM)

Figure 1-53

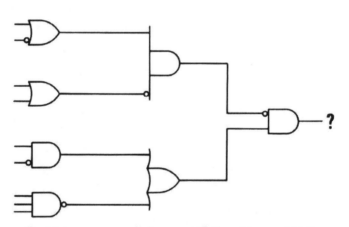

Furnish your own sample inputs - will there be an output ?

Typical Logic Schematic

Figure 1-54 shows a typical landing gear position indicator, Circuit. This circuit utilizes two solid state switches, an inverter, a six input NAND gate, an OR gate and a light circuit, to illuminate the gear down light. A ground is provided to the light circuit through solid state switch "B" with the gear down lock sensor activated (solid state switch "A"). A logic "0" (ground) is inverted and applied as logic "1" at the nand gate input.

As long as all gear sensor inputs to the nand gate are logic "1", the nand gate output will be logic "0". This output is inverted at the "OR" gate to solid state switch "B", thus providing a ground to illuminate the gear down light.

Figure 1-54

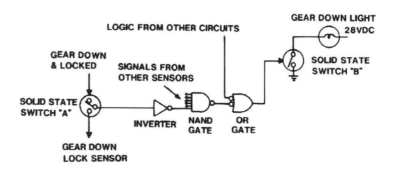

Digital vs. Analog

Our world operates on an analog base. The earth slowly blends day into night. A great deal of our machinery has been built on this method of continuous resolution (Basic Analog). Analog is the most convenient method in our everyday lives, but it is not very efficient or precise when compared to digital in the realm of aircraft avionics.

Analog data is a physical representation of information which bears an exact relationship to the original information. The electrical signals on a telephone line are an analog data representation of the original voice. Figure 1-56 illustrates a sine-wave (analog) signal. It can be measured at various points, such as A, B, C, D and E. The measured values are a direct representation of the continuous values along the sine-wave. Analog data is represented in a continuous form by physical variables, such as voltage, resistance, capacitance, angle of rotation, etc.

The advantage of converting data into a digital representation is that the components are more economical, lighter, and provide a higher degree of resolution. Digital data is also much easier to transmit without loss of data detail. Another advantage is that more than one user can use the same information. By having the data coded, more than one data group can be transmitted by time sharing the line.

Digital data is represented in discrete or discontinuous form. In this case by a series of ones and zeros. A one represents the presence of a value or expression, while a zero represents the absence of that value or expression. Figure 1-57 illustrates the sine-wave measurements from Figure 1-56 being processed by an analog to digital converter. For example, measurement A has a value of + 2.75 volts. The output of the converter consists of a digital word of ten bits (ones and zeros) which represents the analog value of measurement A. A parity bit is added to verify the reliability of data transmission and reception.

Depending upon the usage, coding and required resolution, the length of digital words may be anywhere from 8 to 64 bits. A bit is defined as a single binary decision such as a "1" or "0".

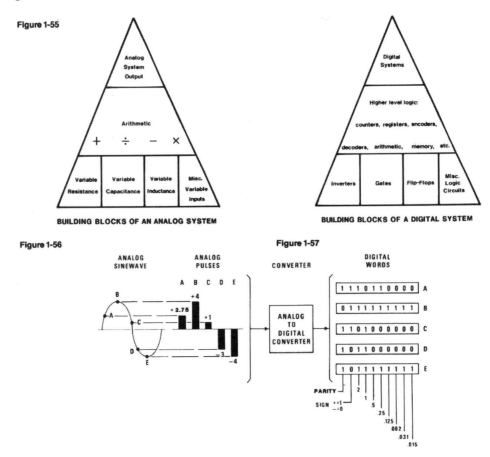

Figure 1-55

BUILDING BLOCKS OF AN ANALOG SYSTEM

BUILDING BLOCKS OF A DIGITAL SYSTEM

Figure 1-56

Figure 1-57

Microprocessors

Microprocessors are regarded as one of the most important devices in our everyday machines called computers. Before we start, we need to understand what exactly microprocessors are and their appropriate implementations. A Microprocessor is an electronic circuit that functions as the central processing unit (CPU) of a computer, providing computational control. Microprocessors are used in electronic systems such as home computer, automobiles, jet airliners and household appliances. Typical microprocessors incorporate arithmetic and logic functional units as well as the associated control logic, instruction processing circuitry.

To understand microprocessor function, first consider a simple light switch. You flip the switch by the door, and the light in the middle of the ceiling comes on. It stays on. When you leave the room, you flip the switch down again, and the light goes out. It stays out. The switch will remain in the position you last left it until you (or someone else) come back and flip it to its other position.

In a sense, it "remembers" what its last command was until you change it, and "overwrite" that command with a new one. In this sense, a light switch represents a sort of rudimentary memory element.

Whether a switch is mechanical, electrical, hydraulic, or something else is irrelevant. What matters is that a switch contains a pattern: On or off, up or down, flow or no flow. To that, this pattern can be assigned a meaning.

In general then, what we call "memory" is an aggregate of switches that will retain a pattern long enough for that pattern to be read and understood by a person or a mechanism. Memory consists of containers for alterable patterns that retain an entered pattern until someone or something alters the pattern.

Computers in the past have been made out of relays, although as you might imagine (with a typical relay being the size of an ice cube) they weren't especially powerful computers.

Fully electronic computers are made out of transistor switches. Modern processors can contain in excess of 10,000,000 transistors. Let's consider a transistor switch a sort of electrical "black box" and describe it in terms of inputs and outputs (Figure 1-57a).

When an electrical current is fed through the base, current flows between emitter and collector. When the current ceases flowing through the base, current ceases to flow between emitter and collector.

In real life, a tiny handful of other components (typically diodes and capacitors) are necessary to make things work smoothly in a computer memory context. These components are not necessarily little gizmos connected by wires to the outside of the transistor (although in early transistorized computers they were), but are now cut from the same silicon crystal the transistor itself is cut from, and occupy almost no space at all.

Taken together, the transistor switch and its support components are called a memory cell. (Figure 1-57b) The electrical complexity of the memory cell is hidden within an appropriate black-box symbol.

Continued on next page ...

Figure 1-57a

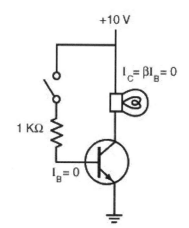

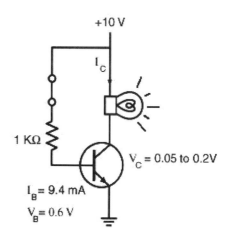

Microprocessors (cont'd.)

The memory cell's circuit is arranged so that if you put a tiny voltage on its input pin and a similar voltage on its select pin, a voltage will appear and remain on its output pin. That output voltage will remain in its set state until you take away the voltage from the cell as a whole, or remove the voltage from the input pin while putting a voltage on the select pin.

The "on" voltage being applied to all of these pins is kept at a consistent level (except, of course, when it is removed entirely). Most memory cells operate at a constant 5 volts. The pattern is binary in nature: you either put a voltage on the input pin or you take away the voltage entirely. Likewise, the output pin either holds a fixed voltage or no voltage at all.

We apply a code to that state of affairs: the presence of voltage indicates a binary 1, and the lack of voltage indicates a binary 0. This is called a bit.

One bit doesn't tell us much. To be useful, we need to bring a lot of memory cells together. In the beginning, one chip held one transistor. In time, the designers crisscrossed the chip into four equal areas, making each area an independent transistor. From there it was an easy jump to adding the other minuscule components needed to turn a transistor into a computer memory cell. These chips are called RAM chips, because they contain random access memory.

Random access works like this: Inside the chip, each bit is stored in its own memory cell, identical to the memory cell in figure 1-57b. Each of the

however-many memory cells has a unique number. This number is a cell's (and hence a bit's) address.

Each chip has a number of pins coming out of it. The bulk of these pins are called address pins. One pin is called a data pin. The address pins are electrical leads that carry a binary address code. Your address is a binary number, expressed in 1s and 0s only. Special circuits inside the RAM chip decode this address to one of the select inputs of the numerous memory cells inside the chip. For any given address applied to the address pins, only one select input will be raised to five volts, thereby selecting that cell.

Depending on whether you intend to read a bit or write a bit, the data pin is switched between the memory cell's input or output, as shown above. But that's all done internally to the chip. Once you've applied the address to the address pins, the data pin will contain a voltage representing the value of the bit you requested. If that bit contained a binary 1, the data pin will contain a 5 volt signal; otherwise, the binary 0 bit will be represented by 0 volts.

From a functional perspective, memory is measured in bytes. A byte is eight bits side-by-side. A binary number eight bits in size can be one of 256 different values, numbered from 0 to 255.

Every byte of memory in the computer has its own unique address, even in computers that process two bytes, or even four bytes, of information at a time.

Continued on next page ...

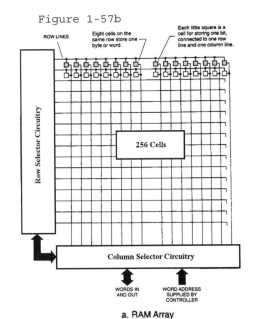

a. RAM Array

Figure 1-57b

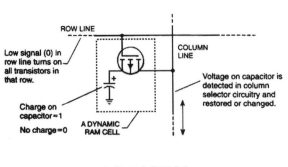

b. Dynamic RAM Cell

Microprocessors (cont'd.)

The CPU chip's most important job is to communicate with the computer's memory system. Like the pins of memory chips, the CPU's pins transfer information encoded as voltage levels. Five volts indicate a binary 1, and zero volts indicate a binary 0.

Like the memory chips (in a memory bank), the CPU chip has a number of pins devoted to memory addresses, and these pins are connected directly to the computer's banks of memory chips Figure 1-57c). When the CPU desires to read a byte from memory, it places the memory address of the byte to be read on its address pins, encoded as a binary number. A split second later, the byte appears (also as a binary number) on the data pins of the memory chips. The CPU chip also has data pins, and it slurps up the byte presented by the memory chips through its own data pins.

The process, of course, also works in reverse: to write a byte into memory, the CPU first places the memory address where it wants to write onto its address pins. A split second later, it places the byte it wishes to write into memory on its data pins. The memory chips then store the byte inside themselves at the requested address.

This give-and-take between the CPU and the memory system represents the bulk of what happens inside your computer. Information flows from memory into the CPU and back again. Information flows in other paths as well. Your computer contains additional devices called peripherals that are either sources or destinations (or both) for information.

Like the CPU and memory, these peripherals are all ultimately electrical devices, having both address pins and data pins. Peripherals "talk" to the CPU (i.e., pass the CPU data or take data from the CPU) and sometimes to one another. These conversations take place across the electrical connections, linking the address pins and data pins that all devices in the computer have in common. These electrical lines are called a data bus, and form a sort of party line linking the CPU with all other parts of the computer.

There is an elaborate system of electrical arbitration that determines when and in what order the different devices can use this party line to talk with one another. But it happens the same way: an address is placed on the bus, followed by a byte (or word or double word) of data. Special signals go out on the bus with the address to indicate whether the address is of a location in memory, or of one of the peripherals attached to the data bus. The address of a peripheral is called an I/O address to differentiate between it and a memory address.

The data bus is the major element in the expansion slots present in most PC-type computers, and most peripherals are boards that plug into these slots. The peripherals talk to the CPU and memory through the data bus connections brought out as electrical pins in the expansion slots.

Continued on next page ...

Figure 1-57c

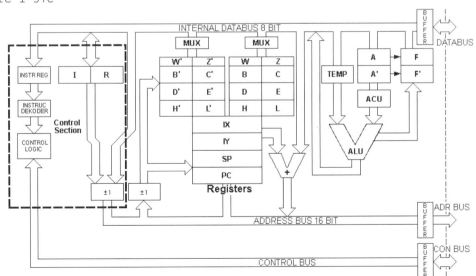

Microprocessors (cont'd.)

Every CPU contains very few data storage cubbyholes called registers (Figure 1-57c). When the CPU needs a place to tuck something away for awhile, an empty register is just the place. The CPU could always store data out in memory, but that takes a lot of time. Because the registers are actually inside the CPU, placing data in a register or reading back again is fast.

But more important, registers are the processor's workbench. When the CPU needs to add two numbers, the easiest and fastest way is to place the numbers in two registers and add the two registers together. The sum (in usual CPU practice) replaces one of the two original numbers that were added, but after that, the sum could then be placed in yet another register, or added to another number in another register, or stored out in memory, or any of a multitude of other operations.

Work involving registers is always fast, because the registers are within the CPU and very little movement of data is necessary.

Like memory cells, and indeed, like the entire CPU, registers are made out of transistors, but rather than having numeric addresses, registers have names like AX or DI. To make matters even more complicated, while all CPU registers have certain common properties, most registers have unique special powers not shared by other registers.

Continuous, furious communication along the data bus between CPU, memory, and peripherals is what accomplishes the work that the computer does. The question then arises: who tells the processor and the rest of the machine what to do? You write a program. Where is the program? It's in memory, along with all the rest of the data stored in memory. In fact, the program is data, and that is the heart of the whole idea of programming as we know it.

The circuitry used to make up the microprocessor system is referred to as the system hardware (Figure 1-57d). Software is the program developed to make the system hardware perform the required functions. When the software instruction codes are programmed into integrated circuit permanent memory, the combined hardware/ software (programmed integrated circuit) is referred to as firmware.

Figure 1-57d

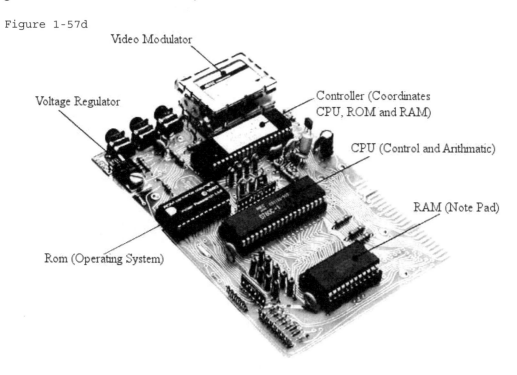

Video Modulator

Voltage Regulator

Controller (Coordinates CPU, ROM and RAM)

CPU (Control and Arithmatic)

RAM (Note Pad)

Rom (Operating System)

Eight Bit Digital Data Bus

The incorporation of a fully integrated digital avionics system requires a digital data bus to provide a two way interface between various navigation sensors, computers and indicators.

The interface between each computer or external device is accomplished via the digital data bus. The bus is made up of a twisted pair of wires which are shielded and jacketed. Data may travel one way or in two directions, depending on the system de-sign. The shielding is grounded at all terminal ends and breakouts to assure a higher degree of immunity to bit distortion and subsequent erroneous data. Shield grounding and high voltage spike protection within the receivers virtually guarantees accurate transmissions from each transmitter to its receiver. Breakouts to specific equipment is normally accomplished at burndy block type modules. Transmission of data within micro-computers and external transmissions between other components is accomplished with 8, 16, 32 or 64 bit digital words.

The serial bus is one on which the data is transmitted sequentially, one word following another word.

It is commonly used for long distance transmissions. The parallel bus interconnects the internal devices of a computer and has enough wires to transmit all bits of the word simultaneously. An eight bit parallel bus would be eight times faster than the serial bus, and a 64 bit parallel bus would be 64 times faster than its equivalent serial bus.

Figure 1-58 illustrates a typical unidirectional bus system used on aircraft in the 1980's. Aircraft being designed for the 1990's will have high speed bidirectional busses using inductive coupling. There is also in the planning stage a fiber optic system which will provide superior performance, along with reduced size and weight.

Figure 1-58

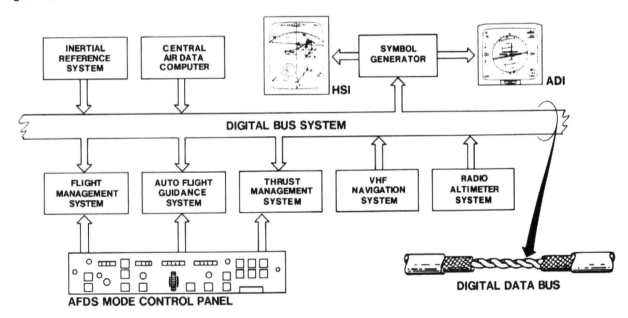

ARINC 429 Data Word

The ARINC 429 digital data word (Figure 1-58a) consists of 32 bits. The data word has five basic parts;

Label
Source/Destination Identifier (SDI)
Data field
Sign Status Matrix (SSM)
Parity bit

LABEL

Bits 1 to 8 comprise the label; this identifies the information contained in the data word (magnetic wind direction, EGT).

SOURCE/DESTINATION IDENTIFIER

Bits 9 and 10 comprise the Source/Destination Identifier (SDI), this is used when it is necessary to indicate the source or where the information is to be directed (1, 2, 3).

DATA FIELD

Bits 11 through 29(or 28) comprise the data field; this contains the specific data assigned to a label.

This information fits certain parameters can be in many different forms of binary codes. The most significant bit (MSB) is at bit 29 and the least significant bit (LSB) is at bit 11. If the data does not require all the available bits, binary 0 or pad bits are used to fill the field.

SIGN STATUS MATRIX

Bits 30 (29) and 31 comprise the sign status matrix (SSM), this identifies the characteristics of a word (+, -, north) and its status (no computed data, test).

PARITY

Bit 32 is the parity bit; the bit value will give odd parity. Odd parity means that the sum of all logic 1s in the data word equals an odd number. This bit is used to check the transmission efficiency, since this is a one way bus.

SYNCHRONIZATION

To enable the receiver to identify the beginning of a transmission, the data word is synchronized by a minimum 4 bit time gap.

Figure 1-58a

Digital data word = 32 bits

SDI-Source/Destination Identifier

SSM-Sign Status Matix

P-Parity

LSB-least significant bit

MSB-most significant bit

ARINC 629 Data Bus
(Typical Boeing 777)

The ARINC 629 data bus is a high speed, serial, bi-directional, multiple terminal, digital data bus which supports data communication between many terminals over a common bus.

ARINC Specification 629 contains the standards for the "Multi-Transmitter Data Bus". Boeing refers to this bus as the "Digital Autonomous Terminal Access Communication" (DATAC) bus.

The bus operates at 2 megabits per second, which allows 100,000 20-bit words to be transmitted each second.

Each component can send a message that contains up to 31 word strings and each word string can have 256 words.

The ARINC 629 data bus cable consists of a pair of twisted (shielded or unshielded) wires up to 160 feet long with a terminator at each end of the cable. The data bus allows any component connected to the bus to transmit and receive on the bus. Up to 60 components can be connected to the bus through inductive bus couplers (Figure 1-58b). Bus cables located inside bus panels are unshielded because the metallic bus panel enclosures provide the shielding benefit. Bus cables that are external to the bus panels are usually shielded.

Figure 1-58b

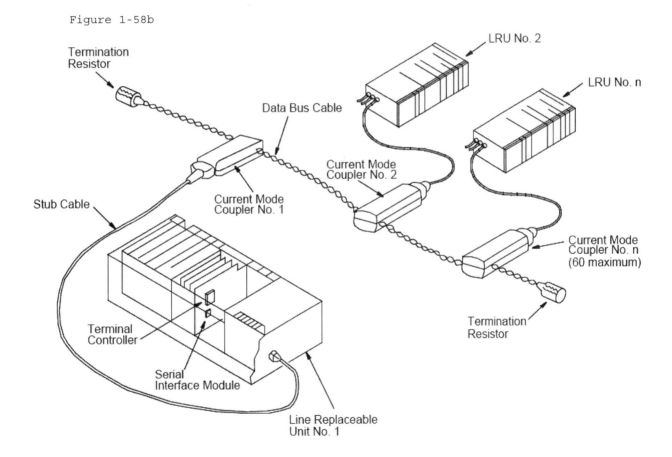

ARINC 629 System

The ARINC 629 communication system (Figure 1-58c) includes the bus cable, controllers, interface modules, stub cables and current mode couplers.

The ARINC 629 communication system has these characteristics:

- Two way transmission
- Multiple transmitters
- Broadcast-type
- Autonomous terminal access
- Time-division multiplex

Each LRU (Line Replaceable Unit) is connected to its coupler by a stub cable. The stub cable contains two sets of shielded twisted pairs of wires. One pair is used for transmitting and the other for receiving. The stub cables can be up to 75 feet long. The LRU's use a coupler and terminal (terminal controller and serial interface module) to communicate with the bus. Each terminal listens to the bus and waits for a quiet period before it transmits. Only one terminal on a bus transmits at a time. After a terminal transmits, three separate timers make sure that it does not

transmit again until all of the other terminals on the bus have had a chance to transmit.

The current-mode coupler (sometimes referred to as bus couplers) connects the bus cable to the LRU via the stub cable. They are referred to as current mode couplers because they use inductive coupling. This is similar to attaching an inductive timing light connector to the spark plug wire when adjusting your car's timing.

These are the physical characteristics of the current-mode coupler:

- It is a two-part assembly for fast installation
- It has a cover with the electronics and the receptacle for the stub cable
- It has a base designed for easy installation on panel
- It has a waterproof housing
- It has a protective wire guide to install the data bus cable

All current-mode couplers are located in bus panels.

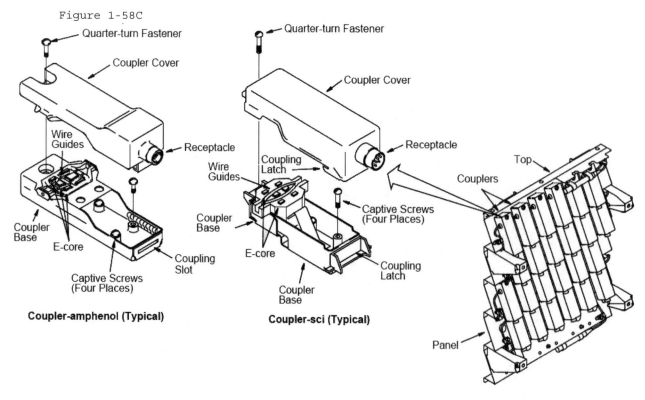

Figure 1-58C

Coupler-amphenol (Typical)

Coupler-sci (Typical)

Bus Panel (Cover Removed)
(Typical)

Airplane Information Management System (Typical Boeing 777)

Newer State of The Art Airplanes have begun migrating away from the "Distributed Architecture" philosophy where the computing functions take place in many different LRUs, to one in which much or all of the computing functions take place in fewer centralized locations. This architecture philosophy is referred to as "Central Architecture". "Federated Architecture" is a compromise of Central and Distributed architecture philosophies.

The airplane information management system (AIMS) is a new system introduced on the Boeing 777 airplane. Advancements in technology microelectronics, fault tolerance, and software permit the development of the highly integrated, digital avionics discussed above. The AIMS integrates the avionics computing functions that require large quantities of data collection, processing and calculations. On other model airplanes, several LRUs handle these avionics computing functions.

The AIMS has two cabinets. Each cabinet has eight line-replaceable units (Figure 1-58d): four input/output LRUs and four core processor LRUs. These cabinets operate as the main computer for eight avionics systems.

The AIMS cabinet integrates the computing functions for the avionics systems. Software partitioning keeps a necessary separation between computing functions. The software partitioning allows the integration of multiple computing functions in a single core processor module.

The AIMS cabinets interface with approximately 130 LRUs, sensors, switches and indicators. The large quantity of interfaces permits the AIMS to integrate the information from a majority of airplane systems in one place. It is efficient to integrate this information for central maintenance computing, flight data recording, airplane condition monitoring and displays.

Several other components are part of the AIMS. The components are:

- EFIS control panel (2)
- Display select panel
- Control display unit (CDU) (3)
- Display switching panels (2)
- Cursor control device (2)

The central maintenance system uses the AIMS cabinets for the computing function. The maintenance crew uses a maintenance access terminal (MAT) to control and access the central maintenance computing system. The MAT is a station with a display module, disk drive module, keyboard, and cursor controller/power supply module. The MAT is at the second observer's position.

Continued on next page...

Figure 1-58d

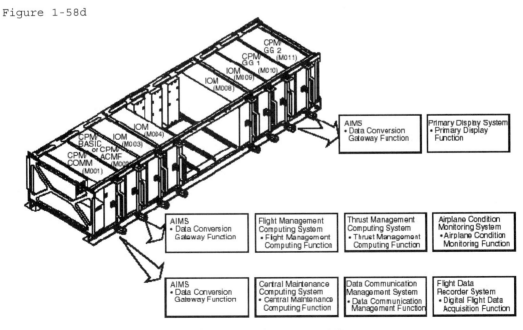

Airplane Information Management System

Airplane Information Management System (cont'd.)

Engineers use the ground based software tool (GBST) to create airline modifiable information (AMI). The AMIs allows the airline to customize information. The AMI software is loaded into these functions: Figure 1-58e illustrates AIMS centralized architecture concept and the integration of many navigation systems into one centralized location. This centralized architecture improves the sharing of data in many aircraft systems improving performance, reliability and efficiency.

- ACMF (airplane condition monitoring function)
- CMCF (central maintenance computing function)

- DCMF (data communication management function)
- FMCF (flight management computing function)
- Airplane information management system
- Data conversion gateway function
- Primary display system
- Flight management computing system
- Thrust management computing system
- Central maintenance computing system
- Maintenance Access Terminal
- Airplane conditioning monitoring system
- Flight data recording system
- Data communication management system

Figure 1-58e

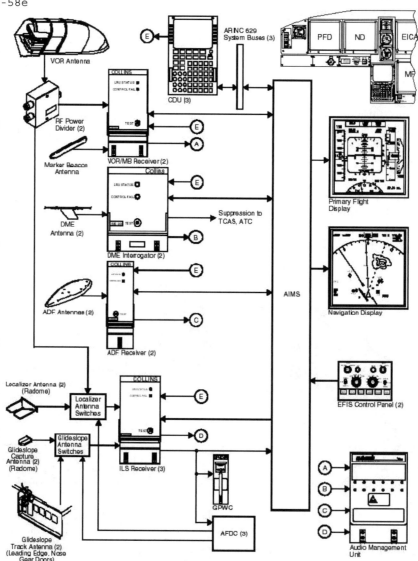

Static Electricity

Static electricity is generated and stored on the surface of non-conductive materials and discharges to a ground source if available. An electrostatic discharge from an object, electrostatic field, or voltage spark can cause over stressing on equipment and produce a failure. Protective measures should be taken to prevent degrading of equipment performance.

Figure 1-59 identifies some of the labels for electrostatic discharge sensitive devices. These labels have black nomenclature on a yellow background. These labels are used on racks and shelves, containing ESD sensitive equipment. Line replaceable units and circuit assemblies that have sensitive circuitry should be labeled.

Figure 1-60: Whenever static electricity can be seen or felt, the electrostatic discharge can usually be measured in thousands of volts.

Even when static electricity cannot be seen or felt, ESD damage can occur. Damage to micro-circuitry can occur at less than thirty volts.

Figure 1-60, a chart of typical electrostatic voltages, illustrates that at 10-20% relative humidity; just by walking across carpet you could generate 35,000 volts.

Figure 1-61, a typical caution label decal D. One caution you would find is to follow the proper handling and grounding procedures to prevent damage to the control card and card file. The control card is a static sensitive device.

Before initiating repairs on any ESD device, be sure to ground your body. Discharge all tools by touching tools to ground. Use proper safety techniques for units under power tests.

Figure 1-59

DECAL	SYMBOL	USAGE
A	**CAUTION** THIS ASSEMBLY CONTAINS ELECTROSTATIC SENSITIVE DEVICES	BLACK BOXES AND ASSEMBLIES
B	**ATTENTION** THIS UNIT CONTAINS STATIC SENSITIVE DEVICES. CONNECT GROUNDING WRIST STRAP TO ELECTROSTATIC GROUND JACK LOCATED AT THE LOWER RIGHT HAND SIDE OF THIS UNIT.	LRU FRONT PANELS
C	STATIC SENSITIVE	CIRCUIT BOARD EXTRACTOR
D	**ATTENTION ELECTROSTATIC GROUND JACK**	GROUNDING JACKS
E	(symbol)	AREAS TOO SMALL TO ACCOMODATE ANY OF THE ABOVE DECALS

CHARACTERS: BLACK
BACKGROUND: YELLOW

Figure 1-60

TYPICAL ELECTROSTATIC VOLTAGES

MEANS OF STATIC GENERATION	ELECTROSTATIC VOLTAGES	
	10 TO 20 PERCENT RELATIVE HUMIDITY	65 TO 90 PERCENT RELATIVE HUMIDITY
WALKING ACROSS CARPET	35,000	1,500
WALKING OVER VINYL FLOOR	12,000	250
WORKER AT BENCH	6,000	100
VINYL ENVELOPES FOR WORK INSTRUCTIONS	7,000	600
COMMON POLY BAG PICKED UP FROM BENCH	20,000	1,200
WORK CHAIR PADDED WITH POLYURETHANE FOAM	18,000	1,500

ELECTROSTATIC DISCHARGE

Figure 1-61

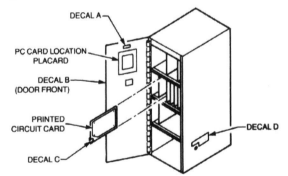

DECAL A

PC CARD LOCATION PLACARD

DECAL B (DOOR FRONT)

PRINTED CIRCUIT CARD

DECAL C

DECAL D

CHAPTER 2 - BRIDGES AND MONITORS

DC Bridges

The simple DC bridges in Figures 2-1 through 2-4 remind you that current and voltage can be reversed in the sensing member by conditions in the legs of a bridge.

Figure 2-1 shows current flowing to the left through the sensing member because of a higher voltage on the right. Figure 2-2 shows equal voltages across the sensing member, and therefore no current flow. Figure 2-3 shows a higher voltage on the right, and therefore current flowing through the ammeter to the left.

Figure 2-4 shows a higher voltage on the left, and therefore current flow to the right through the ammeter.

Remember that current flow in aircraft circuits is almost always more convenient than electron flow because 28 volt DC power is supplied with positive connected to the circuit breakers, and negative to ground. This works out fine when dealing with solid state items because the arrows are already lined up with current flow rather than electron flow.

Figure 2-1

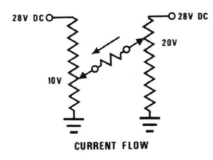

CURRENT FLOW

Figure 2-2

NO CURRENT FLOW

Figure 2-3

CURRENT FLOW

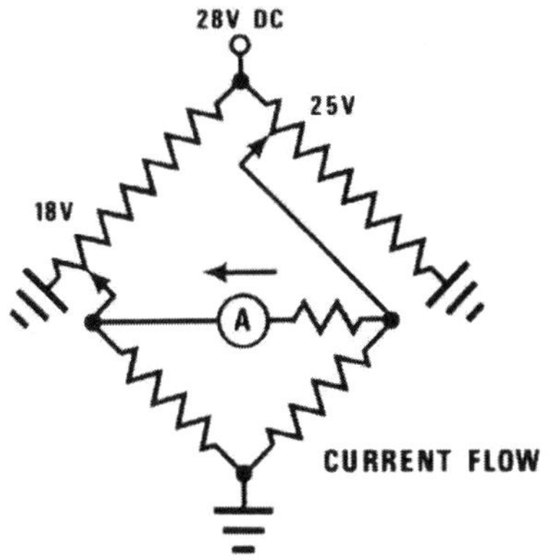

Figure 2-4

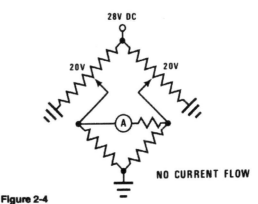

CURRENT FLOW

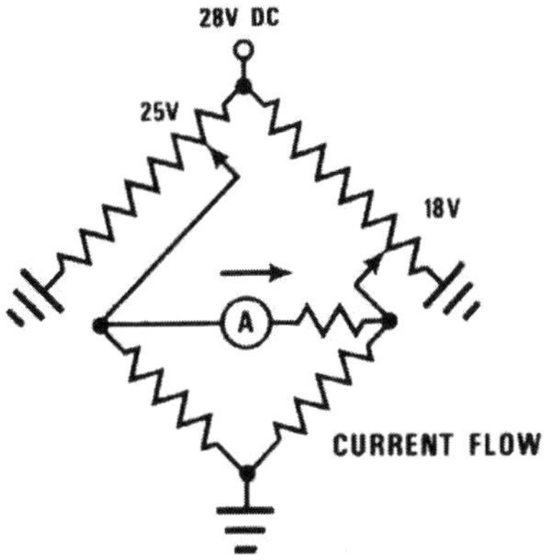

AC Bridges

Phase reversal is an important and useful concept when working with avionics signal circuits.

"Phase reversal" means changing the phase of a voltage from phase angle zero (0°) to phase angle 180 (180°). Phase angle zero is typically the phase of the source voltage. Phase reversal occurs only after a null has been reached and passed (passing through a null).

There are many ways by which phase reversal is accomplished, and a great number of null seeking devices and null situations in which phase is reversed as the signal changes from one side of the null to the other. It is usually the key to the operation of a system; where a particular phase identifies a particular reversible motion or sense of operation, the opposite phase identifies the opposite motion. For example:

if 0° is:	then 180° is:
left	right
up	down
forward	aft
clockwise	counterclockwise
slow	fast
climb	descent
more	less

It is generally desirable to exclude signals which are not phase angle 0° or phase angle 180°. Circuits designed to screen out or correct for these undesired signals are called "quadrantal error" corrections because signals not phase angle 0° or 180° lie somewhere in the "quadrants" of the electrical circle.

AC bridges illustrate one way in which phase reversal and an intervening null are generated. Figure 2-5 is a balanced bridge. It is balanced because the resistance ratios between the left and right members are equal. There is no voltage difference and null exists between the sensing points.

In Figure 2-6 the bridge is unbalanced because the resistance ratios between the right and left members are not equal. There is a voltage difference between the sensing points, with the high voltage on the right sensing point. A phase angle of 180° is arbitrarily assigned this voltage. Depending on your point of reference, a phase angle of 0° could have been assigned (when the little sine wave leads off with a positive loop, it represents phase angle 0°).

If the resistance ratios are reversed as shown in Figure 2-7, the voltage across the sensing points is

Continued on next page ...

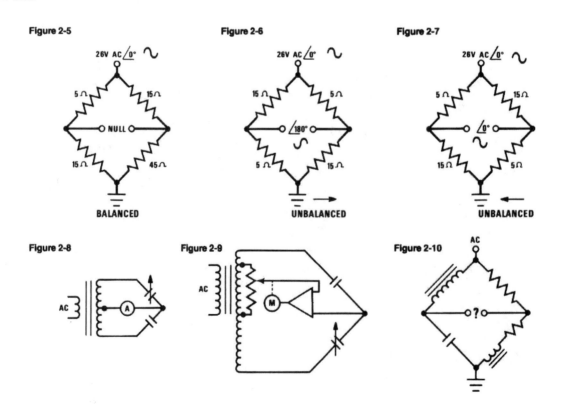

AC Bridges (cont'd.)

of the same amplitude. But since the high voltage is on the left sensing point, the phase is opposite to that in Figure 2-6. These two signals are therefore 180° out of phase. One of the signals is of the same phase as the input signal and the other is opposite in phase to the input signal. Unbalancing the bridge only slightly from one side of null to the other completely reverses the phase of the signal voltage.

The effect of phase reversal in a circuit is to reverse the action in a following circuit. For example, an autopilot moves an aileron up instead of down, a flight director moves a command bar left instead of right, or a compass card moves clockwise instead of counterclockwise.

A null seeking system, discussed later, is one in which a null is sought. Any movement away from this particular null results in a subsequent movement back toward the desired null.

This association of an AC null with a signal phase reversal from one side of the null to the other is very common and very important in avionics circuits. You will see it everywhere.

Figure 2-8 shows an AC bridge with inductances on one side and capacitances on the other. One of the capacitances is variable and, therefore, the bridge can be upset from a null one way or the other by varying its capacity.

Figure 2-9 represents the same circuit with one inductance variable by a potentiometer driven by a two phase servo motor. The other side has a variable capacitance. This is typical of a fuel quantity circuit. The variable capacitor represents a fuel tank element (tubular type capacitor) whose capacity is varied by the amount of fuel in a vertical tubular type capacitor. The phase of the signal causes the motor to drive in the correct direction to the desired null.

Capacitors and inductors develop voltages 180° from each other, so their use in a bridge of this arrangement results in no unusual circumstances.

Figure 2-10 shows a bridge with unsymmetrical components which introduce voltages with phase angles between 0° and 180°.

Without using computations involving vector analysis or j factors, we cannot predict what would be expressed across the sensing points.

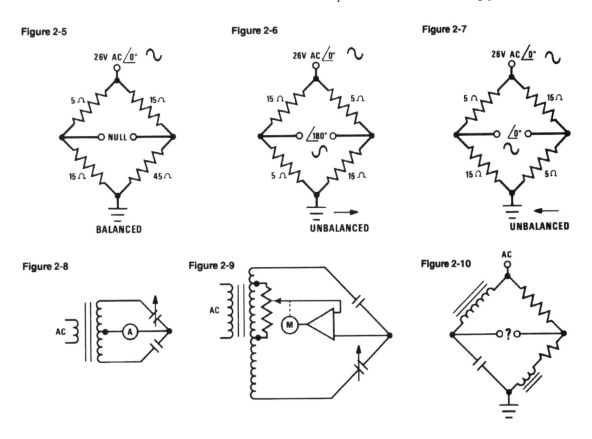

Signal Monitors

Figure 2-11 represents a possible simplified signal presence monitor utilizing a zener diode. Such monitors are used to determine whether signals present are of sufficient strength to be useful in black boxes, such as navigation receivers, autopilots, or flight director computers. They could be monitoring follow-up signals, or any type of signal which must have a certain minimum strength in order to be useful.

If the monitored signal shown in Figure 2-11 does not have sufficient strength to break down the zener, the transistor will not be switched on and the output signal will not appear. Some type of warning or disconnect operation would result.

Sometimes it is desirable, as in null seeking devices, that the monitored signal be kept at a minimum level. If the monitored signal is excessive, it is the result of some type of failure.

In Figure 2-12, if a monitor signal exceeds the preselected value as determined by the zener and its associated components, the transistor switch will be turned on and the indication or disconnect operation of some kind then occurs.

The capacitors ahead of the transistor gates al-low for AC signal monitoring.

Figure 2-11

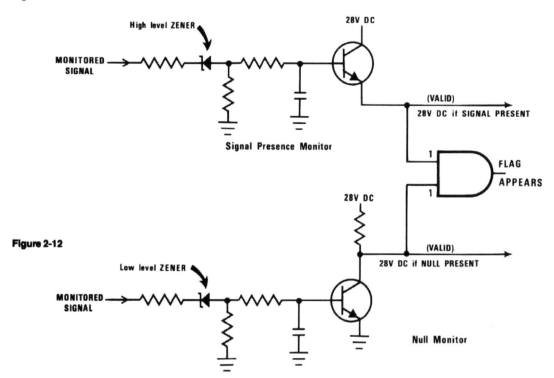

Figure 2-12

Modulator And Demodulator

Sometimes it is desirable to change an AC signal of a certain phase and amplitude into a DC signal of a corresponding amplitude and polarity; and it is sometimes desirable to reverse the process.

Figure 2-13 represents a simplified method of changing a DC signal into an AC signal in that manner. It is called a modulator. In this case the output is a 400 Hertz signal. The polarity of the DC signal determines which one of the transistors will be turned on. Consequently, the 400 Hertz AC signal will use one particular half of the transformer primary. In this manner, the polarity of the signal determines the phase of the output.

In this circuit the transistor is used as an amplifier; therefore, the amplitude of the output is a function of the amplitude of the DC "signal in".

Figure 2-14 shows a demodulator. The transistors in this device serve both as amplifiers and rectifiers.

If the phase of the AC "signal in" is zero degrees, then it is positive at the same time that the lower end of the transformer secondary is positive. The lower transistor is therefore conducting during positive half cycles. The upper transistor does not conduct because, when the upper half of the trans-former secondary is positive, the signal into the upper transistor base is negative. The polarity of the rectified and filtered output is positive on the bottom and negative on the top.

If the phase of the "signal in" is 180°, then the upper transistor conducts during the times that the upper half of the secondary is positive. The polarity of the output signals is then positive on the top and negative on the bottom. The amplitude, in each case, is a function of the amplitude of the "signal in".

These are not real life modulators or demodulators, but they do illustrate the principle of one possible method.

Figure 2-13

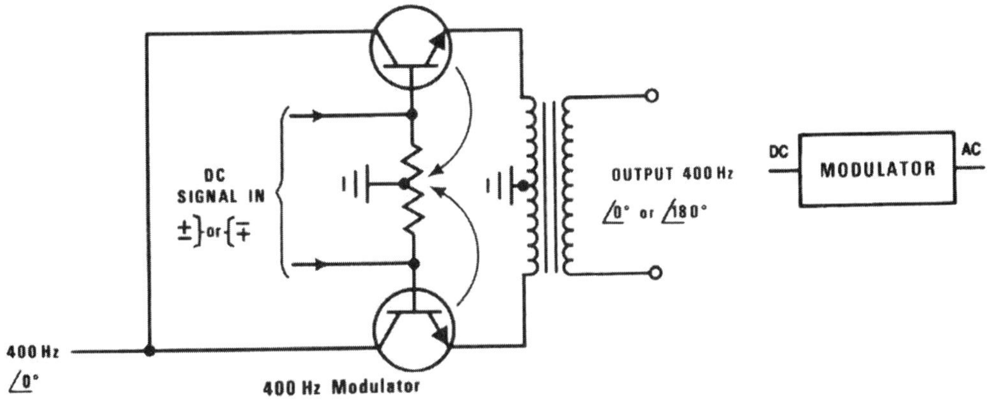

400 Hz Modulator

Figure 2-14

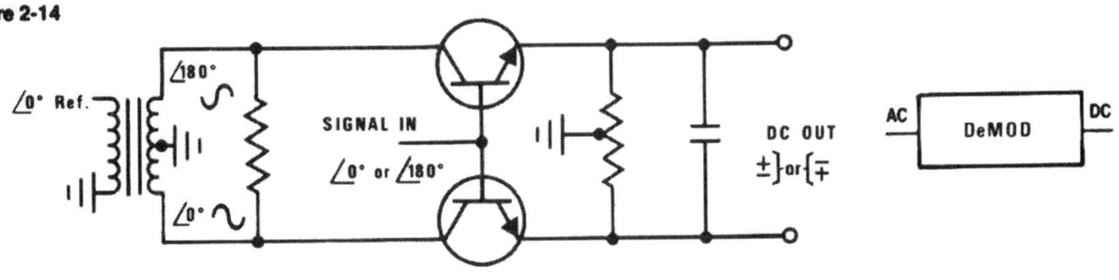

400 Hz Demodulator

Summing Points

Figure 2-15 illustrates the summing point principle. The schematic symbol for a summing point is shown in the center of the illustration.

A summing point is defined as a point into which signals come from as many as three directions, but go out in only one direction. Therefore, no input signal can have any effect on any other input signal.

A summing point might represent any one of the various illustrations around the outside of the symbol, or any combination of them. The important principle is the successful isolation of input signals from each other, even though they have a common output.

Figure 2-15

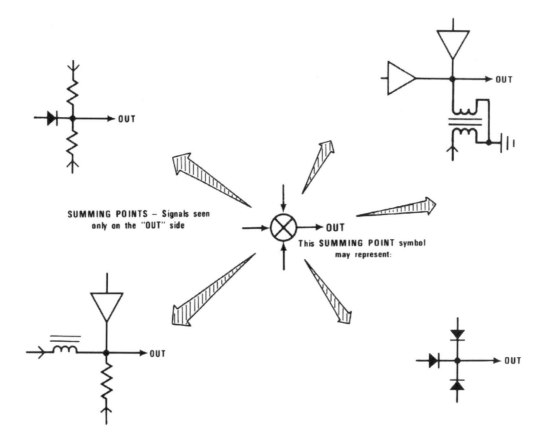

SUMMING POINTS — Signals seen only on the "OUT" side

This SUMMING POINT symbol may represent:

CHAPTER 3 - SYNCHROS

Vectors And Components

This section on vectors has only one objective, which is to show you the concept of vectorial notation of a resultant magnetic field as it applies to synchro devices.

A vector quantity has magnitude and direction. It must have a direction which can actually or arbitrarily be represented by the direction of a line on paper. The length of the line represents the magnitude of the quantity in arbitrary units. For example, electrical quantities are assigned established customary directions in vector notation.

This definition and subsequent examples of how vectors can be added geometrically will help explain the principles of synchro operations, which come later. Magnetic fields do have actual physical magnitude and direction and can be represented vectorially. With the aid of vector symbols and the concept of magnetic fields which they represent, we can visualize what happens to electrical signals as they progress through any type of synchro circuit.

An example of a vector quantity is the velocity of the airplane in Figure 3-1.

By mathematical definition, velocity includes speed and direction. The velocity might be 20 miles per hour in a northwest direction. In this case we can use an arrow at 45° off to the left, as shown, with an arbitrary length representing 20 MPH.

If the plane were going 10 MPH on the same scale, the vector would be only half as long.

To the right of the airplane is a more likely scale showing the same length of arbitrary units. If it were going 160 MPH, the arrow would terminate at the 160 MPH, one-third as long.

The vector has a head and a tail. The head is where the arrow head is, and the tail is the other end.

Any vector quantity can be divided into real or arbitrary components. Figure 3-2 shows a 100 MPH velocity northwest divided into 70.7 MPH west and 70.7 MPH north. This shows that if we are traveling 100 MPH northwest, we are also traveling 70.7 MPH west and 70.7 MPH north. We can do this geometrically by drawing west and north lines, then erecting perpendiculars (dashed) to them from the arrowhead.

Figure 3-3 shows a 100 MPH velocity at a compass heading of 340° divided into west and north components. This shows the magnitude of the west speed and the magnitude of the north speed. These can be measured with a ruler using arbitrary units, or calculated with trigonometry.

In the illustrated case, we are going 34 MPH west and 94 MPH north.

Figure 3-4 shows a 100 MPH velocity with a compass heading of 290°. In this case the western speed is 94 MPH and the northern speed is 34 MPH.

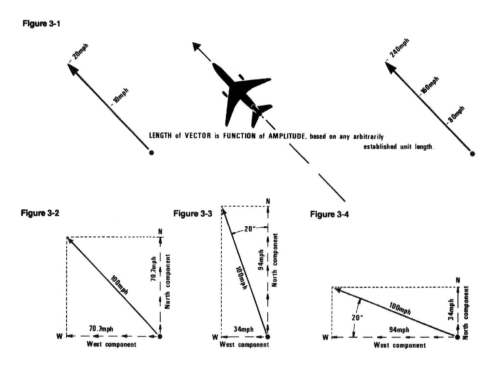

Figure 3-1

Figure 3-2   Figure 3-3   Figure 3-4

Resultant Vectors — Rectangles

Figure 3-5 shows an airplane with a north heading and 200 MPH airspeed. If there were no wind, he would simply be going 200 MPH north. But there is a moving body of air carrying the airplane at 60 MPH west.

Without some kind of calculation we could only guess the direction and speed of travel over the ground. With a simple geometric process using a ruler and protractor we can closely calculate the direction and speed of ground travel. In fact, this is what an airplane pilot can do with his simple little geometric computer that he carries with him. It does the same thing, with moveable sections of transparent ruled plastic that we are doing here on paper.

On paper we lay out a vector pointing north 200 units long. Beginning at the tail of that vector, we lay out another one pointing west 60 units long. Then we complete a parallelogram (in this case also a rectangle) and draw a resultant vector from the tails of the two original vectors to the far corner of the parallelogram.

Measuring the length and direction of this resultant vector gives us a ground speed figure of 209 MPH in a direction of ground travel approximately 342°.

Figure 3-6 is another exercise in adding components. We have the same airplane velocity, but this time the body of air is moving at a speed of 120 MPH west. We lay out the north vector of 200 units and the west vector of 120 units tail to tail. Completing the parallelogram and drawing the resultant from the tails of the two original components reveals a ground speed of 233 MPH heading approximately 326°.

Another method of finding the resultant is to draw one vector and then draw the other vector with its tail at the head of the first. The resultant vector is the third side of the triangle.

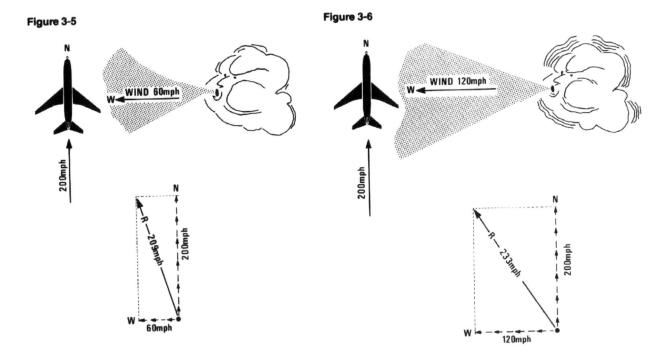

Figure 3-5

Figure 3-6

Resultant Vectors — Parallelograms

To make sure that the vector idea becomes familiar, we have more examples. Figure 3-7 shows our airplane still trying to go north at 200 MPH while the wind is blowing at 100 MPH from the southeast.

Lay out one vector 200 units in a north direction, another vector from its tail of 100 units pointing northwest. Complete the parallelogram and draw the

diagonal. The airplane in this case is traveling 280 MPH with a heading of approximately 345°.

In all of the previous examples the wind has increased airplane speed. In Figure 3-8, a wind is from the northeast at 100 MPH. Lay out the vectors, complete the parallelogram and draw the resultant. This shows the airplane traveling only 150 MPH over the ground in a direction of approximately 328°.

Figure 3-7   **Figure 3-8**

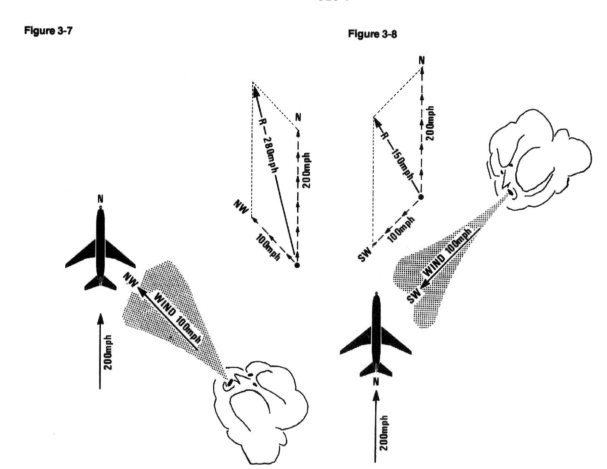

Resultant Magnetic Vectors

In Figure 3-9, we want you to visualize the direction of a magnetic field at a particular point. The fields represented by vectors result from the addition of two fields from two different magnets. The point of consideration is the black dot. It is at the intersection of the centerline extensions of these magnets.

Vector arrows represent the directions of these fields at these particular points. Of course, the fields have other directions at other places, but in these and subsequent drawings, the direction of fields only at the points illustrated is significant.

Figure 3-9

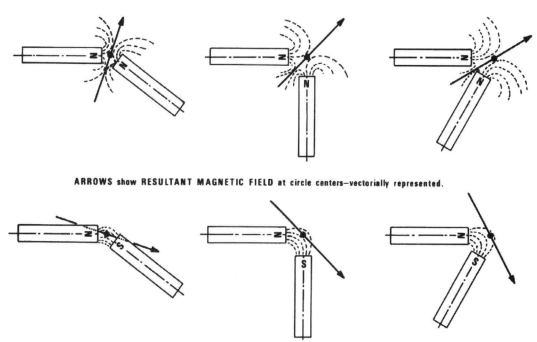

ARROWS show RESULTANT MAGNETIC FIELD at circle centers—vectorially represented.

Two Coil Electromagnetic Field

In Figure 3-10 a permanent magnet, represented by a small arrow with a dot at its center, is pivoted around the dot. The electromagnet on the left has no current and no magnetic field. The one on the bottom has 24 volts applied and a north pole at the top of the magnet. Since only one field is involved, the permanent magnet lines itself up with the single electromagnetic field vector.

In Figure 3-11 the lower magnet has 20 volts applied, and the left magnet, 8 volts. The resultant magnetic field is represented with the vector arrow. The permanent magnet aligns itself with this field.

Figure 3-12 shows the two electromagnets with equal voltages, both generating north poles. The resultant magnetic field is shown by the vector arrow. The permanent magnet aligns itself with this arrow.

In Figure 3-13, the lower electromagnet has no voltage applied. The left one has 24 volts applied with a north pole near the permanent magnet.

The permanent magnet takes up the position shown. We have caused the permanent magnet to rotate through 90° by varying the voltages.

Figure 3-14 shows equal voltages applied to the electromagnets, but the polarity of the lower magnet has been reversed. The resultant electromagnetic field has been rotated 45° away from that shown in Figure 3-13 and the permanent magnet aligns itself with this field.

Figure 3-15 shows the same polarities of voltage applied to the electromagnets, but with different amplitudes. The resulting electromagnetic field has been further shifted clockwise and the permanent magnet has further rotated clockwise.

These illustrations show that by adjusting the voltages and polarities applied to the electromagnets, a resultant field can be positioned in any desired direction.

This concept of the direction of the resultant electromagnetic field is essential to our presentation of synchro operations.

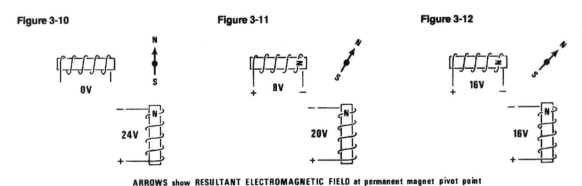

Figure 3-10 **Figure 3-11** **Figure 3-12**

ARROWS show RESULTANT ELECTROMAGNETIC FIELD at permanent magnet pivot point
(Think of MAGNET as a single iron FILING)

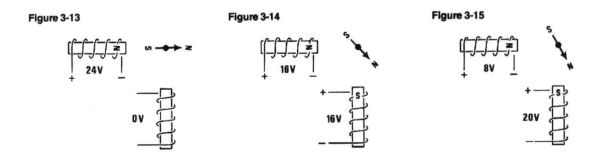

Figure 3-13 **Figure 3-14** **Figure 3-15**

Three Coil Electromagnetic Field

Figures 3-16 through 3-21 present a similar idea, but use three electromagnets.

In Figure 3-16, 6 volts power each of the two lower electromagnets to produce north poles at their tops. Twelve volts power the top electromagnet, producing a south pole at its bottom. Visualize the resultant magnetic field oriented through the center line of the upper electromagnet. The permanent magnet aligns itself as indicated.

In Figure 3-17, the lower right electromagnet has no voltage. Twelve volts are across the lower left electromagnet and 12 volts across the upper electromagnet, with polarities as indicated. You can see the resultant field direction causes the permanent magnet to align itself as shown.

Figure 3-18 shows the lower right electromagnet with 6 volts and the South Pole on top. This condition results in the magnetic field having been rotated 60° from its position in Figure 3-16.

Figure 3-19 removes the voltage from the upper electromagnet, and gives the two lower electromagnets equal power but opposite polarities. This rotates the resultant field clockwise another 30°.

In Figure 3-20, the polarity of the upper electromagnet has been reversed. Again the situation is comparable to Figure 3-19, except that this time the field is rotated 120°.

Figure 3-21 shows the resultant field rotated another 30°.

From these illustrations you should be able to see that by varying the voltages and polarities applied to the electromagnets, we can produce a resultant magnetic field in any desired direction.

The past two pages of illustrations, Figures 3-10 through 3-21, form the beginning of the most useful synchro operations concept. Virtually all synchros are constructed with either two coil fields 90° apart or three coil fields 120° apart.

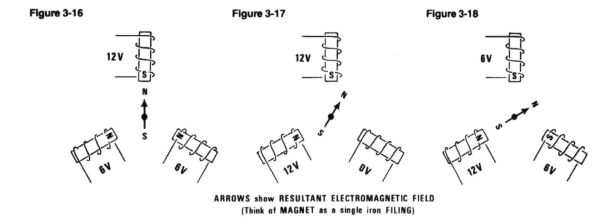

Figure 3-16   **Figure 3-17**   **Figure 3-18**

ARROWS show RESULTANT ELECTROMAGNETIC FIELD
(Think of MAGNET as a single iron FILING)

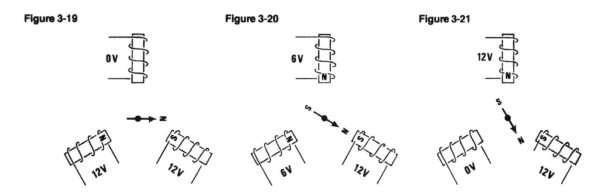

Figure 3-19   **Figure 3-20**   **Figure 3-21**

Synchro Frame Magnetic Schematic

The frame of an ordinary synchro is shown schematically in Figure 3-22. The outer ring with the three pole pieces is not ordinarily movable and is called the stator. The inner bar (with the black dot pivot) is called the rotor.

There are other types of synchro construction, but this is the most common, so we discuss it first.

Permanent magnets are rare in synchros, but are used here to simplify the explanation.

These illustrations are presented to show that by exercising control over the position of the rotor, we can control the direction of the resultant field in the stator.

Figure 3-22

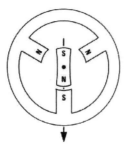 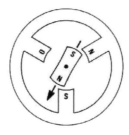 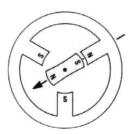

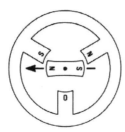 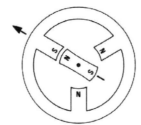 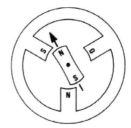

ARROWS show RESULTANT STATOR FIELD— established by rotor

TX And REC Synchro Schematic No. 1

In Figure 3-23 the rotor of the transmit synchro at the top is powered from the circuit breaker through the H connection, and the C connection goes to the ground. Each synchro pole piece winding is connected at one end to a common point. Each of the other winding ends, marked X, Y and Z, is connected to a corresponding point in the receiving synchro stator. These letter labels are typical and common for this type of synchro.

The rotor usually receives 400 Hz power from either a 26 volt AC or 115 volt AC aircraft power source. The upper synchro is labeled "transmitter" because it is there that a signal originates from aircraft power. The significance of a transmit synchro signal is the position of its rotor.

This is a good place to mention a rather common misconception. This device is in no way a three phase device. Only one phase is supplied to the rotor, and the stator windings can never be other than in phase with the supply voltage, or 180° out of phase with the supply voltage. The circumstance of the stator windings being placed 120° apart is the basis for this rather frequent error in understanding.

Previous illustrations demonstrated that a resultant field from electromagnets can be positioned any place throughout 360°, either with two electromagnets at 90° apart or with three electromagnets 120° apart. There is no theoretical difference in the capabilities of the two systems. There are technical and practical considerations which designers take into account, but we are not concerned with these, and from our point of view, either one does the job quite well.

The powered rotor of our transmit synchro functions as the primary of a transformer. The field windings of the stator function as the secondaries. Receiver phases and nulls are discussed in later figures, but are ignored for the moment. A more direct approach, applying to a special type synchro, is used.

Although the field developed by the rotor changes polarity 800 times per second, it can be given a vectorial direction sense related to its phase; for example, placing the vector arrow head down, because when aircraft power is peaked positive, north is down. Reversal of rotor phase would cause north to be up at that instant. The magnetic field of the Z winding is of the same sense.

Continued on next page ....

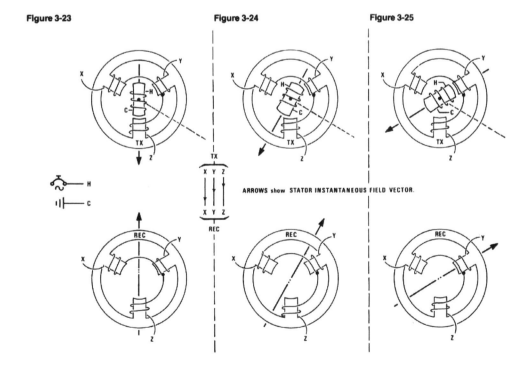

Figure 3-23   Figure 3-24   Figure 3-25

ARROWS show STATOR INSTANTANEOUS FIELD VECTOR.

TX and REC Synchro Schematic No. 1 (cont'd.)

The X and Y windings are at 60° to the vector arrow representing the magnetic field. It is the rotor which establishes the synchro magnetic field. The voltages developed in the X and Y windings are of a smaller magnitude than that in the Z winding because the magnetic coupling is not as great.

If we sum vectorially the three magnetic fields of the windings, they add up in a resultant direction corresponding to the position of the rotor. This is because the strength of the X and Y fields are equal and opposing but sum in the same direction as that of the Z field.

The voltages developed in the transmitter X, Y and Z windings by transformer action cause corresponding current flow in the X, Y and Z windings of the receiver. Since the paired windings are individually connected in series, whatever current is induced in the transmitter Z winding also flows through the Z windings of the receiver. The X and Y circumstances are similar.

Because of the way we are showing the transmitter connected to the receiver, current flow is reversed in windings. Consequently, the instantaneous individual magnetic polarities of the receiver stator pole pieces are the opposite of those in the transmitter. Their magnitudes, however, are the same.

This reversal is represented by a vector pointing "down" in the transmitter and one pointing "up" in the receiver. In Figures 3-26 through 3-31 we recognize this reversal; in later illustrations, however, for convenience and simplicity we will ignore it and show both vectors in the same direction.

Figure 3-24 shows the transmit synchro rotor shifted by 30°. Since it is perpendicular to the X winding there is no transformer action between it and the X winding. Therefore, no voltage is developed in the X winding. The rotor forms the identical angles with both the Y and Z windings. Voltages and currents developed in the Y and Z windings are equal. Their magnetic fields are equal, and when added vectorially, sum along the vector arrow shown (also the direction of the rotor field). The resultant magnetic field developed in the receiver stator is the reverse of that in the transmitter.

Figure 3-25 shows the transmit rotor shifted another 30°, this time lining up with the Y leg. The Y leg currents and magnetic fields are maximum; X and Z fields are equal. When the stator leg fields are added vectorially at the center of the rotor, they align with the rotor.

The sum of the magnetic fields, at the center of the receiver synchro, is also shifted 30° and aligns with the center of the Y leg.

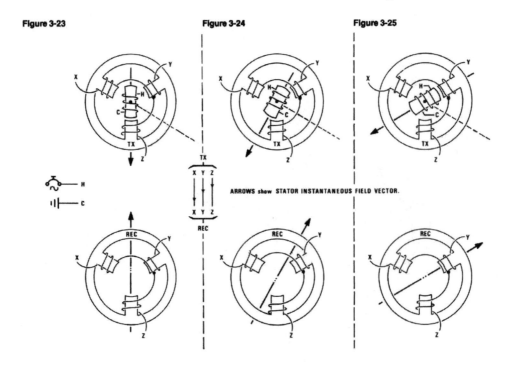

Figure 3-23   Figure 3-24   Figure 3-25

ARROWS show STATOR INSTANTANEOUS FIELD VECTOR.

TX And REC Synchro Schematic No. 2

In Figure 3-26 the transmit rotor has been shifted another 30°, this time perpendicular to the Z leg. Equal voltages induced in the X and Y windings cause equal currents to flow. The magnetic fields developed in their pole pieces when summed at the center of the synchro lie in the line of the vector arrow.

Figure 3-27 shows the transmit synchro rotor lined up with the X leg where it develops the largest voltage and current.

Equal current and magnetic fields are developed on the receiver Y and Z legs. The sum of the Y and Z magnetic fields aligns with the X field. Once again the resultant field in the stator of the receiver is in line with, although directionally opposite to, the resultant field of the transmitter.

Figure 3-28 shows the rotor turned another 30°, with a condition comparable to that of Figures 3-26 and 3-24.

The transmit rotor, the transmit stator fields, and the receiver stator fields have all been shown rotated almost through 180° in steps of 30°. By now you should readily visualize that, regardless of where the transmit synchro rotor is positioned, there is a corresponding resultant field established in the stator of the receiver synchro.

This function of a transmit synchro of establishing, in the stator of a remote receiver synchro, a magnetic field with a position relative to the position of the transmit synchro rotor is the first principle of synchro operations. It is so fundamental that if it is not clear to you, begin with the section on vectors, Figure 3-1, and review until it becomes clear. This concept is not just an analogy; it is valid, demonstrable, and of tremendous usefulness.

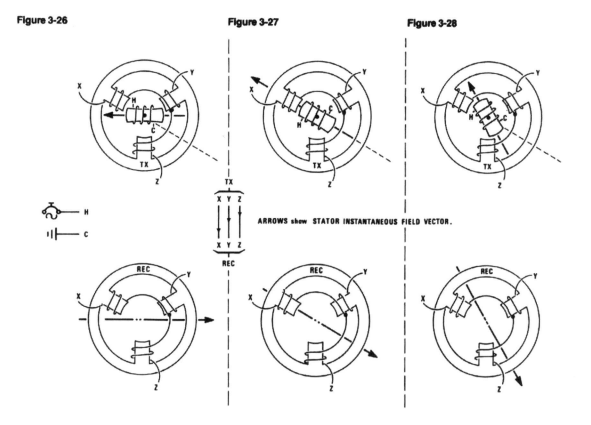

Figure 3-26   Figure 3-27   Figure 3-28

RMI Type TX And REC Synchros No. 1

From Figure 3-29 forward, in order to simplify our presentation and because it causes no real difficulty, we will ignore the actual magnetic reversal of field between the transmitter and receiver. In the transmit synchro at the top, the vector arrow represents both the position of the rotor and the position of the resultant magnetic field of the stator.

In the receiver at the bottom of Figure 3-29, the vector arrow represents the imposed resultant magnetic field in the stator. This receiving synchro is one with a free swinging rotor (usually on jeweled bearings) and typified by a radio magnetic indicator (explained in Chapter 10). It is a dial-type indicator whose rotor is attached to an indicating needle.

Notice that the receiver synchro rotor is powered by the same source which goes to the transmit synchro rotor.

The transmit synchro rotor is mechanically held in a certain position, whereas the receiver synchro rotor is free to turn. Visualize a resultant magnetic field aligned with the Z leg in the receiver synchro stator. The receiver synchro rotor is magnetically powered so it aligns itself with that resultant field.

Figures 3-29, 3-30 and 3-31 are drawn to show how the movement of the transmit synchro rotor controls the position of the free-swinging receiver rotor.

Admittedly, this is not a complete explanation, but it is conceptually valuable and for all practical purposes the truth of the matter. If you force the receiver synchro rotor out of its normal position, the magnetic fields of both receiver and transmitter are displaced. This is, of course, an unnatural situation which would not occur in an operative system.

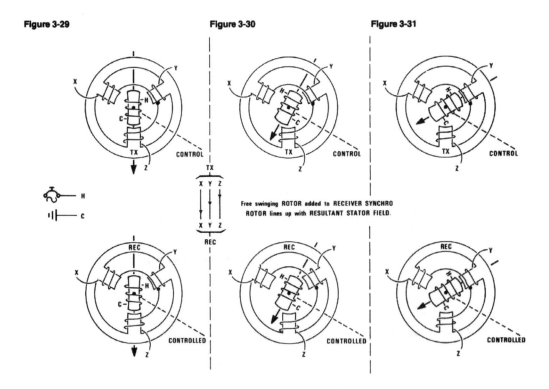

Free swinging ROTOR added to RECEIVER SYNCHRO
ROTOR lines up with RESULTANT STATOR FIELD.

RMI Type TX And REC Synchros No. 2

In Figures 3-32, 3-33 and 3-34 we have further displaced the transmit synchro rotor. The receiver synchro rotor has followed the position of the transmit synchro rotor, and we can remotely read, in an indicator, the position of the transmit synchro rotor.

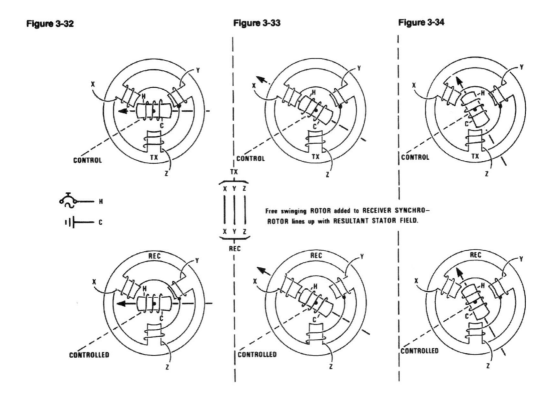

Free swinging ROTOR added to RECEIVER SYNCHRO-ROTOR lines up with RESULTANT STATOR FIELD.

Pivoted Secondary No. 1

In Figures 3-35 through 3-46 we will develop the ideas of a "null" signal and phase reversal. Figure 3-35 shows a fixed position transformer primary and a pivoted separated secondary. In this figure the secondary windings are perpendicular to the primary windings. There is, therefore, no voltage transfer, and we say there is a null signal from the secondary.

The primary winding is shown connected to a voltage source with an arbitrarily designated phase angle 0° (the little sine wave leads off with a positive loop).

In Figure 3-36, the secondary has been rotated clockwise 5°. The two ends of the secondary are differentiated by sectioning one end. In this case there is some net cutting of the windings of the secondary by the magnetic field of the primary.

A two volt potential and a phase angle 0° are assigned to the secondary.

In Figure 3-38 our "100% efficient" transformer secondary is parallel to the primary, and the secondary voltage is therefore 26 volts, and still a phase angle of 0°. Let us say that phase angle 0° in the secondary is the phase which causes a positive voltage to appear on the top wire of the secondary at the same time that a positive voltage appears on the top wire of the primary.

Further rotating of the secondary clockwise as in Figure 3-39 does not change the phase angle, but does diminish the voltage.

Further diminishing of the voltage can be seen in Figure 3-40, without change of phase angle.

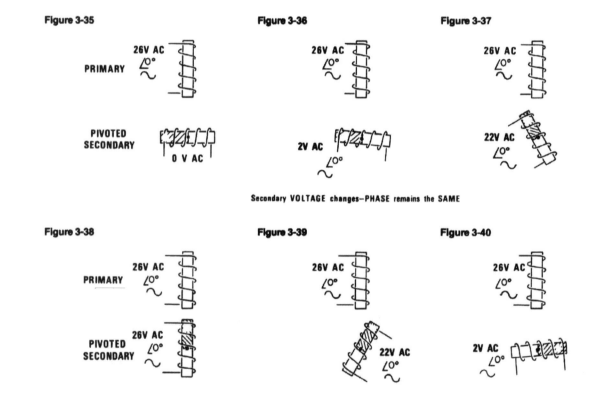

Pivoted Secondary No. 2

In Figure 3-41 the secondary has been rotated until it is perpendicular to the primary. It is now mechanically 180° away from the position shown in Figure 3-35. The shaded portion "A" is on the right side rather than on the left. No voltage is developed in the secondary in this position, and we say it has a "null" signal. Figure 3-35 also shows a null. Its null, however, is distinguished from this null be-cause the secondary has been rotated 180°.

Further clockwise rotation of the secondary as shown in Figure 3-42 develops a small voltage. This voltage is of the same amplitude as in Figure 3-36 because the positions angularly correspond; however, the phase angle has been reversed. It is now labeled phase angle 180°, and the little sine wave is presented with the left half downward. We say that the phase angle has been reversed because the instantaneous polarities of the terminals of the secondary windings are reversed from the corresponding instantaneous polarities of the terminals in Figure 3-36.

An important milestone has been reached here in our presentation of synchro operations because we have shown that when passing through a null condition in these AC circuits, the phase reverses from one side of the null to the other side of the null. This is a fact which you will see not only in synchros but in quite a few other devices throughout this text.

This fact of phase reversal when the signal passes through a null is almost as widely used in aircraft control and indicating systems as are light bulbs in aircraft.

Further rotation of the secondary in Figure 3-43 shows the same phase angle but increased voltage.

Moving the secondary to a position parallel with the primary in Figure 3-44 gives a maximum out-put in this "no-loss" transformer.

Figure 3-38 also has a maximum output. However, it is of the opposite phase. Figures 3-45 and 3-46 show the output voltage reduced without changing the phase angle. Further rotating of the secondary to a position perpendicular to the primary restores the same first null that we saw in Figure 3-35.

The two nulls shown, one in Figure 3-35 and the other in Figure 3-41, are electrically indistinguishable. There is a very important difference, however, that of 180°. Operationally the difference between these two nulls can be discovered by rotating the secondary slightly away from the null and looking at the phase angle developed.

If we rotated the secondary of Figure 3-35 slightly clockwise, we would find a phase angle zero. If we rotated the secondary of Figure 3-41 slightly clockwise, we would find a phase angle of 180°. In operating systems, this is the way in which these two nulls are distinguished.

An important fact can be brought out here. Nulls can be more precisely distinguished than maximums. Since it requires only a very slight movement away from a null position to develop a voltage, it is very easy to discover that precise position. This is in sharp contrast to what happens when at a maximum.

If we move away from a maximum, we cannot distinguish which direction we have moved. Comparing Figure 3-45 with Figure 3-43 illustrates this problem. In each case, we moved away from the maximum. However, the voltage and phase angle are the same for both directions of movement.

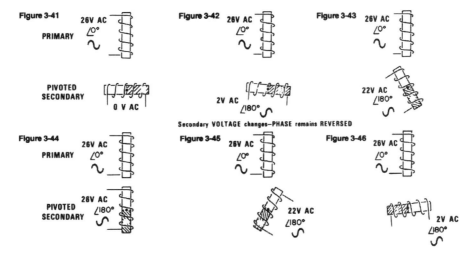

Control Synchro No. 1

Figures 3-47 through 3-52 represent a transmit synchro connected to a control synchro. The control synchro is also sometimes referred to as a control transformer (CT). The control synchro rotor has no voltage input, but instead develops a voltage out-put which is used as a signal input downstream. This arrangement is used in many of the systems described later.

In all of the Figures 3-47 through 3-52, the transmit synchro rotor remains in a single position. Consequently, the resultant field developed in the stator of the receiver synchro remains fixed in all of these figures. Figure 3-47 shows the control synchro rotor displaced by a small amount from perpendicular to the stator field. A small signal voltage is therefore developed to which we have arbitrarily assigned the phase angle 0°.

In Figure 3-48 the control synchro rotor is parallel to the magnetic field of the stator. The signal developed is therefore the maximum signal possible. The phase angle has not changed because the shaded "A" end of the rotor is higher than the other end.

In Figure 3-49 the rotor is again nearly, but not quite, perpendicular to the field. The voltage developed is therefore small. The phase angle and the voltages developed in Figures 3-47 and 3-49 are the same, and the signal user would not be able to distinguish between those two positions of the rotor. In fact, all of the positions of the rotor between the two nulls are ambiguous because each has a corresponding position on the other side of

maximum where the phase angle and voltage is the same.

Even at the maximum position we could not be sure that it was maximum unless we moved the rotor to see whether a higher voltage could be developed. Voltage differences between the maximum position and a position one degree to either side of maximum would be very difficult to distinguish because the voltage change is so small.

If we rotate the rotor of Figure 3-47 slightly counterclockwise, we come to a position where the rotor windings are perpendicular to the field of the stator and the signal developed is therefore a null. Correspondingly, if we rotate the rotor of Figure 3-49 slightly clockwise, we come to a null position. These are two different nulls and the user of the signal can easily distinguish between them be-cause a very small movement (1/4° or less) away from the null develops a discernible voltage whose phase can be discriminated.

For example, if we have the control synchro at a null and do not know which null it is, we move the rotor very slightly — let us say clockwise. If the phase angle is detected to be zero degrees we know that the null is the one almost reached in Figure 3-47.

If the detected phase is 180°, then we know that the null is the one approximated in Figure 3-50.

In practice, sensitivity of 1/4° is not unusual; and in expensive precision equipment, such as inertial navigation systems, sensitivity is much greater.

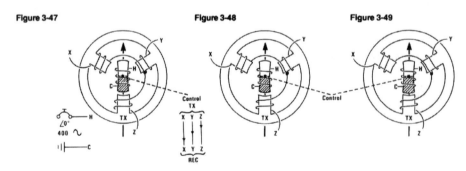

Rec. Rotor VOLTAGE and PHASE depend upon its position relative to RESULTANT STATOR FIELD.

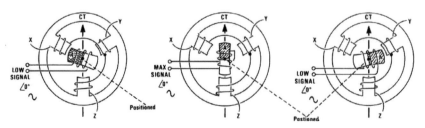

Control Synchro No. 2

In Figure 3-50 the rotor has been moved through a null from its position in Figure 3-49. The phase of the signal has therefore reversed; and the shaded "A" end of the rotor is in the opposite position to that of Figure 3-47.

Figure 3-51 shows the rotor with the shaded "A" end at the bottom. This is the other maximum position, and phase angle is opposite to that of Figure 3-48.

In Figure 3-52 the rotor has been turned through almost 360° and is approaching the original null approximated in Figure 3-47. The phase angle is still 180° in Figure 3-52.

If we turn the rotor through the null clockwise to the position shown in Figure 3-50, the phase angle will once again be reversed and become phase angle 0°.

In null sensing or null seeking systems, a particular one of the nulls is sought. The phase angle detected is the key to whether some motion should be made laterally left or right, circularly clockwise or counterclockwise, or vertically up or down in order to obtain a desired reading, a desired position, a desired action, or a desired indication. This principle of desired null detection is used in many systems covered later.

| Figure 3-50 | Figure 3-51 | Figure 3-52 |

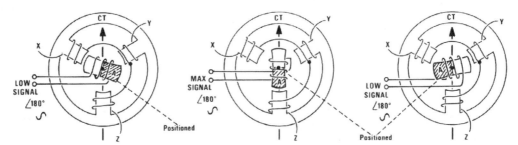

Rec. Rotor VOLTAGE and PHASE depend upon its position relative to RESULTANT STATOR FIELD.

Control Synchro No. 3

Figures 3-53, 3-54 and 3-55 illustrate what happens if we leave the control synchro rotor stationary and move the transmit synchro rotor. These figures show the transmit synchro starting out at about 7 o'clock in Figure 3-53. The resultant field in the stator of the control synchro is in a corresponding position, developing a low level signal of phase angle 0° in its rotor.

In Figure 3-54 the transmit synchro rotor has been moved counterclockwise to 6 o'clock. The control synchro rotor is at a null position with respect to the stator field.

Figure 3-55 shows the transmit synchro rotor moved to about 5 o'clock, developing a low voltage signal of phase angle 180° in the rotor of the control synchro.

Figure 3-53   **Figure 3-54**   **Figure 3-55**

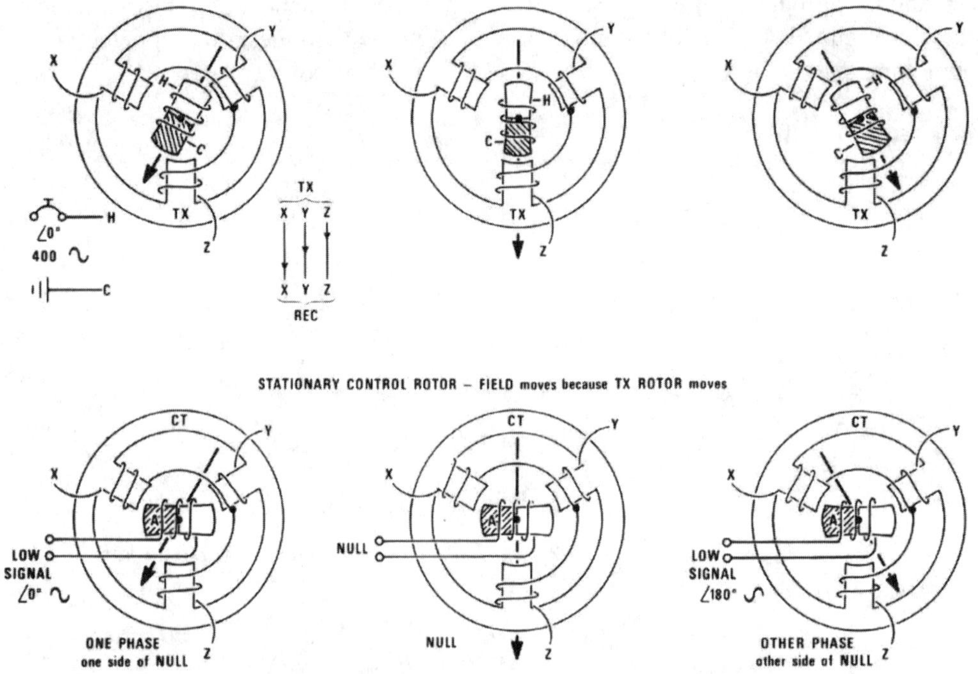

STATIONARY CONTROL ROTOR — FIELD moves because TX ROTOR moves

Control Synchro No. 4

Figures 3-56, 3-57 and 3-58 show the control synchro rotor stationary. The transmit synchro rotor in Figure 3-56 is at 9 o'clock, in Figure 3-57 has been moved to 6 o'clock, and in Figure 3-58 has been moved to 3 o'clock. In the control synchro rotor this develops a maximum signal of phase angle 0° in Figure 3-56 a null in Figure

3-57 and a maximum signal of phase angle 180° in Figure 3-58.

Figures 3-50 through 3-58 show two different ways of changing the signal in the rotor of the control synchro. Both ways are used in actual systems.

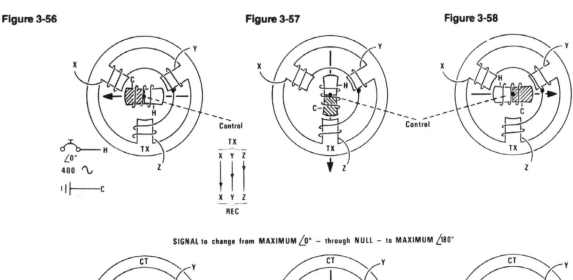

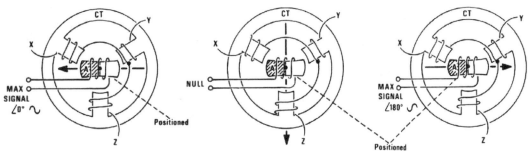

SIGNAL to change from MAXIMUM ∠0° – through NULL – to MAXIMUM ∠180°

Synchro Symbol Equivalents

Figures 3-59, 3-60 and 3-61 show schematic symbol equivalents of the synchro frame that we have been using. The coil-type symbols in the middle of each figure are particularly useful when showing rotor position, or if a change in rotor position is desired to be shown in the schematic. The more compact and simpler circular symbols are used where there is no intention to represent rotor position.

These symbols can be used for any three coil stator/single coil rotor synchro, such as a control synchro. We will use the coil-type symbol because we frequently assign a meaning to the rotor position.

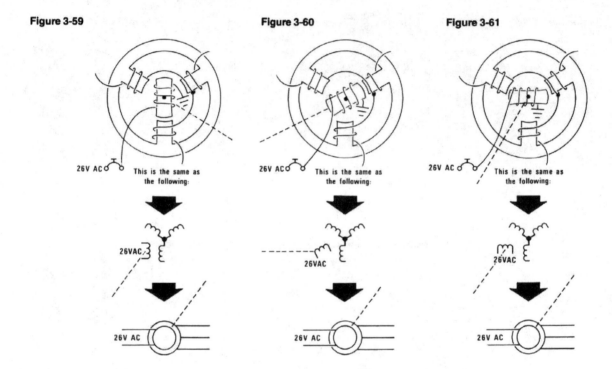

Field Vector Follows Rotor

The figures in 3-62 show our symbol for a three coil stator/single coil rotor transmit synchro, with the rotor in a variety of positions. The vector arrows show a resultant magnetic field being generated in the stator with a direction corresponding to a position of the synchro rotor. The head of the arrow here is related to side "A" of the rotor.

The intention is to show how you should think of these synchros when you see them in a schematic so that you can follow the action through the synchro system. In some schematics, the illustrated position of a synchro rotor has a valid reference to a null or mechanical position situation existing in the synchro system. In other schematics, the rotor position may have no particular significance.

The concept of the field following the rotor can be used by you in any synchro circuit if you actually or mentally assign a field magnetic position in the stator and observe how that field is duplicated in the receiver synchro stator.

This type of synchro usually gives us the position of a particular mechanical device throughout 360° of rotation. For example, position information from a rotating ADF loop, roll and pitch information from a vertical gyro, or azimuth information from a directional gyro.

Figure 3-62

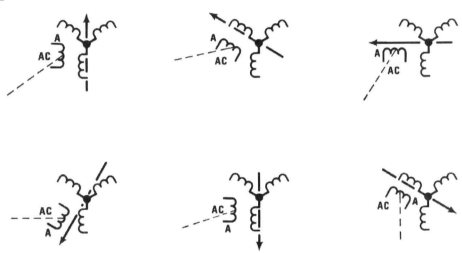

When TRANSMIT SYNCHRO ROTOR is powered – VECTOR DIRECTION in stator field will CORRESPOND to ROTOR POSITION

Control Synchro Signal No. 1

Figures 3-63 through 3-70 show a transmit synchro connected to a control synchro. We have labelled them TX and REC, but in most schematics they will not be so noted. You can identify the transmit synchro by the fact that its rotor has AC power connected. The control synchro rotor passes its signal along to an amplifier or other circuit component using that signal.

Figure 3-63 shows a transmit synchro causing a maximum signal of phase angle 0° to appear in the rotor of the control synchro.

Moving the transmit synchro rotor clockwise (Figure 3-64) rotates the resultant field in the stator of the control synchro, diminishing the amplitude of the control synchro rotor signal.

Figure 3-65 shows the transmit synchro rotor having been moved through 90°. The signal in the control synchro rotor is at one of the two nulls.

Further rotation of the transmit synchro rotor in Figure 3-66 causes a medium amplitude signal of phase angle 180° to appear in the rotor of the control synchro (phase reversal through null).

Figure 3-63

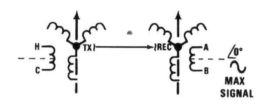

Figure 3-64

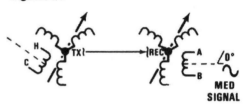

Figure 3-65

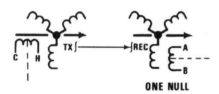

Figure 3-66

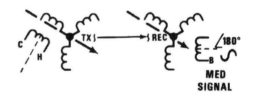

Control Synchro Signal No. 2

Figure 3-67 shows the transmit synchro rotor having been rotated 180° from its position in Figure 3-66, which results in a maximum control synchro rotor signal of phase angle 180°, as compared with the maximum signal of phase angle 0° in Figure 3-63.

Figure 3-68 shows the transmit rotor further shifted clockwise, and therefore a diminished signal in the rotor of the control synchro, but still phase angle 180°.

Figure 3-69 shows the transmit rotor moved so that the field in the stator of the control synchro is perpendicular to its rotor. The rotor now sees the other null from the one shown in Figure 3-65.

Further rotation of the transmit rotor, Figure 3-70, causes a reversed phase to appear in the control synchro rotor; another illustration of phase reversal resulting when a signal passes through a null.

Figure 3-67

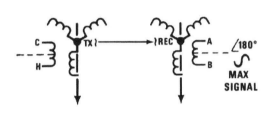

Figure 3-68

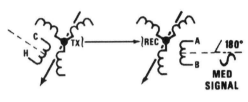

Figure 3-69

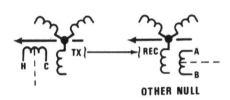

Figure 3-70

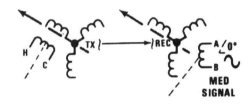

Null Seeking Control Synchro

Figure 3-71 shows a control synchro rotor in a null position. When it is in this desired position, it can tell us remotely the position of some other device to which a transmit synchro is attached. We find this desired null position, because if the rotor is not there (Figures 3-72 and 3-73), its voltage will be of one phase or the other. If it is phase angle 0°, the rotor should be moved clockwise until it reaches a null. If the phase angle is 180°, the rotor should be moved counterclockwise until it reaches the desired null.

Figure 3-74 shows that the mechanical device to which the transmit synchro is attached has moved by 90°. That motion has been followed by the control synchro in observing the principle that if the signal phase angle is 180°, the rotor should be turned counterclockwise. So the field moved counterclockwise, and the rotor followed.

When a control synchro rotor is driven by a servo motor, the movement of the transmit synchro rotor is closely followed as a small voltage is developed in the control synchro rotor. This is why, in an operative system, only small voltages are typically generated in the rotor. In fact, in critical systems, a voltage larger than a predetermined minimum causes an alarm signal.

Figure 3-75 shows the rotor turned too far counterclockwise, developing a phase angle 0° signal. In a servo system this phase would drive the rotor clockwise. The phase angle 180° signal in Figure 3-76 would drive the rotor counterclockwise.

Figure 3-71   **Figure 3-72**   **Figure 3-73**

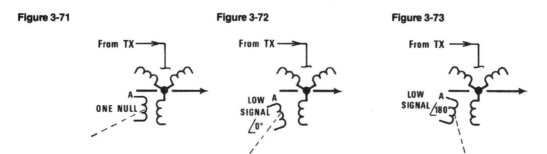

FUNCTION of CONTROL SYNCHRO usually is to detect a PARTICULAR NULL – SIGNAL PHASE also shows direction toward that particular NULL.

Figure 3-74   **Figure 3-75** 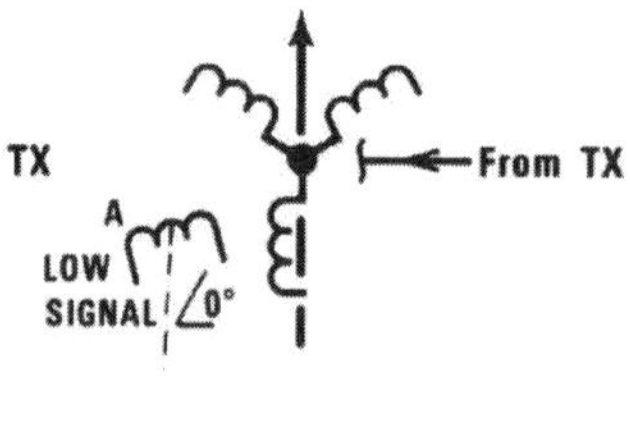   **Figure 3-76**

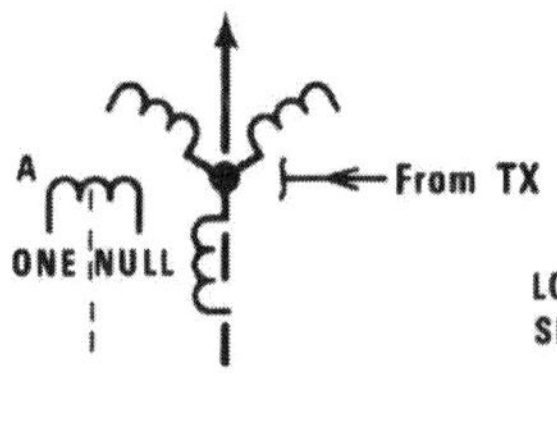

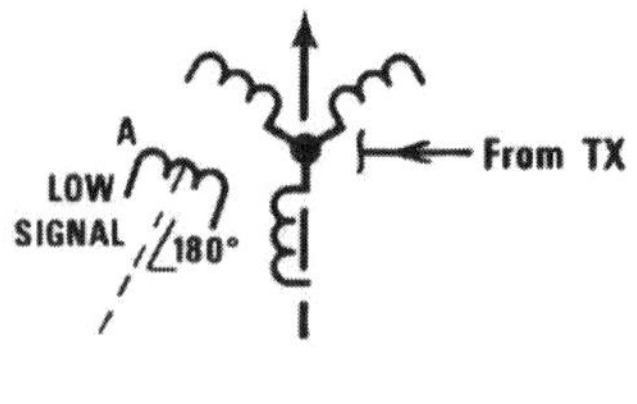

Combination Control Synchro

The synchro shown in Figures 3-77, 3-78 and 3-79 is a special application of a three coil stator/single coil rotor synchro (used in a flight director system). It is presented here to provide practice in visualizing voltage and phase angle relationships. Since it is more difficult than most, it makes a good example.

In normal use the rotor is not moved more than about 20° maximum away from the position shown in Figure 3-78. So the positions in Figures 3-77 and 3-79 are extreme. The voltages at X and Z are referenced to ground; X is a follow-up voltage to the servo which drives the rotor, and Z is a signal presence monitor voltage.

In Figure 3-78 the voltages developed at Y and X are equal, and are of the same phase. The voltage appearing at X is ground potential in Figure 3-78. Ground potential is a null.

Another way of looking at it is to consider instantaneous DC potentials. Consider the voltage generated in the Y leg as 10 volts positive at the center. The voltage generated in the X leg is the same since the angle formed with the rotor is the same. (Remember that the rotor actually sits in the center of these three coil stator windings.) Since the DC drops across the X and Y legs are the same, the voltage at the X terminal must, therefore, be ground potential.

Moving the rotor away from the position shown in Figure 3-78 causes a voltage to appear at X which is not ground potential, and which is of one phase or the other (passing through a null).

In Figure 3-77 we have moved the rotor (beyond its normal limits) and assigned phase angle 180° to the voltage appearing at X. Converting to instantaneous DC potentials, if we say that the DC voltage at the center of the stator generated across leg Y equals a positive 20 volts, the same voltage must appear at X also, since the rotor is perpendicular to the X leg and generating no voltage across it.

In Figure 3-79 the rotor has been moved so that it is now perpendicular to the Y leg. Since no voltage is generated in the Y leg, the center of the stator must be ground potential. At the same reference instant as previously considered, the voltage generated across the X leg is a positive 20 volts with the positive end at the center of the stator. Since the center of the stator is now at ground potential, the X point must be a negative 20 volts. It therefore is of the opposite phase from that illustrated in Figure 3-77.

To sum it up, the X voltage will be of one phase if the rotor is moved counterclockwise from its null position in Figure 3-78, and of the opposite phase if it is moved clockwise from that position.

Figure 3-77 shows the voltage present at Z to be a maximum. This is true because the voltages across Y and Z legs are added at point Z.

In Figure 3-78 the voltage at Z is high, but not as high as in Figure 3-77.

In Figure 3-79 the voltage at Z is lower, but still substantial. It is in fact .866 times the maximum across an individual leg. Therefore the voltage at Z, in normal use, is always high enough to indicate that a signal is present.

Figure 3-77   **Figure 3-78**   **Figure 3-79**

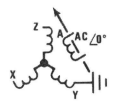

SIGNAL at 'Z' is MAXIMUM – SIGNAL at 'X' is PHASE ANGLE 180° (Same as top of 'Y' COIL)

SIGNAL at 'Z' is HIGH – SIGNAL at 'X' is NULL (Equals SIGNAL at 'Y')

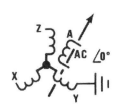

SIGNAL at 'Z' is LOWER – SIGNAL at 'X' is PHASE ANGLE 0° (Has passed thru NULL)

SYNCHRO used to DEVELOP FOLLOW-UP SIGNAL on 'X' LEG, and a SIGNAL PRESENCE MONITOR VOLTAGE on 'Z' LEG.

All VOLTAGES referenced to GROUND.

Differential Synchro

Figures 3-80, 3-81 and 3-82 illustrate a differential synchro. The top simplified illustrations of stator frames are the same as you have seen in past discussions. The rotor is different because it is a three-legged rotor with three coil windings. The bottom illustrations are the schematic symbols for differential synchros. The stator and rotor are represented by the same coil type symbol. The rotor can be identified by the dashed line from its center to its mechanical connection source. Dashed lines are commonly used in schematics to indicate mechanical connections.

If a stator is receiving a signal from the transmit synchro, the stator establishes the resultant magnetic field of the synchro. The rotor is then the transformer secondary, and the signals it develops are transmitted in turn to a receiving synchro somewhere else in the circuit.

The rotor could be receiving the transmit synchro signal, in which case the rotor would establish the magnetic field of the synchro and the stator would be a secondary, transmitting its signal to a receiving synchro. As far as we are concerned, it would not make any difference.

In Figure 3-80, if we say that the rotor position shown represents the portion where the rotor X leg is lined up with the stator X leg, and so on, then so far as the receiving synchro is concerned there might as well not be a differential synchro in the circuit.

But if we move the synchro rotor (Figure 3-81), then the effect on the receiving synchro is the same as if we had held the differential synchro rotor still (Figure 3-80), and had instead moved the transmit synchro rotor a corresponding direction and distance.

Figure 3-82 shows the synchro rotor turned 120° from its first position. The receiving synchro has no way of knowing whether it was the differential synchro rotor that moved 120° or the transmit synchro rotor that moved 120°.

Since there is only one resultant magnetic field in the synchro, the schematic symbols in these figures would be misleading if we did not understand them. The stator and the rotor have the same mechanical center, but are separated in the schematic symbol. If we show a resultant field arrow, we must show it twice, once for the rotor and once for the stator; and, of course, these arrows must be parallel.

The differential synchro has the capability of changing the received position of a transmit synchro rotor by any amount up to 360°. These synchros are used where such a function is useful, as in some later compass systems, or where it is desired to combine two mechanical positions into one signal.

An example is a follow-up signal in an autopilot which combines elevator position and stabilizer position information.

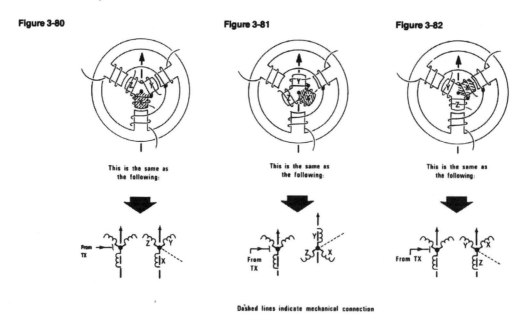

Figure 3-80   **Figure 3-81**   **Figure 3-82**

This is the same as the following:

Dashed lines indicate mechanical connection

Either ROTOR or STATOR can receive a SIGNAL from a TRANSMIT SYNCHRO — The other (ROTOR or STATOR) will then be a SECONDARY and forward a SIGNAL to the next device in the circuit.

Resolver

Figures 3-83 and 3-84 illustrate the simplified framework and the schematic symbol equivalent for a resolver. Understanding this versatile synchro in any particular application is not difficult when we use the resultant magnetic field concept. The four pole windings are paired off so that they constitute two separate stator fields perpendicular to each other. Therefore, we show them schematically as stators, physically at 90° to each other. The rotors are also actually and schematically 90° to each other.

As demonstrated earlier, we can generate a resultant magnetic field of any position throughout 360° by varying the amplitudes and polarities of the voltages of the two fields. There is only one resultant field in a resolver, established either in the stator or in the rotor. Whichever one is receiving a signal from the outside is the primary; the other is the secondary. It is useful to say that either the rotor or the stator establishes the field and the other looks at it.

Figure 3-83   **Figure 3-84**

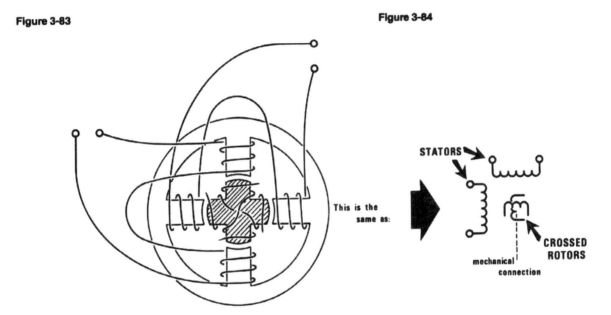

Transmit And Receiver Resolvers

Figures 3-85 and 3-86 show a transmit resolver connected to a receiving resolver. The transmit resolver has one unused rotor winding, shorted out so that it does not affect the rest of the resolver. AC power is connected across the other rotor. This powered rotor establishes the direction of the magnetic field in the stator. Since the transmit stator is connected in parallel to the receiver stator, the direction of the field in the transmitter is duplicated in the field of the receiver.

The receiver resolver rotors see the magnetic field from two different positions, since there are two rotors 90° apart. The ability to develop receiver signals 90° apart is useful, for example, in VOR systems, ADF systems using a non-rotating loop, in autopilots, and in flight directors. The information from one receiver rotor can be used to find a null position. When that rotor is at a null, the other is developing a maximum signal.

Figure 3-85   **Figure 3-86**

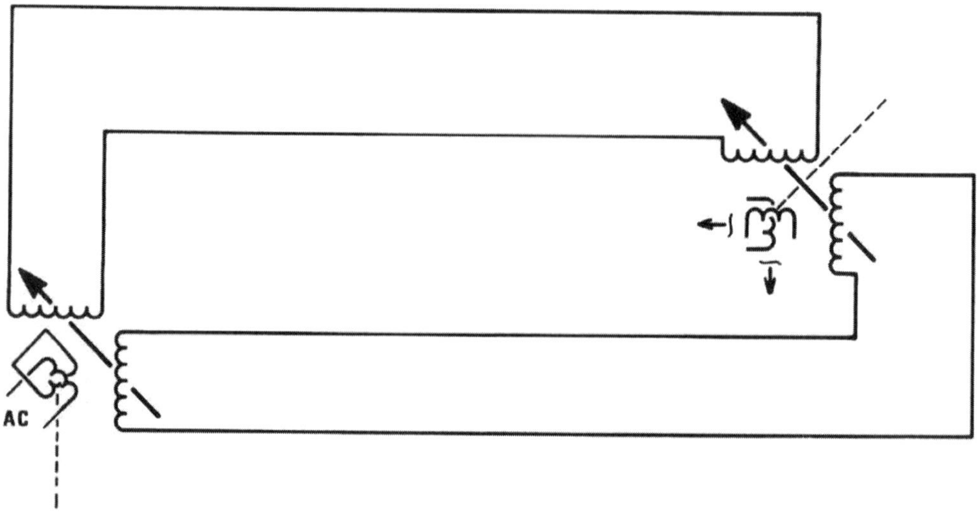

TRANSMITTER RESOLVER

STATOR VECTOR follows
powered ROTOR

RECEIVER RESOLVER

STATOR VECTOR position duplicated
from TRANSMITTER STATOR. The two
ROTORS see the field from different positions.

Resolver Connections

Figure 3-87 shows a resolver whose rotors are receiving an incoming signal, and Figure 3-88 shows a resolver whose stators are receiving an incoming signal. So far as we are concerned, there is no difference in their operation. A resolver duplicates, in its secondary windings, the information present in the primary windings.

Resolvers can function in two-legged circuits the same as a differential synchro functions in three-legged circuits.

The resolver connected as in Figure 3-89 usually gives information as to when the rotor has reached a particular null.

The resolver symbol shown in Figure 3-90 is functionally the same as in Figure 3-87. It does not matter schematically whether two moving rotors look at a fixed magnetic field, or whether two fixed stators look at a moving magnetic field.

Figure 3-87

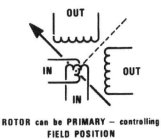

ROTOR can be PRIMARY — controlling
FIELD POSITION

Figure 3-88

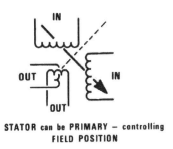

STATOR can be PRIMARY — controlling
FIELD POSITION

(RESOLVERS could function similar to DIFFERENTIAL SYNCHROS)

Figure 3-89

Figure 3-90

Sine And Cosine Signals

The intent of this section is to familiarize you with sine and cosine synchro signals. "Sine" and "cosine" are trigonometric terms for functions of angles. We will not attempt to teach trigonometry, but to explain terminology and concepts that will be useful to you.

Sines and cosines are referenced to and can be considered the property of angles. Any angle found anywhere, regardless of the length of its sides, has sine and cosine values depending only upon the value of the angular spread between the sides of the angle.

In order to illustrate sines and cosines, a right angle triangle (one with a 90° angle) is always constructed. Since sines and cosines are ratios (fractions expressed as decimals), the actual size of the triangle is of no consequence. The ratio is one side of the triangle divided by the hypotenuse (the side opposite the 90° angle), and that ratio is constant regardless of the size of the triangle.

To be specific, the sine of an angle is the side opposite that angle in a right triangle divided by the hypotenuse. In Figure 3-91, the side opposite 30°, when divided by the hypotenuse, is the numerical value of the sine, if you value the hypotenuse at 1 (unity). If you measure the side opposite 30° in Figure 3-91, you will find it is r/2 the hypotenuse. The sine of 30° is therefore 0.5.

Similarly, the cosine of 30° (Figure 3-91) is the side adjacent to the 30° angle divided by the hypotenuse, if the hypotenuse is valued at 1 (unity). Measuring that side will show the cosine of 30° to be equal to 0.866 of the hypotenuse. The cosine of 30° is therefore 0.866.

Figure 3-92 illustrates the sine and cosine of the angle 45°. Measuring these sides in units equal to the length of the hypotenuse will show that both the sine and the cosine equal 0.707.

Measuring the sine and cosine sides of the triangle in Figure 3-93 in a similar manner will show the sine ratio to be 0.866 and the cosine ratio to be 0.5.

As the size of the angle increases toward 90°, the sine ratio increases toward 1.000 and the cosine ratio de-creases toward 0.000. From here we go to Resolvers.

Resolvers have sine and cosine windings which are so named because the voltage values developed are proportional to the sine and cosine values of the angle of rotor position as measured counterclockwise from parallel to the cosine winding. For example, in Figure 3-91 the voltage developed in the sine winding is 0.5 and the voltage developed in the cosine winding is 0.866 if the maximum voltage (rotor parallel to stator winding) is 1.000.

These are the respective values of the sine and cosine of the angle 30°, the angle through which the rotor moved away from parallel to the cosine winding.

These actual voltage amplitudes are not usually important to us, but the general idea of how the voltage is changed in these windings is important.

Bear in mind that every time a voltage passes through a null the phase changes. We see that there is a phase angle change for every 90° of rotation of the rotor in either sine or cosine winding.

If we assign plus values to the phase angles in the sine and cosine windings during the first 90° of rotation (the first quadrant), then the signs of these values are trigonometrically valid through the other three quadrants as well.

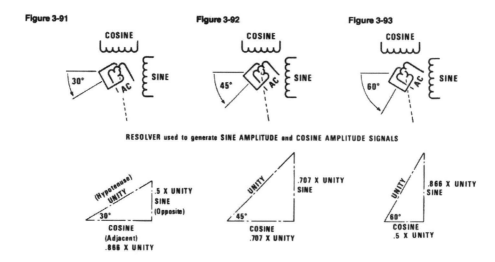

Figure 3-91   Figure 3-92   Figure 3-93

RESOLVER used to generate SINE AMPLITUDE and COSINE AMPLITUDE SIGNALS

Phase Shifting

Figures 3-94, 3-95, 3-96 and 3-97 show how the combination of two voltages, one of phase angle 0°, and one of phase angle 90°, can be varied in relative amplitude so that the result of their combination is a voltage of any desired phase angle between 0° and 90°

If the phase angle 0° voltage is maximum and the phase angle 90° voltage is zero, the result of adding the two, of course, is phase angle 0°. Adding a little phase angle 90° voltage to the 0° voltage will shift the resultant toward 90° a little way. Further increasing the amplitude of the 0° voltage causes the resultant to be shifted farther towards 90°.

Continuing this until the 0° voltage is quite large will shift the resultant voltage nearly all the way to 90°. When the amplitude of the phase angle 0° voltage becomes zero and the 90° voltage be-comes maximum, then, of course, the resultant is a voltage of 90°.

The amplitudes of the 0° and 90° voltages in these four figures are adjusted in such a manner that the resultant is always the same amplitude. These amplitudes of the phase angle 0° and phase angle 90° voltages are adjusted in accordance with the cosine and sine values respectively, beginning with the angle of 0° (not shown).

At angle 0° the cosine value would be maximum and the sine value would be zero.

In Figure 3-94 the amplitude of phase angle zero voltage is the cosine value of the angle 22.5°, and

the amplitude of the phase angle 90° voltage is the value of the sine of the angle 22.5. Looking at the instantaneous values of the two input voltages at the point marked 22.5, you can see that their addition results in the peak value shown on the heavy resultant wave form. The phase angle of the resultant wave form is therefore 22.5°.

In Figure 3-95 the sine and cosine values of angle 45° are given to the two input voltages. In this case they are equal. You can see that at the point marked 45° in the graph, the sum of the instantaneous values of the two input voltages is the graph value assigned to the heavy resultant wave form, with a peak at 45°. The phase angle, there-fore, of the resultant wave form is 45°.

In Figure 3-96 the input voltages have been assigned the peak amplitude values of the sine and cosine values of the angle 67.5°, and the instantaneous values of the additions of these two input voltages peaks out at 67.5°.

In Figure 3-97 the 0° input voltage has been diminished to zero, and the resultant wave form is a voltage of phase angle 90°.

These graphs should make it possible for you to visualize how we can achieve a voltage of any desired phase angle between 0° and 90° by adding two voltages, phase angle 0° and phase angle 90° respectfully. This is the basic principle of operation of a phase shifter (next pages).

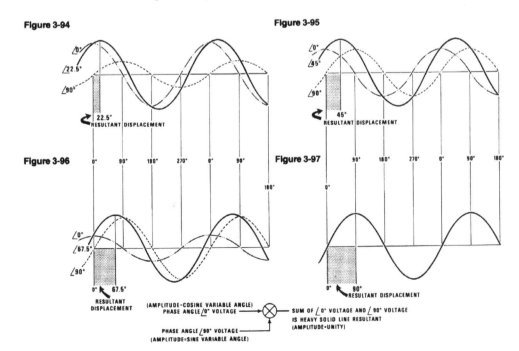

Phase Shifter No. 1

Figures 3-98 through 3-114 show how a phase shifter can shift the phase of any given voltage throughout the range of 360°. Mechanical phase shifters such as this are used in VOR systems, and fixed loop ADF systems. Understanding this phase shifter will make VOR and ADF systems much less mysterious. We use a resolver with one rotor powered and have omitted drawing the other unpowered rotor. The stators are marked "A" and "B" for reference convenience.

The AC power input to the rotor is assigned the phase angle of 0°. The rotor position in Figure 3-98 is assigned the physical angle 0°. Stator A is the sine winding and has no output. Stator B is the cosine winding and has a maximum output. Stator B is connected in such a manner that its output has a phase angle of 0°, the same as the rotor. Therefore the output of the summing point in Figure 3-98 is also of phase angle 0°.

In Figure 3-99 the rotor has moved 22.5° clockwise, and stator A output of phase angle 0° is shifted to phase angle 90° by the condenser. At the summing point, therefore, we have a small voltage of phase angle 90° combined with a large voltage of

phase angle 0°, and the output of the summing point is a voltage of phase angle 22.5°. This is the situation illustrated in Figure 3-94.

Figure 3-100 shows the rotor at 45°. The amplitudes of the stator voltages are therefore equal and, added at the summing point, give a resultant with a phase angle 45° (Figure 3-95).

The rotor position in Figure 3-101 is 67.5°, the phase angle 0° voltage is small and the phase angle 90° voltage is large. The resultant from the summing point is a voltage of phase angle 67.5 (Figure 3-96).

In Figure 3-102 the rotor has been moved clockwise 90° from its original position. Stator A output is maximum and the stator B output is null. The output of the summing point is a voltage of phase angle 90° (Figure 3-97).

Figure 3-103 shows the rotor moved through an angle of 112.5°. Since the stator B output has passed through a null, its phase angle has been reversed. At the summing point, we are adding voltages of phase angle 90° and phase angle 180°. In fact it is 112.5°. It is plain by now that the phase angle of the resultant voltage will be the same as the physical angle of the resolver rotor.

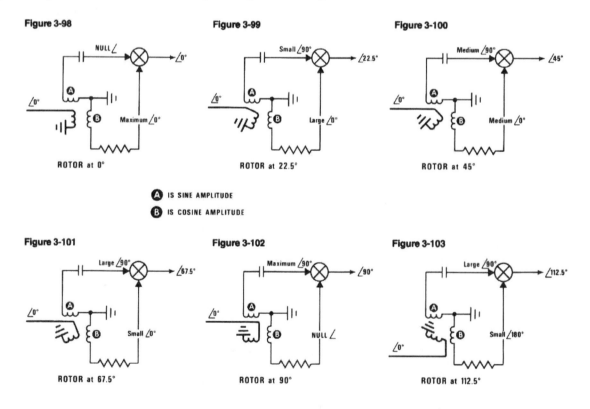

Figure 3-98   Figure 3-99   Figure 3-100

ROTOR at 0°   ROTOR at 22.5°   ROTOR at 45°

Ⓐ IS SINE AMPLITUDE
Ⓑ IS COSINE AMPLITUDE

Figure 3-101   Figure 3-102   Figure 3-103

ROTOR at 67.5°   ROTOR at 90°   ROTOR at 112.5°

Phase Shifter No. 2

The rotor position in Figure 3-104 is 135°. The amplitudes of the stator A and stator B outputs are equal, so the resultant phase angle is halfway between, or 135°.

In Figure 3-105 the rotor has been moved an-other 22.5 to 157.5°, developing a large 180° voltage and a small 90° voltage. Their resultant has the phase angle 157.5°.

Figure 3-106 shows the rotor at 180°. Stator A output is a null and stator B output is maximum. Therefore, the phase angle of the resultant voltage is 180°.

Further clockwise movement of the rotor causes the stator A output to pass through a null, and the phase of its output is therefore reversed to 270°. The rotor position in Figure 3-107 is 202.5°. Stator A output is a small 270° voltage, stator B is a large 180° voltage; the resultant voltage is closer to 180°, namely 202.5°.

In Figure 3-108 the rotor has been moved to 225°. Both stator outputs are equal. Adding equal amplitudes of 180° voltage and 270° voltage gives a resultant voltage of 225°.

Figure 3-109 moves the rotor to 247.5°. This diminishes the amplitude of the 180° voltage and in-creases the amplitude of the 270° voltage, shifting the resultant to 247.5°.

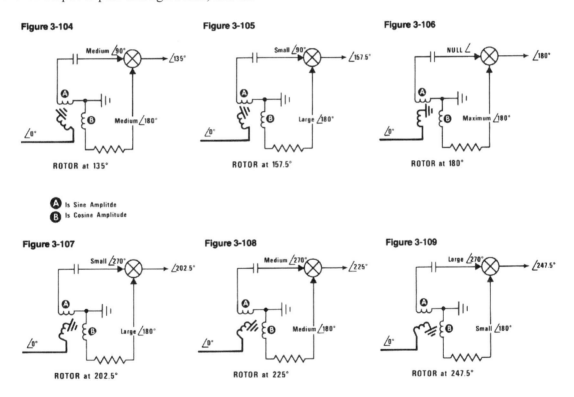

Figure 3-104   Figure 3-105   Figure 3-106

ROTOR at 135°   ROTOR at 157.5°   ROTOR at 180°

A Is Sine Amplitde
B Is Cosine Amplitude

Figure 3-107   Figure 3-108   Figure 3-109

ROTOR at 202.5°   ROTOR at 225°   ROTOR at 247.5°

Phase Shifter No. 3

The rotor position in Figure 3-110 is 270°. The voltage from stator B is a null and the voltage from stator A is a maximum. The phase angle of the resultant is therefore 270°.

Moving the rotor clockwise away from 270° causes the stator B output to reverse in phase.

In Figure 3-111, the rotor has been moved to 292.5°; the phase angle of the resultant voltage is also 292.5°.

Figure 3-112 shows equal amplitude stator outputs. The phase angle of the resultant is halfway between, 315°.

The rotor position in Figure 3-113 is 337.5°. We have gone almost through 360° of phase angle change in the resultant voltage.

In Figure 3-114 the rotor position, 360°, is the same as in Figure 3-98. A null voltage from stator A and a maximum voltage from stator B cause the resultant to have a phase angle of 0°.

These figures demonstrate that the phase angle of the output voltage from the summing point is a direct function of the position of the resolver rotor. Any movement of the resolver rotor causes a corresponding shift in the phase angle of the output voltage degree for degree.

Usually the rotors of phase shifter resolvers are servo motor driven, but in one type of VOR system the resolver rotor is moved manually in the process of selecting a desired course.

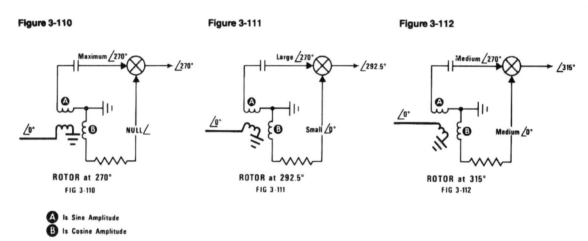

Figure 3-110   Figure 3-111   Figure 3-112

ROTOR at 270°   ROTOR at 292.5°   ROTOR at 315°
FIG 3-110   FIG 3-111   FIG 3-112

Ⓐ Is Sine Amplitude
Ⓑ Is Cosine Amplitude

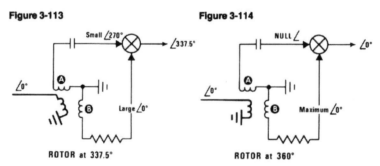

Figure 3-113   Figure 3-114

ROTOR at 337.5°   ROTOR at 360°

Differential Resolver

Figures 3-115 and 3-116 show schematically a differential resolver. The term "Differential Resolver" is an arbitrarily selected name and is not necessarily descriptive of its use. Basically, the differential resolver is a synchro with a two-legged stator and a three-legged rotor.

Figure 3-115 shows it converting a two-legged input signal into a three-legged output signal. Figure 3-116 shows it converting a three-legged input signal into a two-legged output signal. The input signal always establishes the direction of the magnetic field in the synchro and the other section,

either rotor or stator, becomes the secondary looking at the field established by the input signal.

The needs of a particular circuit determine the way a differential resolver may be used. For example, as shown in Figure 3-116, one of the stator windings could be used for a null-seeking signal, while the other stator winding is used as a signal presence monitor. If desired, the differential resolver could also be used in a manner similar to the differential synchro; that is, to provide a second mechanical input to a sensing circuit.

Figure 3-115

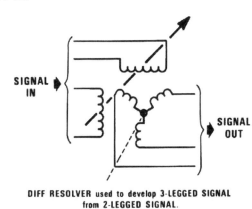

DIFF RESOLVER used to develop 3-LEGGED SIGNAL
from 2-LEGGED SIGNAL.

Figure 3-116

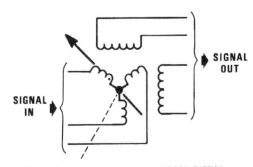

DIFF RESOLVER used to develop 2-LEGGED SIGNAL
from 3-LEGGED SIGNAL.

Transolver

Figures 3-117 and 3-118 illustrate schematically another type of synchro called a transolver. It is similar to the differential resolver except that it has a three-legged stator and a two-legged rotor. The transolver also has the capability of converting two-legged signals to three-legged signals or vice versa.

The reasons why one synchro is chosen in preference to another does not usually concern us.

What really matters is that we can visualize the operations these synchros perform on the signals which they receive and send.

By utilizing the concept of a resultant magnetic field and visualizing its effect upon whichever section of the synchro is the secondary, we can understand the operation of any synchro system.

Figure 3-117   **Figure 3-118**

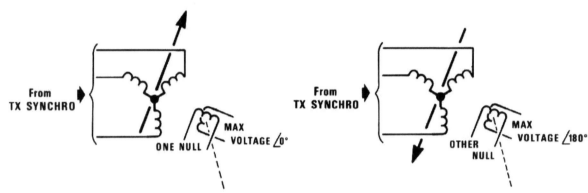

Synchro Signal Practice

The circuit of Figure 3-119 is a hypothetical circuit, given as an exercise in following signals through the various types of synchros we have covered.

At (1) is a three-legged transmit synchro generating a three-legged signal telling the position of its rotor. This signal is given to the rotor of a differential resolver at (2) which transforms it into a two-legged signal, which is then given to the rotor of a resolver at (3). The resolver passes it on as a two-legged signal to the rotor of a transolver at (4) which changes it to a three-legged signal. This signal is then passed through a differential synchro at (5). It shows up on the stator of a control synchro at (6). There are five different places (the rotors of the first five synchros) where the original position signal can be changed by mechanical inputs.

Notice that the signal source is basic AC power. Visualize the original signal as a certain position of the transmit rotor. Its magnetic field is felt in the stator and transferred to the rotor at (2). This position is electrically duplicated in the stator of (2) and given to the rotor of (3). The stator of (3) sees that position and passes it to the rotor of (4). The stator of (4) duplicates he signal in the stator of (5). The rotor of (5) looks at it and passes on its version to the control synchro stator.

The control synchro stator signal is the sum of the positions of the five previous rotors.

If we move the rotor of (1) 10° clockwise, the resultant fields all the way down the line are moved 10° clockwise. If we then move the rotor of (2) 10° clock-wise, all of the following fields are shifted an additional 10°, making a total field movement of 20° at the end of the line. Moving the rotor of (3) still another 10° clockwise shifts all the following fields 10° more.

If we move the rotor of (4) still another 10° clock-wise, the total field movement is 40° in the stator of (6). Now, if we move the rotor of (5) 40° counterclockwise, the field in the stator of (6) is restored to the same position it had prior to moving any of the rotors.

Another example is to move the rotor of (2) 30° counterclockwise while simultaneously moving the rotor of (5) 30° clockwise at the same speed. This causes no change in the field position of the stator of (6).

If these operations are clear to you, then you should not have any problem with synchro signals wherever you find them.

Figure 3-119

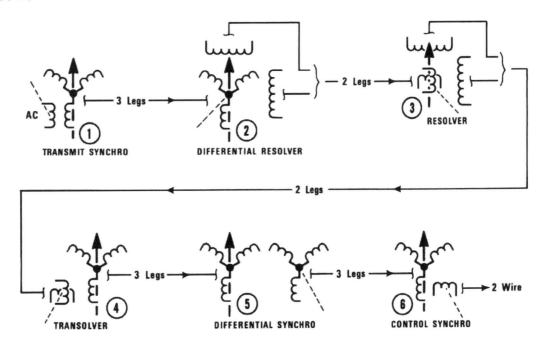

Shorted And Open Leads

The symptoms for shorted and for open synchro leads are such that it is difficult to know which is causing the trouble. It is usually not significant anyway because it involves component change or troubleshooting the wiring of the airplane. The important thing is to be able to recognize the probability of a short, or an open. A dead short between two leads or a complete open of one lead will cause much the same symptom to appear. The symptom is masked to the extent that the short is not a dead short, or the open is not a complete open.

Figure 3-120 shows two leads shorted together with no transmitted error as long as the transmit rotor is in the position shown. If the leads were not shorted together, the voltage that is developed on the two upper legs would have been equal anyway, so shorting them together with the rotor in this position makes no difference.

If we think of the two upper legs as being effectively out of business so far as generating voltages are concerned because they are shorted together, the controlling factor of the transmitted signal is whatever current is developed in the lower leg and passed through the lower leg of the receiving synchro.

For example, if we move the transmit synchro rotor 30° either way we diminish the amplitude of the signal in the lower leg. But, moving the position of the transmit synchro rotor less than 90° does not change the position of the resultant magnetic field in the transmitter stator or in the receiver stator, only its amplitude.

In theory, this is what happens if we have perfect inductances without capacitance or resistance in the stator windings. In the real world we can't have such things, so there is always some movement of the magnetic field. The amount of movement is not the same in high resistance synchros as in low resistance synchros; in high resistance synchros you can expect more movement.

In any case, you will find that when the transmit rotor is moved 10° or 20° to either side across the position where it is parallel to the non-shorted leg, the resultant field movement is small. When the rotor is perpendicular to the non-shorted leg it is at a null, and the other two legs are shorted out.

Moving the synchro rotor 10° or 20° to either side of the null position causes large swings in the resultant field. It goes from one phase on one side of the null to the other phase on the other side of the null, causing the field position to reverse itself. In some synchro systems, moving the transmit rotor 15° to one side of its null causes the field to move 75°. In contrast, moving the rotor 75° from its position parallel to the nonshorted leg causes only 15° of field movement.

Remember that whatever field is developed in the transmit synchro is duplicated in the receiving synchro.

Continued on next page ...

Figure 3-120

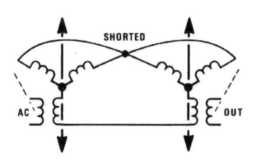

Shorted And Open Leads (cont'd.)

Following this reasoning a little further in Figure 3-121: The rotor of the transmit synchro has been left in the same position, but two different leads are shorted out. Consider any voltages and currents developed in the two shorted leads to be circulating only between themselves, and there-fore of equal magnitude.

The two magnetic fields generated in the two shorted coils add vectorially to be in line with the field in the non-shorted coil. In our perfect inductances, any resultant field developed has to be of one direction or the other in line with the non-shorted coil.

In Figure 3-122 one lead has been opened up so that only two coils are functioning in each synchro. Any currents developed are of equal amplitudes in all four operative windings since they are all in series. Since the magnetic fields of the two operative windings in each synchro are equal, their vectorial sum always has the same position, although the direction can be reversed. This position is perpendicular to the open leg (in a synchro without capacitance or resistance).

Figure 3-123 shows a different leg open and the resultant field perpendicular to that leg.

In a resolver the situation is simpler and easier to visualize. Figure 3-124 shows a resolver with what should have been an input signal causing a field to appear in some position other than parallel to one or the other rotors. Shorting out or opening up either rotor leaves only the other rotor to supply the field; that field is always parallel to the non-shorted coil.

In Figure 3-125 one of the resolver stator fields is opened up; that leaves only one possible position for the stator field to transmit to its receiver

Figure 3-121

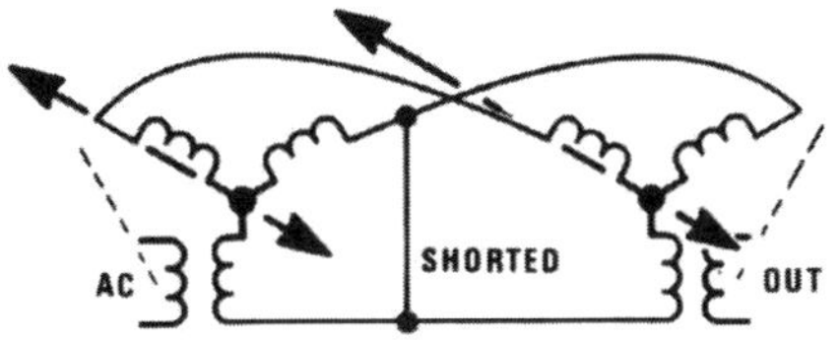

Figure 3-122

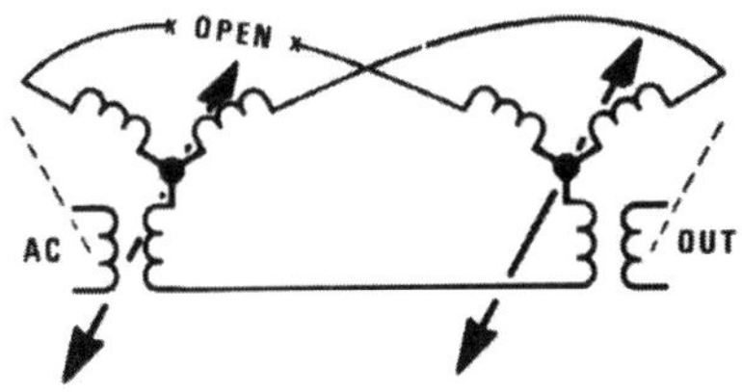

Figure 3-123

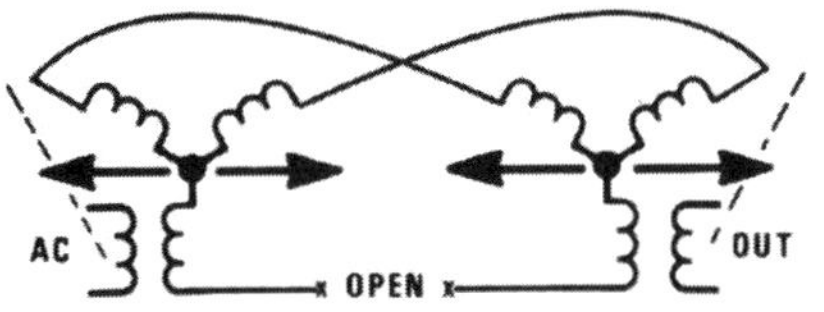

Figure 3-124

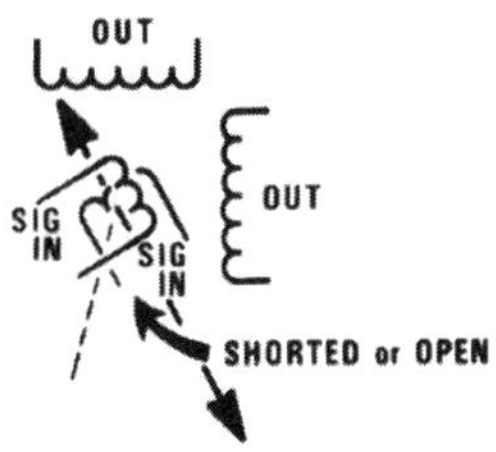

Figure 3-125

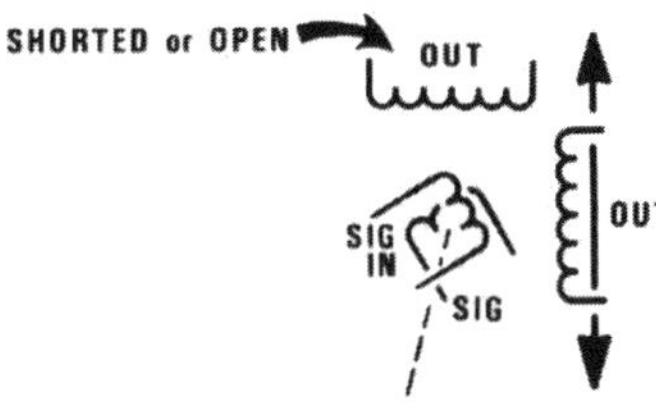

Coil And Circle Synchro Symbols

Figures 3-126 through 3-131 show two different types of synchro symbols. In the circular symbols the inner circle is always the rotor. The circles are more convenient, but the user of the symbol must know exactly what is represented, otherwise the operation becomes mysterious. The coil type symbols are preferred where it is desired to make the synchro operation more apparent.

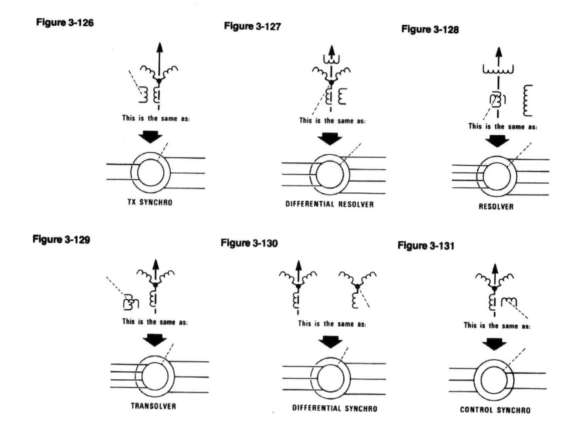

Figure 3-126

Figure 3-127

Figure 3-128

This is the same as:

This is the same as:

This is the same as:

TX SYNCHRO

DIFFERENTIAL RESOLVER

RESOLVER

Figure 3-129

Figure 3-130

Figure 3-131

This is the same as:

This is the same as:

This is the same as:

TRANSOLVER

DIFFERENTIAL SYNCHRO

CONTROL SYNCHRO

CHAPTER 4 - SERVO MOTORS AND TACH RATE GENERATORS

DC Servo Motor Control Principle

Understanding the operating principles of servo motor systems is necessary for complete understanding of avionics systems in general. Figure 4-1 shows a DC servo motor loop. It is called a loop because of the closed nature of the system operation. Whether or not an operated item is attached to the servo motor is of no consequence to the operation of the servo motor loop. The operated item could be anything ranging from an indicator readout to a flight control surface of an airplane. Servo motors are also commonly used inside avionics black boxes.

The 28 volt DC source is connected to the variable control potentiometer and to the follow-up potentiometer. As long as the voltages seen by the wiper arms of the potentiometers are equal, the servo motor does not run. If the control potentiometer wiper arm is moved downward, there is a more positive voltage on the right motor brush than on the left motor brush, and the motor runs so as to cause its potentiometer also to move downward until wiper arm voltages are equal and the motor stops running.

Moving the control potentiometer wiper arm upward would cause a more positive voltage to appear on the left motor brush than on the right, and the motor would run in the opposite direction, causing its potentiometer wiper arm also to move upward until the voltages are equal.

Figure 4-2 indicates another arrangement for a DC servo motor loop. As in the previous example, the servo motor is operating the wiper arm of a potentiometer connected to a positive 28 volts DC.

The motor is connected so that if there is no control signal in, the potentiometer wiper arm is driven to zero potential so that there is no signal into the amplifier. In this case, control signals are always negative DC.

If we supply a negative 14 volt DC control signal to the amplifier, the servo motor will run until its potentiometer wiper arm sees a positive 14 volts DC. At that time the amplifier has no net input and the servo motor stops running.

Any particular control signal voltage (down to a negative 28 volts DC) results in the servo motor wiper arm being driven to a particular position. The operated item moves to a particular position in response to a particular control signal.

Figure 4-1

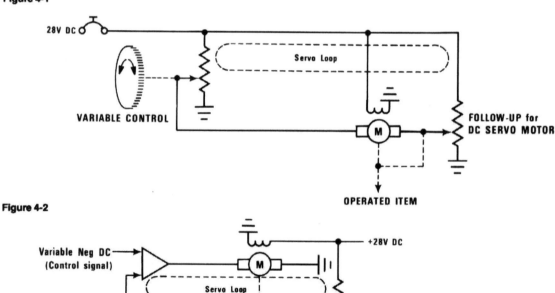

Figure 4-2

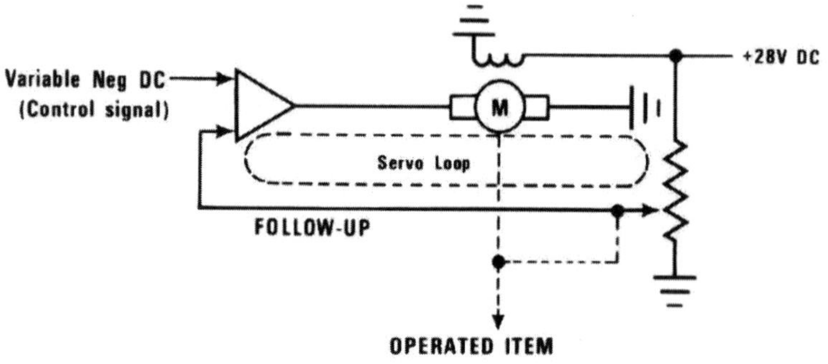

Two-Phase Servo Motor

Figure 4-3 shows a schematic symbol for what is called a two-phase inductance motor. This is a typical AC servo motor. It may be very small, on a computer card, or it may be quite powerful. The construction of the stator is similar to the construction of the resolver stator. It has four poles, paired, so that there are actually two fields. These two fields are represented symbolically by two coils drawn at right angles to each other, because in the motor itself the two fields do exist at right angles to each other.

If a 400 Hertz voltage of phase angle 0° is connected to one field and a 400 Hertz voltage of phase angle 90° (a capacitor in series with the field can shift the voltage 90°) to the other field, the instantaneous field, resulting from the addition of these two, rotates at 400 revolutions per second. You can visualize this rotation if you consider the instantaneous DC voltages present across these fields.

For example, begin with the phase angle 0° voltage at a null, but rising. The phase angle 90° voltage is at a positive peak, so the magnetic field is parallel to the fixed winding. As the fixed winding voltage falls, the variable winding voltage rises.

Electrically 45° later, they will be equal and both positive. The magnetic fields will then be of equal strength with a resultant that has rotated through 45°. If you pursue these instantaneous DC voltages and the resulting magnetic fields through one complete electrical cycle, you will see that the magnetic field has rotated through 360°.

Rotation of the magnetic field in the motor tends to drag the rotor after it in the same direction. How fast the motor moves depends upon its load and the strength of the magnetic field, which effectively is dependent upon the strength of the variable signal.

In Figure 4-4 the phase of the variable signal has been reversed. This reverses the direction of rotation of the resultant field.

The direction of motor rotation depends upon the phase of the variable signal, and the speed of rotation depends upon its amplitude. Often times it is desirable to apply an electrical brake to a two-phase servo motor (Figure 4-5). This can be done by disconnecting either the variable field or the fixed field. If only one field is left operative, the motor does not rotate because the field does not rotate. This tends to hold the rotor of the motor in a fixed position.

Figure 4-6 shows a two-phase servo motor loop using a control synchro input. It is typical of many such loops used throughout aircraft systems. Whenever the amplifier sees a signal of a particular phase; it drives the motor in a particular direction until the synchro rotor comes to a particular null. A signal of opposite phase from the synchro rotor drives the servo motor in the opposite direction. The synchro rotor therefore always is driven to a particular null.

The operating signal will come from some remote source whose mechanical position we want to duplicate in the operated item. For example, the remote source could be a directional gyro and the operated item could be a compass indicator.

Let's continue with this example and say that the synchro rotor has been driven to the desired null. The compass indicator, therefore, shows the correct airplane heading. If the airplane changes heading, the signal from the directional gyro transmits synchro changes, moving the magnetic field in the stator of the control synchro. The rotor therefore is no longer at a null. The phase of the signal to the amplifier is such as to cause the servo motor to drive the synchro rotor to the new correct null position. If the field moves clockwise, the rotor will be driven clockwise, and vice versa.

When the new null position is reached, the compass indicator shows the new aircraft heading.

Since the amplifier is quite sensitive, the rotor will never be very far away from its null, which means that the indicated heading will never be very far away from the actual heading of the airplane. In actual compass systems, the indicated heading probably lags not more than a couple of degrees.

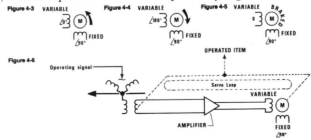

Tachometer/Rate Generator

Tachometer generator and rate generator are two names commonly given to the same device. The schematic symbol is shown in Figures 4-7 and 4-8. It is similar to the symbol for a two-phase servo motor except for the letters in the symbol.

Aircraft power supply generators supply three phases of output on separate wires. These phases are called A, B and C. Any single avionics box (or system) uses only one of these. So far as that particular avionics system is concerned, its power input phase is phase angle 0°. It's internally derived opposite is phase angle 180°.

The desired output from a tachometer generator is phase angle 0° or phase angle 180°. In order for the output to be one of the desired phases, it is common for the fixed phase of the tachometer generator to be of phase angle 90°. This phase for the fixed field, as in the two-phase servo motor, is frequently achieved by placing a capacitor in series with the aircraft power supply to the fixed field.

The rate generator principle of operation can be explained as follows: Figure 4-7 shows a motor driving the rotor of the rate generator in a clock-wise direction. Consider the tachometer at rest. The powered fixed field is perpendicular to the variable field, which has no power. Since the magnetic field is perpendicular to the variable field there is no transformer action and no signal out of the variable field winding.

However, if the rotor begins to turn, the magnetic field does not remain perpendicular to the output winding.

You can think of the rotor as tending to drag the magnetic field along with it. This effect is caused by eddy currents generated in the rotor. A slow rate of movement of the rotor does not bend the magnetic field very much, whereas a greater rate of motion moves the field farther. The generated volt-ages which cause eddy currents in the rotor in-crease with increases in speed of the rotor. So picture the fixed field being bent farther away from perpendicular to the variable field as rotor speed increases.

A low rotor speed moves the magnetic field away from perpendicular to the output winding only a small amount. The small transformer coupling results in a small output voltage. In Figure 4-7 we have assigned phase angle 180° to the output. This is purely arbitrary and it could have been phase angle 0° if we had connected it up the other way.

The output is a 400 Hertz phase angle 180° volt-age of small amplitude. Speeding up the rotor displaces the magnetic field farther away from perpendicular, causing a larger voltage to appear at the output.

The amplitude of the output is a direct function of the rotor speed, so high speeds can give a high output.

Figure 4-8 shows the rotor being turned in the opposite direction. The field has been moved away from perpendicular in the opposite direction, and therefore the phase of the output is opposite to that in Figure 4-7. This is another instance of signal phase reversal as the signal passes through a null.

One of the greatest uses of the rate generator is to provide inverse feedback signals in servo motor systems for speed limiting and smoothing functions. Another use is to provide rate signals.

Figure 4-7

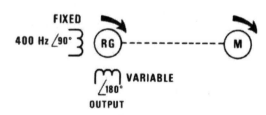

Figure 4-8

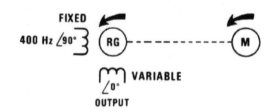

AC Servo Motor No. 1

Figures 4-9 and 4-10 show AC servo motor loops with two different types of follow-up signals. If the potentiometer wiper arm in Figure 4-9 is at its midpoint, its voltage, referenced to ground, is zero. This is sometimes called a phantom ground be-cause there is no direct connection to ground. If the wiper arm is not at its midpoint, there is a voltage at its contact. On one side of the mid position, the voltage is of one phase. On the other side, the voltage is of the opposite phase (because opposite ends of the transformer have opposite phase voltages).

With no control signal input, the servo motor will have driven the follow-up potentiometer to a null. If an input control signal appears at the servo amplifier, it will be a voltage of one phase or the other. The servo motor will run one direction or the other, driving the wiper in the direction which will give a follow-up voltage of the phase opposite to that of the input signal.

The distance through which the wiper arm has to move is a function of the amplitude of the input signal. The larger the amplitude of the input signal, the farther the wiper arm has to be driven to cancel the input and the greater the movement of the operated item.

Cancellation occurs when control signal and follow-up voltages are equal in amplitude. If the control signal becomes a null, the follow-up signal drives the servo motor until the follow-up signal is also a null.

Follow-up voltage and signal presence monitor voltage are developed from the same synchro in Figure 4-10. This synchro is explained in Figure 3-77. When the synchro rotor is in the position shown, the voltages across the X leg and Y leg are equal. Since they are, there is no voltage difference between these X and Y points.

The Y point is grounded; therefore, the X point also has a ground potential (phantom ground). Movement of the rotor in either direction from this position develops a voltage (measured to ground) from point X of one phase or the other.

A synchro connected in this way is never intended to have its rotor moved more than 20° or so. An input signal of phase angle 0° causes the servo motor to run in the direction which will develop an X leg output of phase angle 180°. It runs until the X leg voltage equals the control signal input. If the amplitude of the input voltage is then decreased or eliminated, the servo motor reverses because the follow-up signal exceeds the control signal. It runs until the follow-up signal equals the control signal amplitude.

As long as the synchro rotor remains within its normal limits of movement, the amplitude of the voltage on the Z leg is always sufficient to satisfy the signal presence monitor (refer to Figure 2-11).

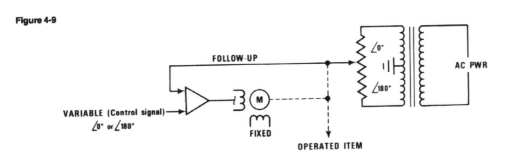

Figure 4-9

Figure 4-10

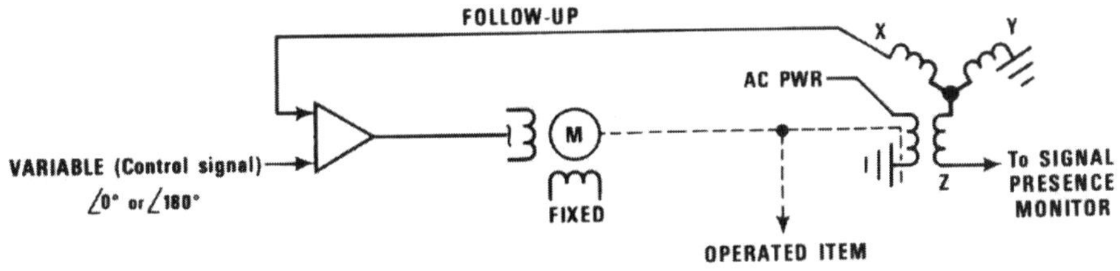

AC Servo Motor No. 2

The servo system in Figure 4-11 is much more common than either of the two previous examples. The servo loop is shown nulled out (synchro rotor perpendicular to its field).

Moving the rotor of the transmit synchro with the knob moves the field in the control synchro, causing the servo motor to drive its rotor to a new null position. These servo motor loops always drive the synchro rotor to a particular null, even if it has to drive the rotor 179.5°.

Figure 4-12 illustrates a servo loop controlled by inputs from two different mechanical devices. The transmit synchro on the left provides the original signal, which is fed through the differential synchro to the control synchro in the servo loop. Movement of the rotor of the transmit synchro by its knob

rotates the field in the differential synchro and in the stator of the control synchro. This causes the servo motor to drive the control synchro rotor to a new null position.

Movement of the differential synchro rotor by itself also moves the field in the stator of the control synchro, causing the servo motor to drive. If the transmit synchro rotor is moved 30° clockwise, and the differential synchro rotor is also moved 30° clockwise, the servo motor will have to drive its control synchro rotor 60° to find a new null.

If we could move the transmit synchro rotor 30° clockwise and simultaneously, at the same rate, move the differential synchro rotor 30° counter-clockwise, the servo motor would not run because the control synchro field would not move.

Figure 4-11

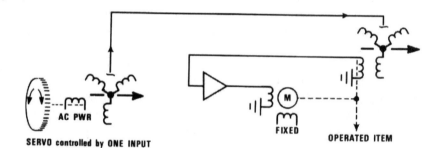

SERVO controlled by ONE INPUT

Figure 4-12

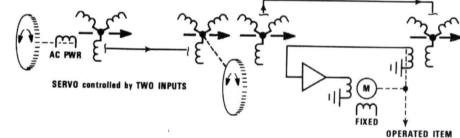

SERVO controlled by TWO INPUTS

Rate Generator And Servo

In Figure 4-13 we show a servo motor loop whose servo drives a tachometer generator. The tachometer generator output is fed into the servo amplifier. This illustrates a typical and very common use of tachometer generators in servo motor systems.

Sometimes the tachometer generator is not drawn in the schematic, even though it exists and is used in the servo system. Other times the fixed field portion of the tachometer generator is omitted and only the output winding shown.

A tachometer generator is always connected to the servo amplifier so that when the motor runs, the generator supplies an opposition signal. It can never stop the servo motor, of course, because as soon as the servo motor stops, the tachometer generator has no output. But, the faster the motor runs, the greater the opposition. In this way it provides inverse feedback for speed limiting and smoothing.

Figure 4-13

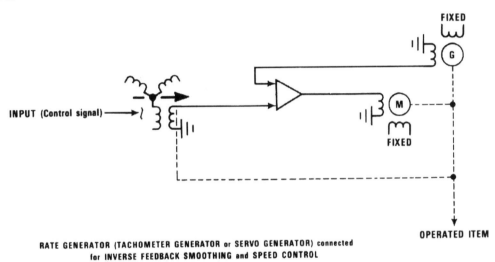

RATE GENERATOR (TACHOMETER GENERATOR or SERVO GENERATOR) connected
for INVERSE FEEDBACK SMOOTHING and SPEED CONTROL

Coarse And Fine Synchro System

An example of a "coarse and fine" synchro system is given in Figure 4-14. This illustration shows a Central Air Data Computer (CADC) transmitting highly accurate, very sensitive information to a remote altimeter indicator.

In a coarse and fine synchro pair, like the one in the CADC, the coarse synchro is used as the name implies. It gives unambiguous information throughout the entire range of operation of the system. It never turns a full 360°, so every position represents a definite altitude. Measuring that automatically imposes a limit on the sensitivity, and therefore accuracy, of the information from the coarse synchro.

The fine synchro turns twelve times as fast as the coarse synchro and is twelve times as sensitive. Since it makes several revolutions on the way to cruise altitude it is highly ambiguous and could not by itself give altitude readings. For example, if the coarse synchro rotor is turned 270° from the minimum to maximum altimeter indication, the fine synchro rotor is turned through nine complete revolutions.

Assuming that each synchro has the same definition of sensitivity, the fine synchro output is twelve times as sensitive as the coarse synchro output. The fine synchro information by itself, however, would not have specific altitude significance.

Any given rotor position in the fine synchro might, for example, be repeated nine times between sea level and 70,000 feet.

In the altimeter indicator, the output of the coarse synchro rotor is fed to a level detector which operates the servo output switches to their triangular contacts if the rotor is more than 10° away from its null. This error signal will cause the servo motor to reposition the coarse synchro until the error is less than 10°. This error signal will cause the servo motor to reposition the coarse synchro until the error is less than 10°. Control then reverts to the fine synchro.

This ensures that the fine synchro signal will not be used unless its rotor is within 120° of its correct null for the correct altitude. Such a large error is not ordinarily present. The error probably does not exceed about 3° in the fine synchro rotor, even when altitude is changing rapidly.

It is not unusual for two synchros to be paired in this way in order to benefit from the greatly increased sensitivity. The fine and coarse synchro pair in the receiving device must be driven by a gear train which has the same gear ratio as the one in the transmitting device. In addition, when these systems are assembled, the synchro rotors, like all synchro rotors, must be carefully positioned to a predetermined spot.

Figure 4-14

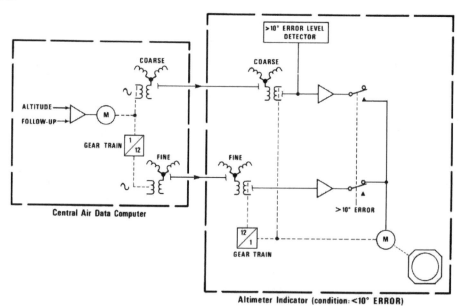

Central Air Data Computer

Altimeter Indicator (condition: <10° ERROR)

CHAPTER 5 - INERTIAL REFERENCE

Direction Of Gyro Precession

This section deals with the use of gyros in aircraft systems. Some of them are so important that no commercial aircraft would be permitted to leave the ground without them. They are fascinating to study and a great deal of material is available on them. For the most part, we will be concerned with only two of the properties of spinning gyros. The first is the tendency of a spinning gyro to remain fixed in space if it is not acted upon by outside forces such as bearing friction.

The other property of a spinning gyro that concerns us is its right angle obstinacy. It never goes in the direction that you push it, but off to one side. Figure 5-1 illustrates this obstinate characteristic. The rules for anticipating the actual direction of motion from a given applied force arc not useful to us. However, the action of this right angle precession is involved in later presentations.

Figure 5-1

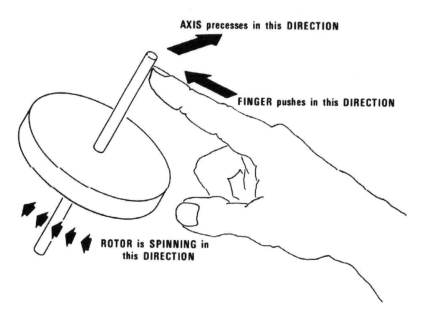

When a FORCE or PUSH is applied to the AXIS of a GYROSCOPE, the AXIS does not MOVE in the DIRECTION of the APPLIED FORCE, but moves instead at RIGHT ANGLES to the DIRECTION of that FORCE.

Behavior Of A Perfect Free Gyro

Figure 5-2 illustrates the behavior of a perfect gyro. A perfect gyro would be one without any external forces acting upon it, mounted in a perfect suspension system that would give it complete freedom of movement in all three axes. All the gyros in this figure are perfect gyros.

Only four gyms are represented — A, B, C, and D. The other gyro symbols shown illustrate the various positions of B, C, and D as the earth rotates.

Gyro A has its spin axis parallel with the spin axis of the earth, sitting on top of the North Pole. It could maintain that position indefinitely.

Gyro B has its spin axis parallel to the earth's spin axis, and is located above the equator. The other gyms in its group represent Gyro B as it would appear at different times of the day. If we were to look at Gyro B sitting on a table in front of us, we would see that the upper end of its spin axis is pointing off toward the North Star. As time goes on and the earth turns 360°, we would not see any change in its attitude on the table. Its spin axis would always point toward the North Star.

Gyro C is situated on the equator. The other gyros in its group represent Gyro C as it would appear at different times of the day.

Let's say that we have the Gyro C in front of us on a table. Its spin axis is parallel to the earth's surface. As time goes on and the earth rotates, we would see its spin axis gradually tilting upward at one end until, six hours later (90° of earth rotation); we would see it perpendicular to the earth's surface, illustrated by the gyro shown to the right of the earth. Six hours later (behind the earth out of sight in this drawing) the spin axis would once again be parallel to the earth, but with the end which was first pointing east now pointing west.

Another six hours later, the spin axis would once again be perpendicular, but this time the opposite end of the axis would be another six hours later. When we get to the same time of day at which we started, the gyro will again be occupying its original position.

Gyro D and its group illustrate another changing aspect of a gyro, in different positions as viewed from the earth's surface at different times of day.

These perfect gyros illustrate what any gyro tries to do but cannot because of its orientation of the spin axis — always in the same direction in space.

Figure 5-2

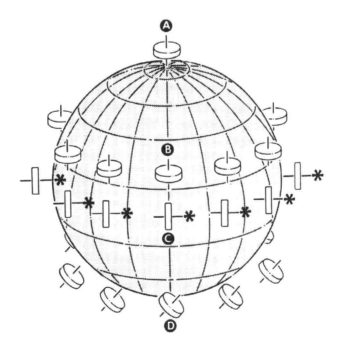

Regardless of the EARTH'S MOTION — A PERFECT GYRO'S AXIS would maintain a fixed position relative to space.

Behavior Of Vertical Gyro

The outer ring of gyros in Figure 5-3 demonstrates that a completely free gyro in an airplane circling the earth would be perpendicular to the earth's surface at only two points.

The gyros drawn in the airplanes are continuously being corrected to a vertical position as the airplane moves around the surface of the earth. The corrections are gentle and slow, since the amount of correction needed in a ten minute period, for example, is small. The gyro is relatively very stable during the pitch and roll maneuvers of the airplane.

The gyro's stable position with respect to the movements of the airplane makes it possible for the pilot to know the actual attitude of his air-plane, nose up or down, and wings level or not. This is quite important to him when all he can see out of the window is a gray fog.

The airplane attitude information derived from the gyros is also used by such systems as the autopilot, radar antenna stabilization, flight recorders and flight directors.

Figure 5-3

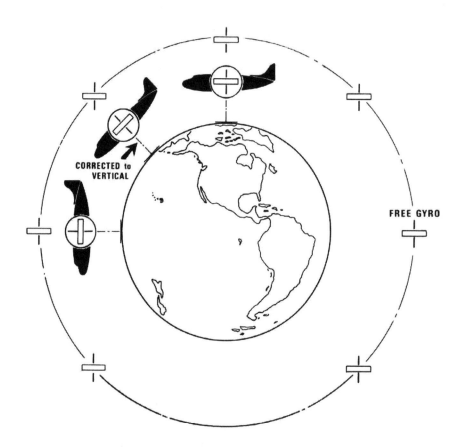

Gyros Inside Indicators

Figures 5-4 and 5-5 illustrate how gyros can be used inside indicators mounted directly on the flight instrument panel. Figure 5-4 shows a gyro mounted with its spin axis vertical. Bearings and gimbals give it freedom of motion in three axes.

The bearings at the top of the illustration, along the fore and aft line of the airplane, give it freedom along the X axis, therefore banking the airplane does not affect it. The bearings on the Y axis prevent it from being affected by the pitching of the airplane, within the limits of the stops provided.

The bearings on the rotor axis itself prevent it from being affected by the turning of the airplane. A horizon indicator coupled to the inner gimbal shows the pitch and roll attitudes of the airplane.

The gyro in Figure 5-5 has its spin axis horizontal. It also has freedom of motion in three axes. A compass indicator card is attached to the vertical gimbal. As long as the spin axis has the correct end pointed north, the indication read by the pilot is the compass heading of the airplane.

This type of gyro needs repeated manual heading correction, since it has no automatic correction system. It is generally useful for periods up to 15 or 20 minutes. During periods of level flight the pilot must manually correct it using information from his magnetic compass.

The vertical gyro has a gravity operated erection system which keeps its spin axis vertical, and the compass gyro has a similar system maintaining its gyro axis horizontal.

The uncorrected gyro compass indicators are no longer used in commercial airplanes. Instead, large expensive gyros are mounted not too far from the center of gravity and control remote indicators on the flight instrument panels.

Figure 5-4

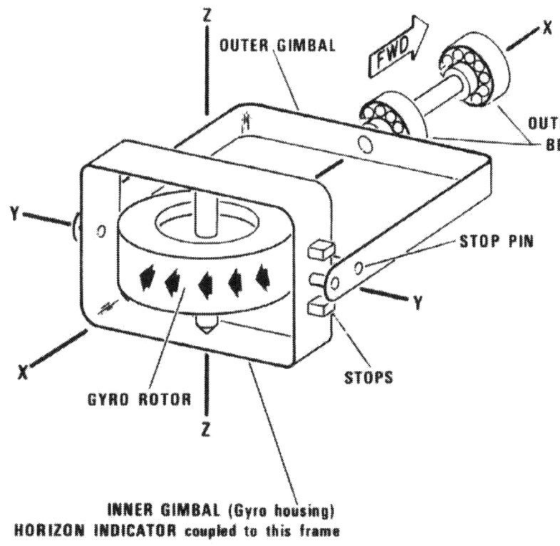

Figure 5-5

VERTICAL GYRO for HORIZON INDICATION

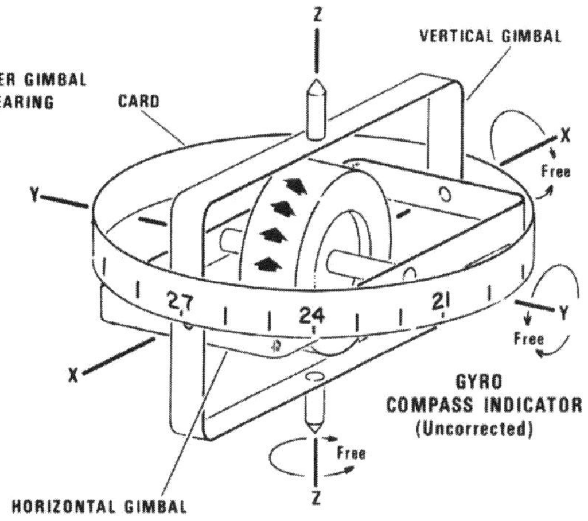

Gyro Horizon Indicator No. 1

Figures 5-6 through 5-14 illustrate the indications given by a gyro horizon indicator for a variety of airplane attitudes. The inner portion of this indicator (and most others) is a movable part of a sphere. A heavy black line represents the horizon, and separates the light colored upper half from the dark colored lower half The "W" with a dot in the center and horizontal bars extending from the sides represents the airplane. This symbol is in a fixed position in front of the sphere.

Figure 5-6 shows the airplane flying wings level ("W" bars parallel with the horizon), and 6° nose down. Pitch attitude is read on the scale drawn on the sphere, at the point where the dot in the middle of the "W" is showing. In this case it is at the 6° position below the horizon.

Figure 5-7 shows the airplane flying wings level ("W" bars parallel with the horizon), and nose on the horizon ("W" dot on horizon line).

Figure 5-8 shows the airplane flying wings level ("W" bars parallel to the horizon), and 6° nose up ("W" dot at the 6° position above the horizon).

Figure 5-6

Figure 5-7

Figure 5-8

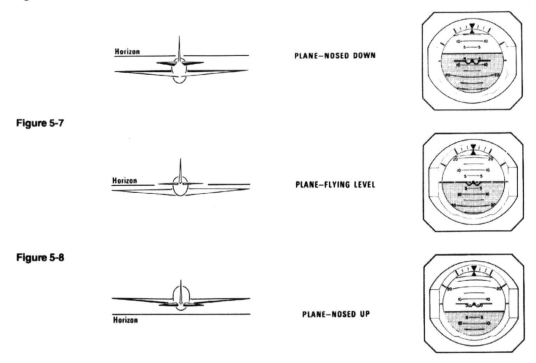

Gyro Horizon Indicator No. 2

In Figures 5-9 and 5-10 the case of the indicator has been banked along with the airplane to show how it would look to us if we could see through the airplane body to the instrument panel.

In Figures 5-11 through 5-14 the indicator is drawn to show how it appears to the pilot in the airplane.

In Figure 5-9 the pilot sees that his left wing is lower than his right wing and that the nose of the airplane is 6° below the horizon. The pilot reads his bank angle at the top of the indicator. The index at the top of the indicator is three spaces off to the right of center. Since each space represents 10° he is in a 30° left bank.

In Figure 5-10 the airplane is 6° nose down, while in a 30° right bank.

In Figure 5-11, the airplane nose is on the horizon, and the left wing is down 30°.

In Figure 5-12, the airplane nose is also on the horizon, but the right wing is down 30°.

In Figure 5-13, the airplane nose is above the horizon by 6°, while in a 30° left bank.

In Figure 5-14, the airplane nose is above the horizon 6°, while in a 30° right bank.

Figure 5-9

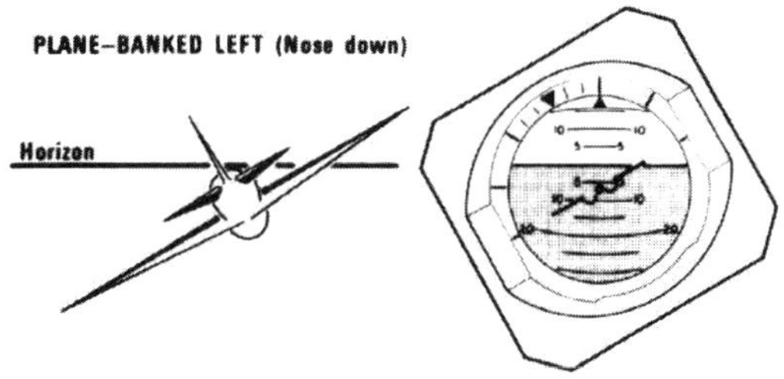

PLANE–BANKED LEFT (Nose down)

Horizon

Figure 5-10

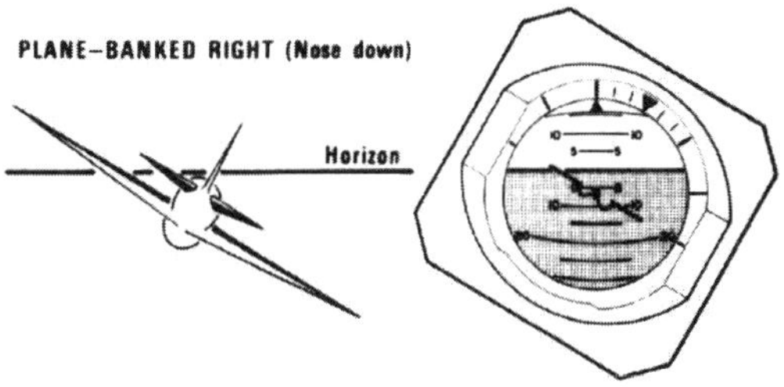

PLANE–BANKED RIGHT (Nose down)

Horizon

Figure 5-11

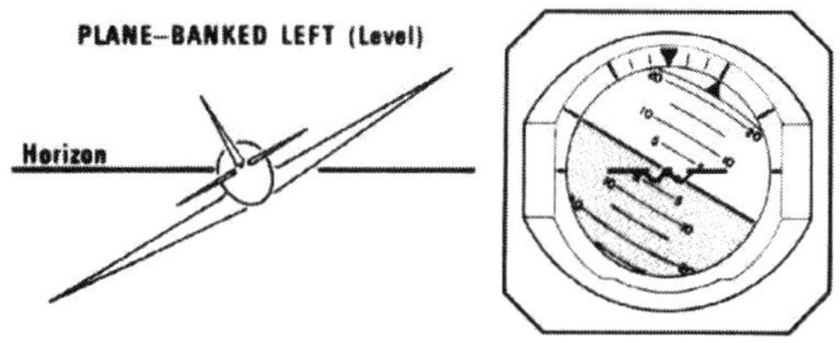

PLANE–BANKED LEFT (Level)

Horizon

Figure 5-12

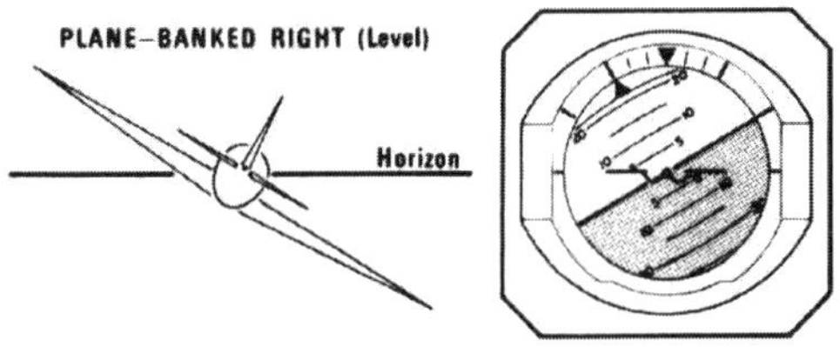

PLANE–BANKED RIGHT (Level)

Horizon

Figure 5-13

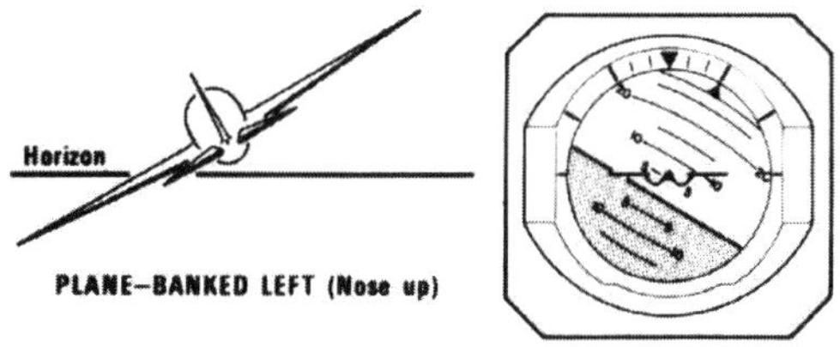

Horizon

PLANE–BANKED LEFT (Nose up)

Figure 5-14

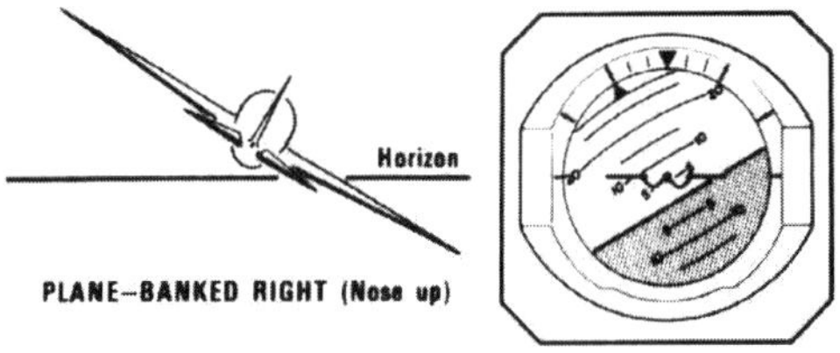

Horizon

PLANE–BANKED RIGHT (Nose up)

Remote Gyro Horizon

Figures 5-15 through 5-18 are photographs of a late model gyro horizon indicator. It is an integrated instrument showing not only airplane pitch and roll attitude, but other indications as well.

At the extreme bottom is a slip indicator. This is a simple device, with a ball enclosed in a curved tube filled with a damping fluid, and a center position indicated. If the ball is not centered, the aircraft is slipping or skidding to one side or the other, a situation which is infrequent in commercial aircraft.

Just above that is a rate of turn indicator so calibrated that if the needle is directly below one of the outside dots, the airplane is turning at a rate of 180° per minute in the direction indicated.

On the right is a glideslope deviation indicator.

On the left is a speed error indicator showing whether the airplane is traveling faster or slower than a preselected speed.

The upper right corner has a decision light showing when a preselected radio altitude above the runway has been reached on descent.

On the left is a light which indicates whether a flight director is turned on. The flight director command hart also part of this instrument, are retracted from view in these pictures.

There is provided a failure warning flag for each indication.

At the lower left is the push-to-test button.

Figure 5-15

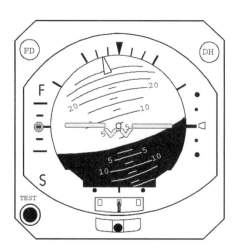

PLANE—NOSE UP (Right wing down)

Figure 5-16

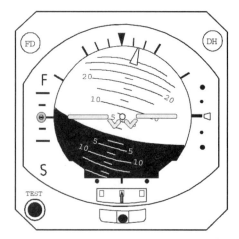

PLANE—NOSE UP (Left wing down)

Figure 5-17

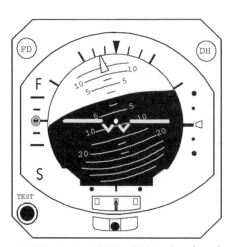

PLANE—NOSE DOWN (Right wing down)

Figure 5-18

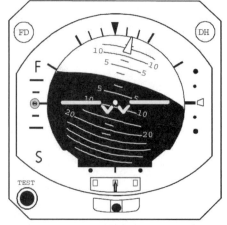

PLANE—NOSE DOWN (Left wing down)

Remote Vertical Gyro Schematic

Figure 5-19 is a remote vertical gyro. It is called a vertical gyro because its spin axis is automatically maintained vertically by gravity sensing pitch and roll erection systems. The torquers and synchros are not drawn, only their position is indicated.

The black arrow pointing forward indicates that the inner gimbal bearing shafts must be aligned with the fore and aft line of the airplane. This is so that the roll synchro will not be affected by pitch attitude changes, and the pitch synchro will not be affected by roll attitude changes.

In order to maintain the spin axis vertically, two erection systems must be used, one in the roll axis and one in the pitch axis. The obstinate reaction of a gyro to an applied force (Figure 5-1) makes it necessary to use erection forces at right angles to the desired direction of motion. This accounts for the pitch erection torquer mounted in the roll axis, and the roll erection torquer mounted in the pitch axis.

A torquer is a frustrated motor. It never gets to turn anything, not even itself; but when called upon to do so, will try. A gravity sensing liquid switch, constructed on the principle of a carpenter's level, provides power to the torquer when the switch is not level. The torquer then provides the force to erect the spin axis vertically in one axis.

Roll erection torquing is cut off when the bank angle exceeds about 6° to eliminate the tendency to erect to a false sense of vertical.

Two types of signals are developed from the roll and pitch transmit synchros. The three legged signal is unambiguous in that any attitude through 360°

could not be mistaken for any other attitude. These signals are always used for controlling gyro horizon indicators. They will also be used by any other system where an unambiguous signal may be desired. The two legged error signals will be nulls when the airplane is wings level and has its nose on the horizon. For example, the roll transmit synchro is positioned in the gyro so that when the airplane is wings level, the voltages developed on the two upper legs of the roll synchro are equal. There is, therefore, a null signal between them.

An isolation transformer is used in the output to lessen possible bad effects on the three legged signal. If the airplane is banked to the right, these voltages are no longer equal, and the difference will be of one phase. If the airplane is banked to the left, the voltages are no longer equal and the difference will be of the other phase. Pitch error signals are developed similarly.

These error signals are ambiguous. For example, the phase and amplitude of a 5° right bank signal is the same as the phase and amplitude of a 175° right bank signal. Also, the phase and amplitude of an 85° right bank signal is the same as the phase and amplitude of a 95° right bank signal.

Every signal developed has a twin at another attitude. However, within the first 90° of bank angle or pitch attitude, there is no ambiguity. Since commercial aircraft never bank or pitch beyond 90°, some less critical systems can use these error signals, which are frequently more convenient.

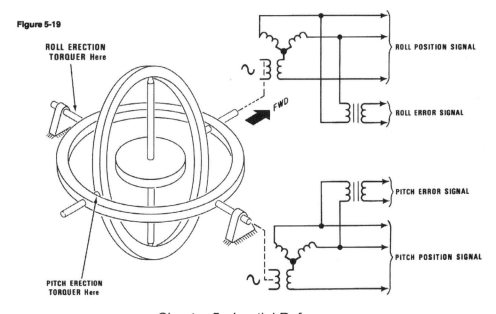

Figure 5-19

ROLL ERECTION TORQUER Here

FWD

ROLL POSITION SIGNAL

ROLL ERROR SIGNAL

PITCH ERROR SIGNAL

PITCH POSITION SIGNAL

PITCH ERECTION TORQUER Here

Remote Horizon System

Figure 5-20 is a remote gyro horizon indicating system comprised of vertical gyro, "instruments modules" and horizon indicator. The airplane attitude signals given by the gyro could be furnished by an Inertial Navigation System (discussed in another section). The items in the "instruments modules" could be in a radio rack black box or in separate black boxes, and the servo amplifiers could be in the horizon indicator itself The horizon indicator could be one of several types, for example, the one illustrated in Figure 5-6. Regardless of location and types of parts, the operation principle will be the same.

The "monitor" box in the gyro names some of the conditions monitored. If the monitor sees normal conditions, there is a positive output to the "attitude monitor". The "gyro" flag, to which the output of the attitude monitor is connected, is a meter movement spring-loaded into view, but powered out of view when the attitude monitor is satisfied.

The two other inputs to the attitude monitor are through NOT circuits. These NOT circuits are a convenient way of representing null monitors, an exception to our rule that logic circuit inputs should be positive voltages. If both servo loops in the indicator have nulls (discussed later), and the gyro monitor is satisfied, the attitude monitor will power the gyro flag out of view.

The horizon indicator has two servo motor systems. One controls the indicated airplane pitch attitude and the other controls the indicated roll attitude. Since both function alike, we will describe the roll operation.

The roll attitude of the airplane, as seen by the vertical gyro, appears in the stator of the roll attitude control synchro as a magnetic field of a particular direction. If the control synchro rotor is not perpendicular to the field in the stator, a signal to the servo amplifier causes the servo motor to drive until the synchro rotor is perpendicular to its field. The servo motor fixed field windings are not illustrated. When the servo motor has driven its control rotor to a null position, the roll attitude indicated is the same as the roll attitude sensed by the vertical gyro.

In normal operation, the servo motor loop keeps the control synchro rotor very close to the correct indication. The rotor does not normally move more than 3° away from perpendicular to its field. In fact, if the signal appearing in the synchro rotor is 3° away from null, the null monitor (represented by the NOT circuit in the attitude monitor) will not have an output to the AND circuit. The AND circuit then will not have an output, and the gyro flag will appear. The operation and monitoring of the pitch attitude indication is accomplished in the same manner as for roll.

Continued on next page ...

Remote Horizon System (cont'd.)

The push-to-test button is on the front of the indicator and, when pushed, introduces a test signal into both the roll circuit and the pitch circuit. In a parked airplane, for example, this causes the gyro horizon indication to show right wing down and nose up, and the gyro warning flag to appear. The amount of motion caused by this test operation is not significant. It is the readiness and sureness of operation of the horizon indicator that counts.

It is a test of the effectiveness of the servo amplifiers and of the servo motors.

In a parked airplane, if the attitude indication is not normal, this test would indicate whether the trouble is probably in the vertical gyro, or in the servo amplifier, or in the horizon indicator. Late model horizon indicators have the servo amplifiers and their null monitors included in their indicator cases.

Figure 5-20

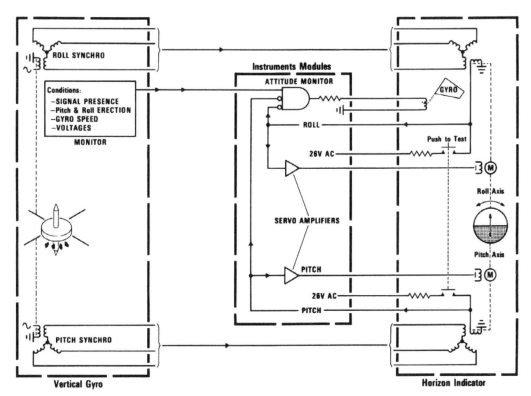

Remote Directional Gyro Schematic

A remote directional gyro, represented in Figure 5-21, is made large and remote for the same reasons as a remote vertical gyro. The spin axis of a directional gyro is maintained in a horizontal position by an automatic leveling system consisting of a torquer motor and level sensing device similar to those used in a vertical gyro.

Only one synchro is shown connected to the gyro. Usually there are two — one at the top and one at the bottom — both giving the same information. The only information supplied by a directional gyro is azimuth (around the horizon) information.

Because of the way the directional gyro is used in a remote reading compass, the function utilized is its stability, and it is just as stable whether the spin axis is pointing north, northwest, southeast or any other direction. Therefore, in all but a very few old remote directional gyro systems, the orientation of the spin axis is of no consequence at the time it is erected.

Later compass systems do not slave the directional gyro, but permit it to drift. Slaving is an operation which, in earlier but still very common systems, maintains the spin axis oriented toward whatever magnetic compass heading it had at the time it was erected.

For example, if it happened to be erected at a compass heading of 187°, then the slaving system would maintain the spin axis in that particular direction as long as the flight continued. In those compass systems that do slave the remote gyro, slaving provides two principal corrections. One corrects for precessing resulting from relative movement between magnetic north and geographic north as an airplane moves across the face of the earth.

In the present systems which permit the directional gyro to drift, the two corrections are made by utilizing a differential synchro operation. Gyro slaving and the differential synchro operation are covered in more detail later.

Figure 5-21

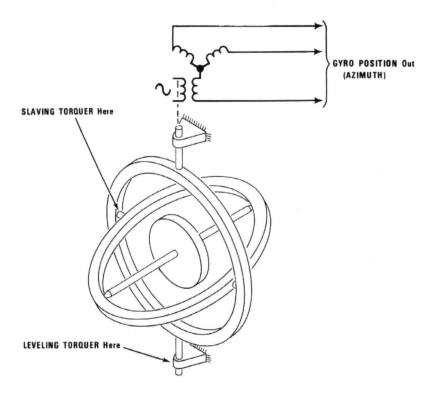

GYRO POSITION Out
(AZIMUTH)

SLAVING TORQUER Here

LEVELING TORQUER Here

Rate Gyro Inside Turn Indicator

The rate gyro principle of operation is shown in Figure 5-22. This particular rate gyro is shown as it could be used inside a panel mounted rate-of-turn indicator. It is, therefore, a yaw rate gyro. Rate gyros are also used by autopilots, particularly in the roll and pitch axes.

Rate gyros are a different class of gyros which do not have complete freedom in all three axes. The freedom of movement in one axis is carefully limited. This gyro has limited movement around the X axis. The pointed shaft ends, pointing forward and aft on the gimbal, fit into bearings in the housing of the turn indicator. The gimbal is capable of moving around the X axis, but its motion is limited by the spring.

If the airplane turns to the left, it pushes on the gyro spin axis trying, with limited success, to rotate the gyro to the left. However, the gyro doesn't like this kind of treatment and resists by trying to roll over; that is, trying to rotate around the X axis in a direction at 90° to the applied force.

The airplane rate of turn determines the amount of force applied to the gyro, which in turn deter-mines how far that spring is going to be stretched as the gyro rotates around the X axis. The indicator rate-of-turn needle is coupled to the gimbal, and as the gimbal rotates around the X axis, the rate of turn is indicated by the amount of motion of the needle.

The rate-of-turn indicators with gyros inside are no longer required in commercial aircraft. If a rate-of-turn indication is given, it is obtained from a remote rate gyro.

Figure 5-22

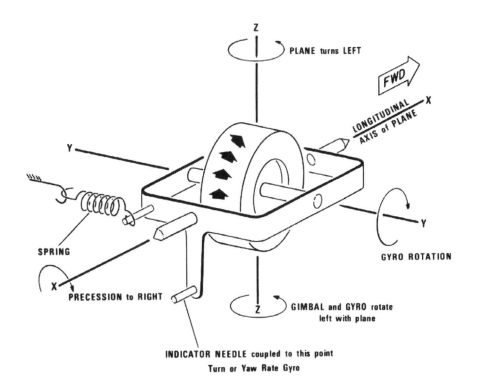

Turn or Yaw Rate Gyro

Turn And Slip Indication

Figures 5-23, 5-24 and 5-25 are examples of turn and slip indications. In these illustrations the indicator becomes an airplane fuselage with wings extending outboard. If you imagine yourself seated in the cockpit looking at this instrument, you will have to give yourself the same bank angle the airplane has.

Let's start with the three illustrations of Figure 5-24 because the airplane is not turning. Since it is not turning, the rate of turn indicator needle is centered. In the top illustration with the left wing down, the ball slip indicator has fallen off to the left side, due to the force of gravity.

In the middle illustration, with the right wing down the ball has fallen off to the right side.

The bottom illustration shows the wings level and the ball centered.

Figure 5-23 illustrates three different kinds of left turn at the rate of 180° per minute. In the top illustration there is too much bank angle and the airplane is slipping off towards the left. The 180° rate of turn is shown by the turn indicator needle centered just below the left dot. The ball indicator shows that the airplane is banked too much and is slipping to the left. The center illustration shows the ball centered. This constitutes a coordinated turn. The force of gravity combined with the centrifugal force of turning keeps the ball centered in its tube.

If you are riding in an airplane during a coordinated turn, which practically all commercial air-craft turns are, you do not feel any force pushing you to the left or the right in your seat, and that glass of champagne on the table in front of you looks the same as it would sitting on your coffee table at home. If you, however, are seated in an airplane slipping to the left (top illustration in Figure 5-23), you would feel pushed against the left side of the seat and your champagne would slop out of the left side of the glass.

The bottom illustration shows a left turn with too little bank. The airplane is skidding off to the right. In this case the champagne is going to be sloshing out of the right side of the glass.

Figure 5-25 illustrates the indications for a slipping, a coordinated, and a skidding turn to the right.

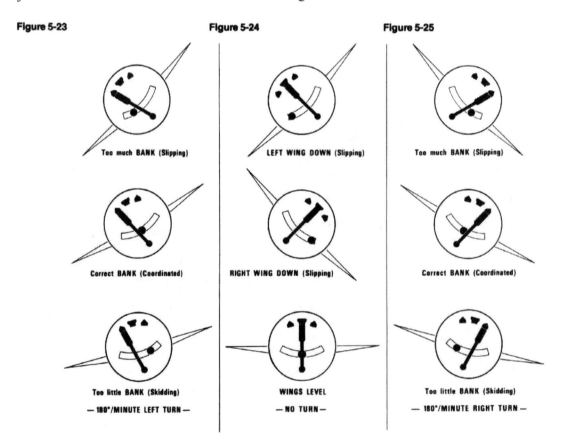

Figure 5-23   Figure 5-24   Figure 5-25

Too much BANK (Slipping)   LEFT WING DOWN (Slipping)   Too much BANK (Slipping)

Correct BANK (Coordinated)   RIGHT WING DOWN (Slipping)   Correct BANK (Coordinated)

Too little BANK (Skidding)   WINGS LEVEL   Too little BANK (Skidding)

— 180°/MINUTE LEFT TURN —   — NO TURN —   — 180°/MINUTE RIGHT TURN —

Basic Rate Gyro

Figure 5-26 schematically shows the construction of a basic remote rate gyro. Instead of a spring restricting the motion in one axis, there is a torsion bar. On the side opposite the torsion bar is a transmit synchro whose rotor is attached to the gimbal.

When the gimbal is at rest, there is no force on the torsion bar, and the output from the synchro is a null signal. This rate gyro is shown drawn as it would be positioned in the aircraft for a rate-of-turn gyro. The rate-of-turn signal is of one phase or the other depending upon the direction of turn. The rate of turn determines the amplitude.

Figure 5-26

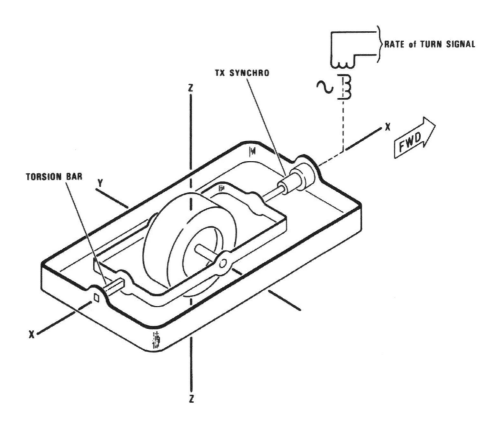

Ring Laser Gyro

Figure 5-27 illustrates the ring laser gyro, which uses two laser light beams to measure angular rotation. A laser is a sophisticated device with four working parts.

The Excitation Mechanism: High voltage is applied between the cathode and the anode, which ionizes the helium-neon gas and produces two beams of laser light around the gyro, in opposite directions.

The Gain: The gain helps to overcome inherent natural losses of any system. Gain is accomplished by ionizing low pressure helium-neon gas with high voltage to produce glow discharge.

The Feedback Mechanism: This is where the glow discharge light fills a triangular cavity, and reflects around the cavity by mirrors mounted in each corner.

The Output Coupler: One of three corners of a triangle; contains a prism to allow two beams to mix together to form a readout detector.

If the gyro is stationary, the frequencies of both light beams remain the same, as does the fringe pattern. The fringe pattern corresponds to gyro rotations, by movements to left and right. Photo diodes located in the readout detector convert fringe pattern movements into electrical pulses.

The two light beams get coupled together at low rotation rates and have a condition called laser lock-in.

A dither motor helps prevent the loss of information at these low rates by vibrating the gyro assembly. Through the lock-in region, the dither motor vibrations can be felt on the IRU case and makes a humming sound.

Figure 5-27

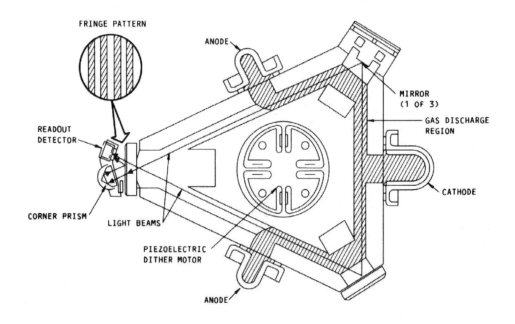

Inertial Reference Unit

The main function of each inertial reference unit (Figure 5-27a) is to sense and compute linear accelerations and angular turning rates about the airplane's pitch, roll and yaw axes. This data is used for pitch and roll displays and navigational computations.

Each IRU provides guidance reference to the airplane by sensing angular rate and linear acceleration. A digital computer is in each inertial reference unit to perform calculations for the inertial reference system.

Each IRU contains three laser gyros, three accelerometers, fourteen circuit cards, power supply and chassis. These sense angular rates and linear accelerations respectively. This sensed data and local vertical coordinates, combined with air data inputs, serve to compute the following: position (latitude and longitude), attitude (pitch, roll and yaw), true and magnetic heading, wind speed and direction, velocity, accelerations, angular rate data and altitude.

When airplane power fails, the IRU automatically switches to battery power. The left and center systems stay on battery power for a maximum of five minutes. The right systems stay on battery power indefinitely. All three systems stay on power during autoland.

In the inertial reference system, the gyros and accelerometers provide respectively: precise angular rate information about the airplane's pitch, roll, and yaw axes; and linear accelerations along these three axes. The computer performs all computations required for navigation, plus sensor systematic error compensation for known fixed biases: misalignments and thermal effects.

Using the same sensor data, the computer also provides attitude, ground speed and track angle, wind speed and direction, and true magnetic heading info to automatic flight control.

There are no illustrations as this is a black box.

Figure 5-27a

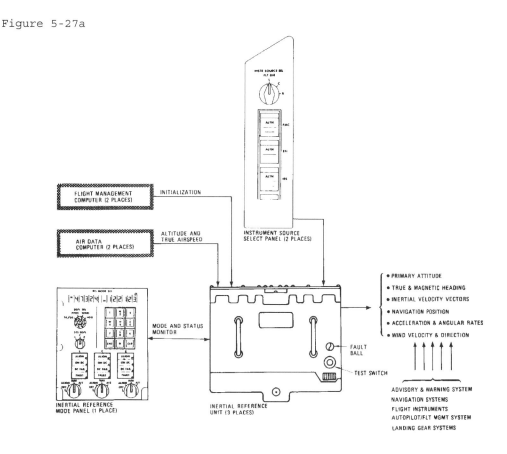

Air Data Inertial Reference Unit
(Typical Boeing 777)

The Air Data Inertial Reference System (ADIRS) (Figure 5-27b) consists of:

- One Air Data Inertial Reference Unit (ADIRU)
- One Secondary attitude Air data Reference Unit (SAARU)
- Eight air data modules (ADMs)
- A standby attitude indicator
- The air data sensors

The ADIRS function the same as a system of three IRUs, two ADCs and its pressure, temperature, and angle of attack sensors. The ADIRS sends primary, secondary, and standby air data and inertial reference information to the flight deck displays, flight controls, autopilot system, and other airplane systems.

The ADIRU uses six ring laser gyro sensors, six linear accelerometer sensors, four processors, three power supplies and three dual-channel ARINC 629 interfaces

The ring laser gyro and linear accelerometers are along six non-parallel, symmetrically skewed axes. This orientation gives a highly fault-tolerant system.

The ADIRU uses ring laser gyros and accelerometers to sense angular rates and linear accelerations. This data is calculated with air data inputs to give:

- Attitude (pitch, roll, and yaw)
- Position (latitude and longitude)
- True heading
- Magnetic heading
- Inertial velocity vectors
- Linear accelerations
- Angular rates
- Track angle
- Wind speed and direction
- Inertial altitude
- Vertical speed

If a failure occurs, all IRU functions are available with three gyros and accelerometers, one power supply, one processor and a single ARINC 629 interface.

Continued on next page ...

Figure 5-27b

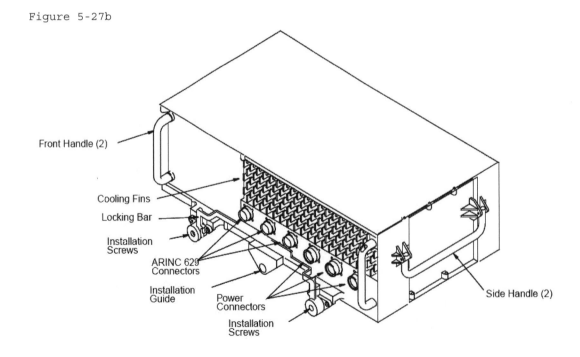

Air Data Inertial Reference System (cont'd.)

The Secondary Attitude Air Data Reference Unit (SAARU) (Figure 5-27c) supplies pitch and roll attitude to the standby attitude indicators. It is also the secondary source of inertial navigation and air data for the Primary Flight Displays (PFDs), primary flight controls system (PFCS), autopilot flight director system (AFDS) and other airplane systems.

During a catastrophic failure of the ADIRU, the SAARU gives the AFDS reduced navigation data.

IRS AND AIR DATA FUNCTIONS

The SAARU uses:

- Four fiber optic rate gyros
- Four analog linear accelerometers
- Two processors
- Three separate ARINC 629 interfaces

Three of the gyro and accelerometers sensors are in the pitch roll and yaw axes, and the fourth sensor pair is in a skewed position.

The ADIRU and the SAARU send inertial reference data and air data to the:

- Airplane Information Management System (AIMS)
- Autopilot Flight Director System (AFDS)
- Primary Flight Control System (PFCS)
- Control Display Units (CDUs)

AIMS sends the inertial reference data and air data to many other airplane systems and systems components. These include:

- GPS
- WES
- TCAS
- EEC
- APU

Figure 5-27c

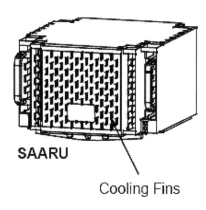

SAARU

Cooling Fins

Gimbal vs. Strapdown

Fig. 5-28, The Gimbal System, isolates the acceleration sensing devices from rotational motion of the airplane, and maintains the platform parallel to the earth's rotation. The gimbal system effectively de-couples the inertial platform from the airframe and provides an all-attitude freedom of motion.

The platform is stabilized by use of three single degree of freedom gyros mounted on the platform in the gimbal system. The gyros are mounted with their sense (input) axes mutually perpendicular. Any minute tilt or rotation of platform produces an input torque about the sense axes of one or more of the gyros. This induces gyroscopic precession, which causes the gyro gimbal to rotate about the gyro output axis and generate an output signal from the gyro pickoff, which represents the angular displacement of the gimbal.

This amplified signal is used to drive the plat-form gimbal torquer through a corresponding angle of rotation to maintain the gyro at a null, and maintain correct platform orientation.

The Strapdown System is said to be more reliable than a gimbal system (Figure 5-29). A strap-down

system's accelerometers and gyros are mounted on, or "strapped down" to, the body of the vehicle, in contrast to the gimbal system, where the accelerometers are mounted on a platform, and the gimbaled gyros allow it to move independently of the airplane's movement.

Many Strapdown systems use laser rate sensors, aligned with the airplane's axes, to get data that is used to interface with the navigation function.

The laser rate sensors have no moving parts, are not body rate limited, and have very little drift. Gyro and accelerometer readings are fed in digital form to a computer which calculates the actual altitude velocity and position of the airplane, compares to desired values, and initiates whatever correction may be necessary.

The total number of sensors in each system is twelve (six gyros, six accelerometers). If any sensor in the gimbal system fails, the entire system is useless (navigational failure).

The Strapdown system remains operational, even after three gyros and three accelerometers have failed.

Figure 5-28

Figure 5-29

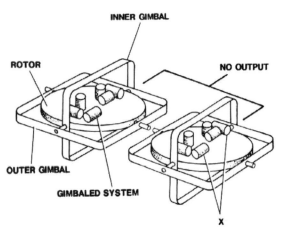

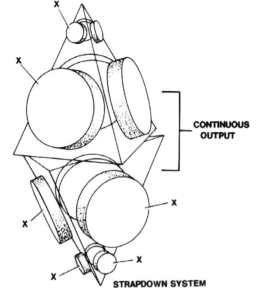

CHAPTER 6 - COMPASS SYSTEMS

Earth's Magnetic Field

In order to understand the functioning of an electrical compass system, we need to know something about the earth's magnetic field. Figure 6-1 illustrates the approximate directions of the earth's field at different places around the earth. It can be visualized as something like that developed by a huge cylindrical bar magnet buried in the interior of the earth. The field lines are in the pattern you would expect from iron filings scattered around a cylindrical bar magnet.

The direction of the earth's field at the north and south magnetic poles is vertical. At the magnetic equator, it is horizontal and toward the poles.

The only information that we want from a compass is the horizontal direction toward the north magnetic pole. But everywhere, excepting at the magnetic equator, the lines of force point not only toward the pole, but also downward toward the surface of the earth. If you consider only the

horizontal component, vectorially derived, of the earth's field, you see that as we get closer to one of the magnetic poles, the horizontal component becomes smaller. At the poles it is zero.

Since the horizontal component is all that we are interested in, the magnetic fields near the poles are not of much use to us, not only because they weaken, but because their direction changes rapidly. This becomes more obvious in Figure 6-2.

When we are walking around on the earth carrying a pocket compass, it is easy enough to hold the compass approximately horizontal, and therefore read on it only the horizontal component. But in an airplane a magnetic compass tips when the airplane tips.

A later diagram indicates the difficulty that arises with an airplane magnetic compass when pitched or rolled away from the level position.

Figure 6-1

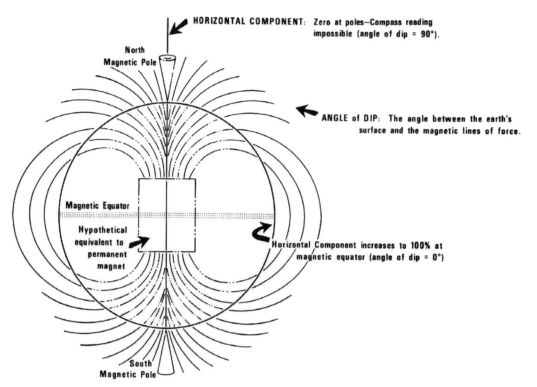

Magnetic Variation

Figure 6-2 shows the difference between geographic north and the north magnetic pole. Since our compass is not going to tell us about the north geographic pole, we have to know for any particular location the difference in heading between magnetic north and true north.

The difference at JFK is that magnetic north is 11° west of true north. If we happen to be in Fort Wayne, we are lucky because magnetic north corresponds with true north. West of Fort Wayne, the sense of the difference reverses. At San Francisco magnetic north is 171/2° east of true north. So even though we have a perfect, no drift directional gyro, reading correctly at JFK, it still must be torqued through 281/2° during the flight from JFK to San Francisco to correct for this continual change in variation.

Putting it another way, if we maintain the same free space direction with our perfect directional gyro all the way across the country from JFK to San Francisco, our gyro would have introduced an error of 281/2° by the time we arrive in San Francisco.

Earlier remote reading electrical compass systems use a slaving system to provide the torquing force, keeping the position of the gyro constantly corrected for changes in position on the earth's surface, and for drift errors resulting from gyro imperfection. Later compass systems accomplish the correction with a servo system in a black box.

Figure 6-2

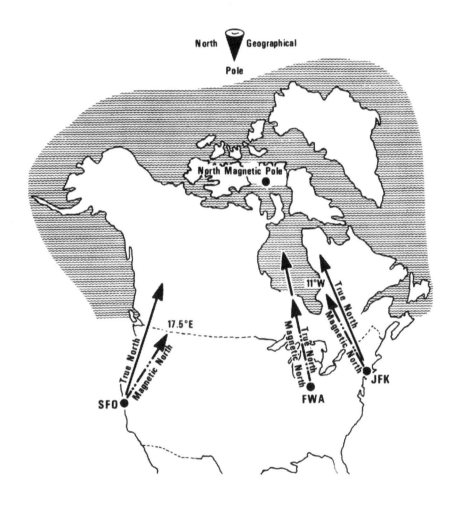

Magnetic Compass In Level Flight

Figure 6-3 represents a typical aircraft magnetic compass. The side view is rotated to show how it would appear to the pilot in the airplane in Figure 6-4.

The compass card assembly includes a pair of needle magnets attached below it. The weight of the needle magnets and the fact that the suspension point is above them keep the compass card level when the airplane is in level flight. When the bar magnets are level, they are influenced only by the horizontal component of the earth's field, and the compass gives a reliable, though unsteady, indication of the heading of the airplane.

Examining the top view of the compass card in Figure 6-3, you notice that magnetic north is toward the top of the page or at the point marked "S" at the compass card. The pilot's view of the compass card, however, is from our side, so the letter "N" appears at the bottom.

If the airplane is headed north, the letter "N" will appear at the lubber line fixed to the case in the pilot's view. Figure 6-4 illustrates our side view of a magnetic compass in the level airplane heading north. The lubber line is at the letter "N" (out of view). The compass needle magnets see only the horizontal component of the earth's field. If the airplane is not wings level, the compass cannot very well be horizontal. This introduces an error called "northerly turning error," discussed next in Figure 6-5.

Figure 6-3

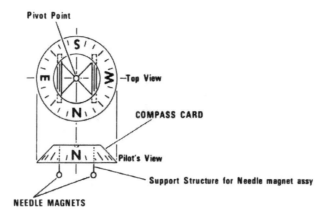

Figure 6-4

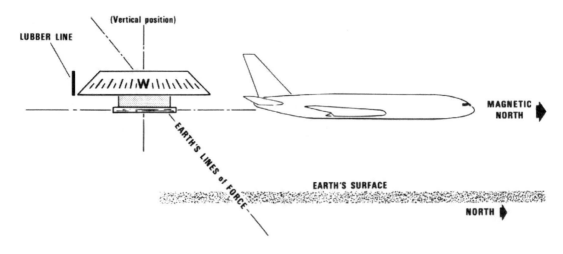

Magnetic Compass When Banked

In Figure 6-5 the airplane is headed north and banked right. In the United States the earth's lines of force are tipped toward the horizon ahead of the airplane. The compass, no longer restricted to the horizontal component, has tipped and turned in the direction of the arrows in order to line itself up with the earth's field.

Even though we have not changed the heading of the airplane, but merely banked it, the compass card has swung around, indicating that we have turned toward the west, when in fact we are about to turn east. This is shown by the letter "W" having been moved closer to the lubber line than it was in Figure 6-4. If such a maneuver is very sudden, the compass card may be given enough turning momentum to spin it completely around.

The disturbance to the magnetic compass by the airplane banking is sufficient to make the compass card completely useless in most maneuvers. The first improvement over a plain magnetic compass for airplanes was the gyro compass mounted in a panel indicator (Figure 5-4). This gave a reliable heading indication for about fifteen minutes. Prior to making a maneuver, the pilot would manually set the gyro compass to agree with the magnetic compass in level flight. Then he could forget the magnetic compass for a few minutes.

The gyro compass became obsolete when electrical remote reading, gyro stabilized compass systems were introduced.

Figure 6-5

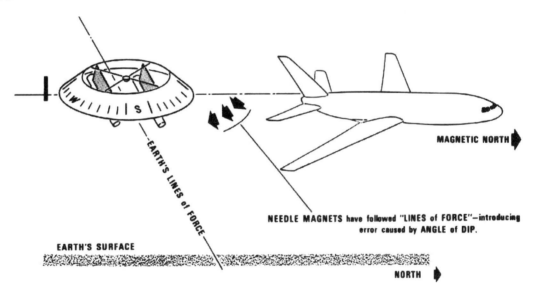

EARTH'S LINES of FORCE

EARTH'S SURFACE

NEEDLE MAGNETS have followed "LINES of FORCE"—introducing error caused by ANGLE of DIP.

MAGNETIC NORTH

NORTH

Flux Valve Frame

Figure 6-6 shows the top and sectional views of a flux valve frame. It electronically "valves" the earth's magnetic field into and out of its frame 800 times each second in order to sense the direction of magnetic north. It is located in the airplane where it will be as free as possible from magnetic influences other than the earth's field. The theory of operation is explained in the next several pages.

The frame is composed of two pieces of sheet metal formed together across a center post. The spokes from the center post to the rim are called spider legs. The excitation coil is wound around the center post, and the signal sensing coils are wound around the spider legs.

A 400 Hertz voltage is fed into the excitation coil. This voltage produces enough ampere turns at its peaks (800 per second) to develop a magnetic field strong enough to practically saturate the spider legs. The design of the frame makes it possible to bring the spider legs very close to magnetic saturation without magnetic spillover into the surrounding air between the upper and lower pieces of sheet metal. The rim is important in this because magnetic spillover which might otherwise spread into the air near the signal sensing coils moves into the rim section. This careful designing makes possible the pulsing of strong magnetic fields in and out of the legs with little effect on the signal coils.

The magnetic fields from the exciter have a minimum effect on the signal sensing coils for three reasons. First, the excitation coil is perpendicular to all three of the signal sensing coils. Second, the design of the flux valve frame keeps the excitation magnetic fields within the metal of the frame. Third, the flux valve signals (explained later) are 800 Hertz because the spider legs are saturated 800 times per second, and the unwanted residual 400 Hertz voltages are filtered out of the signal winding outputs.

Figure 6-6

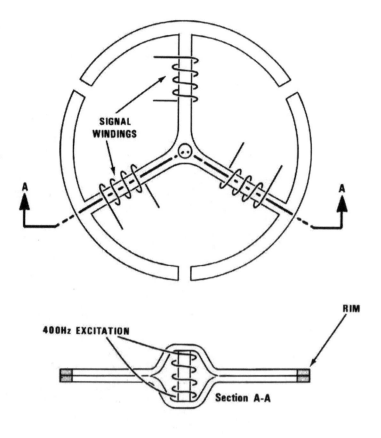

Flux Valve Cross Sectional View

Figure 6-7 is a schematic sectional view of a flux valve. Two hermetically sealed cases are shown. The inner case contains the electrical elements. It is suspended from its top by a universal joint. The universal joint permits the sensing element to swing back and forth in any direction, within limits, with no rotation inside the outer mounting case.

The space between the two cases is filled with a damping fluid which prevents the sensing element from becoming an active pendulum.

The universal joint keeps the flux valve horizontal over the long term when the airplane is flying wings level, but not nose on the horizon.

Since airplanes seldom fly with their noses exactly on the horizon, if this sensing element were rigidly mounted, it would be reading some of the vertical components of the earth's field. It also keeps the flux valve horizontal when the airplane is on the ground, even though the airplane's nose might be pitched down, as in the DC-8; or one wing down because of a lopsided fuel load or unevenly inflated struts. Without the universal joint, accurate compass swings or ground readings would be unlikely.

The fact that the flux valve is not horizontal during turns or during acceleration is not significant because the primary heading reference (direct control of the cockpit indication) is the directional gyro, which is quite stable.

The flux valve function is to correct, very slowly (3° per minute), gyro errors over the long term (slaving). The short term errors in the flux valve signal when the airplane is banked can influence the gyro only at the rate of 3° per minute, and average

themselves out in extended turns. In short turns, only a small gyro error could be introduced, which would subsequently be corrected in level flight at the same rate.

Even if we could hold an error from the flux valve in flight for as long as one minute, the resulting error in the compass indications could not be more than 3°. This error would be eliminated during the next minute of level flight, or by a compensating error in the other direction during the turn.

The mounting flange has elongated mounting holes making it possible to turn the flux valve in its aircraft mounting if necessary for proper alignment. The alignment of a flux valve is critical and must be carefully accomplished under controlled conditions. If the flux valve should be installed rotated a few degrees away from its correct position, the compass indication is in error by that same number of degrees. This is the only error that will show up on all aircraft headings in the same sense.

All other errors, such as magnetic material too close to the flux valve, or wiring problems, average themselves out (if designated plus and minus errors) through 360° of aircraft heading.

The flux valves are now usually mounted on indexing brackets from which they should not be removed. The shop carefully positions the flux valve on its bracket, and the mating bracket in the airplane is carefully aligned with the airplane's fore and aft line. If you have to change a flux valve without the benefit of an indexing bracket, you should be careful to position the new flux valve in exactly the same alignment as the old.

Figure 6-7

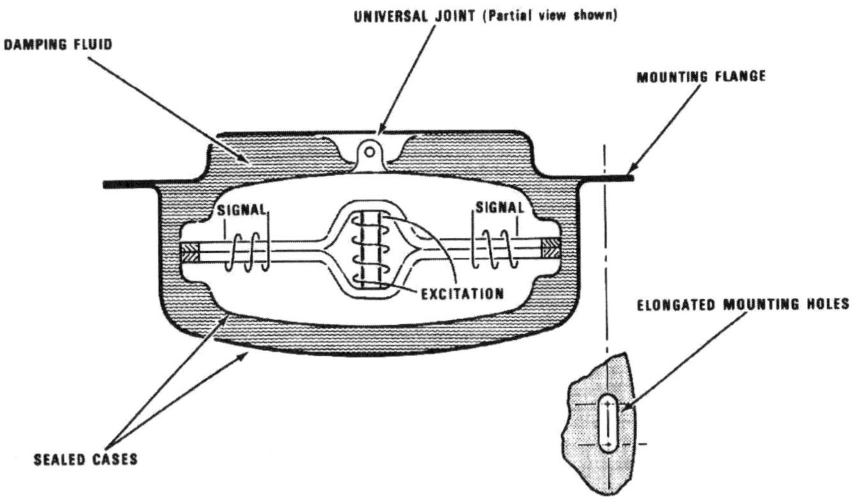

Pulsing Earth's Field

Figures 6-8 through 6-13 are sectional views of a spider leg of the flux valve frame, including the center post and the exciter winding. The rim is omitted. The top and bottom sections of the spider leg are shown slightly separated to help explain the operation. In the actual flux valve frame, these sections are tightly pressed together.

Figure 6-8 illustrates the condition at the instant when the exciter winding current is at maximum, magnetically saturating the spider leg. The earth's magnetic field is shown surrounding the spider leg in an undistorted pattern. Since the spider leg is already saturated with magnetic current generated by the exciter winding, the earth's field is no more distorted than it would be by passing through air or glass.

In Figure 6-9 the exciter current has reached a null, and the spider legs have no magnetic flux from the exciter current. The earth's field sees this leg as a path of less reluctance and is drawn into the spider leg.

In Figure 6-10 the exciter current has again reached maximum and has driven the earth's field out of the leg.

In figure 6-11 a signal coil is shown wrapped around the spider leg and the condition is the same as in figure 6-8, with none of the earth's field in the spider leg.

In Figure 6-12, the exciter current has reached a null. While the earth's field was moving into the leg it was cutting the windings of the signal coil. This is the same action performed in a transformer or generator to produce a voltage. While the field was cutting the windings of the signal coil, it generated a voltage of a particular polarity because the direction of movement of the earth's field was inward.

Figure 6-13 shows the exciter current at maximum and the spider leg saturated. This time the earth's field moved outward, so the polarity of the generated voltage was reversed.

The exciter winding uses a voltage derived from aircraft power with its frequency of 400 Hertz. The effect on the earth's field movement is the same whether the spider leg is saturated with magnetic current of one direction or the other.

Since the exciter current reaches a maximum twice during each cycle and a null twice during each cycle, the frequency of the earth's field-induced signal in the signal coil is double the frequency of the exciter voltage. This is an advantage in the use of the signal coil voltages since any unwanted 400 Hertz signal can be filtered out, making sure that only voltages generated by the movement of the earth's field are used in the circuit.

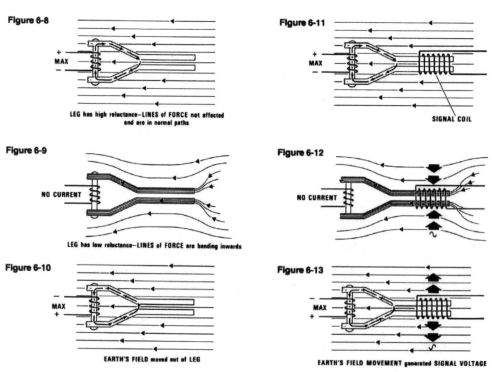

Figure 6-8
LEG has high reluctance—LINES of FORCE not affected and are in normal paths

Figure 6-9
LEG has low reluctance—LINES of FORCE are bending inwards

Figure 6-10
EARTH'S FIELD moved out of LEG

Figure 6-11
SIGNAL COIL

Figure 6-12
NO CURRENT

Figure 6-13
EARTH'S FIELD MOVEMENT generated SIGNAL VOLTAGE

Flux Valve Heading Voltages

Figures 6-14 and 6-15 are illustrations of how the earth's field would pass through the frame of a flux valve in two different positions. The signal coil windings have been omitted in order to show this more clearly. From these illustrations you can see how the different signal coils will generate signals which depend upon their individual positions relative to the direction of the earth's field.

In Figure 6-14 the leg of coil No. 1 is parallel to the earth's field; therefore, the greatest amount of earth's field is being pulsed in and out of that leg, developing a maximum voltage. The coils on legs No. 2 and No. 3 have less of the earth's field pulsing in and out, and their voltages are less, but equal. The signals developed by these three coils are comparable to those of a three-legged transmit synchro with its signal coil rotor parallel to leg No. 1.

In Figure 6-15, the airplane and flux valve are turned 90° to the right. With the flux valve in this position there is no pulsing of the earth's field in and out of leg No. 1, at least not in a manner which would cut the windings on the coil and generate a signal. Legs No. 2 and No. 3, however, have considerable earth's field pulsing in and out; therefore their voltages are fairly strong. These are strong voltages only comparatively speaking. Actual signals from flux valve windings are of a low level.

In Figure 6-15, the signals generated by the flux valve's three signal coils are comparable to those of the stator of a three-legged transmit synchro whose rotor is perpendicular to leg No.1. By extending this analogy we can see that the signals generated in the three legs of the flux valve are comparable to those in the stator of a three-legged transmit synchro whose rotor is always held in a fixed position, relative to the fore and aft line of the airplane.

The signals of the flux valve signal coils are, in fact, used just as if the flux valve were a transmit synchro. The flux valve signal output is connected to a receiving synchro in a heading indicator or compass coupler box where the position of the resultant magnetic field in the stator of the receiving synchro is a direct function of the aircraft heading.

The flux valve is more stable than a magnetic compass because it is not subject to the momentum forces which cause a magnetic compass to overshoot, then swing back toward the correct indication, and overshoot again. However, it is still subject to the northerly turning error which results when it is not horizontal. Its signal is consequently not used for direct indication in the cockpit compass indicator. The cockpit indication comes from a stabilized source, a gyro or an inertial navigation azimuth head in signal. The function of the flux valve is to correct, as necessary over the long term, the stabilized source information.

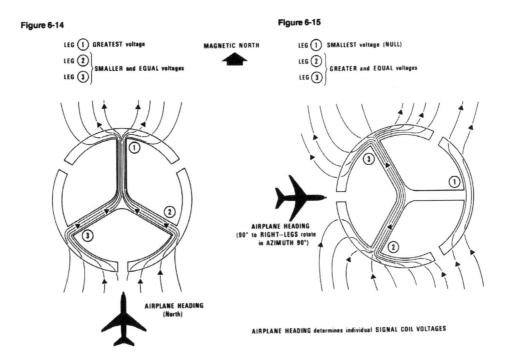

Figure 6-14

LEG ① GREATEST voltage
LEG ② } SMALLER and EQUAL voltages
LEG ③

MAGNETIC NORTH

Figure 6-15

LEG ① SMALLEST voltage (NULL)
LEG ② } GREATER and EQUAL voltages
LEG ③

AIRPLANE HEADING
(90° to RIGHT—LEGS rotate
in AZIMUTH 90°)

AIRPLANE HEADING
(North)

AIRPLANE HEADING determines individual SIGNAL COIL VOLTAGES

Flux Valve System Schematic

Figure 6-16 shows a typical compass system using a directional gyro slaved by information from the flux valve. The latest airplanes have a different version using a compass coupler, discussed later. However, there are still thousands of airplanes using this basic compass system. The compass coupler arrangement will be easier to understand if you first understand this one.

The item called "master indicator" gives several indications besides compass heading, which leads to its having different names (reference Figure 6-17). Sperry Corporation designates it "MHR-4" (Master Heading Radio-4). To be consistent and more descriptive, we will use the title "Heading and Radio Pictorial Indicator". In this system, the title "Master Indicator" is given because it is the location of centralized control.

"Directional gyro" has been divided horizontally into three sections to emphasize the separateness of the three electrical circuits shown. The interconnection between the circuits is strictly mechanical.

We will discuss circuit No. 2 first, since it is the one which determines the reading on the master heading indicator. This statement is essential to the explanation of the operation of this system. It also brings up a point about which there is a great deal of confusion. Notice that the flux valve is not in this circuit. The flux valve, therefore, is not the immediate determining factor in the compass card indication.

The directional gyro determines the compass card reading in this system. The signal source in circuit No. 2 is the transmit synchro in the directional gyro. This synchro gives the position of the directional gyro relative to aircraft heading. As the aircraft heading changes, the rotor of the synchro holds its position in space because it is connected to the spin axis of the gyro, which holds its own position in space.

The outside case of the gyro carries the stator of this transmit synchro, and moves in azimuth as the airplane changes heading. Consequently, during heading changes, the rotor of the transmit synchro moves with respect to the stator, which moves the field in the stator of the control synchro in the master indicator.

The rotor of this control synchro is driven by a servo motor so that it follows up on heading changes, a typical servo loop operation. The servo motor drives the compass card and all three synchro rotors simultaneously degree for degree, so rotor positions represent the heading indicated on the compass card.

As long as: a) the directional gyro is operational,

b) the transmit synchro generates a valid signal, and c) the master indicator servo loop is functional, any aircraft heading change results in a corresponding change of the compass indication, whether the flux valve circuit is operative or not. It makes no difference whether that indication is correct or incorrect — this system functions as described.

Circuit No. 1 is the flux valve circuit. The flux valve circuit makes it possible for the compass indication to be correct, and also for the pilot to know whether the indication is correct in a normal system.

The annunciator is a zero centering DC meter movement whose indication (through a triangular opening) is a blank (null) if the compass heading is correct. If it is not correct, the annunciator shows a cross or a dot. The function of the flux valve circuit is to torque the directional gyro in azimuth in one direction or the other (slaving). It does this if the compass card indication disagrees with the flux valve information.

The slaving torquer in the directional gyro is a low powered torquer which cannot move the gyro around any faster than 3° per minute. Consequently, northerly turning errors which the flux valve sees when the airplane is in a bank have little effect on the directional gyro rotor position. If the airplane remains in a bank long enough to make a 180° turn, the northerly turning error that the flux valve sees during this time is reversed during the turn; consequently, the net effect on the directional gyro is null.

If the northerly turning error is not averaged out to a null during the turn, the small error introduced in the directional gyro rotor position will be corrected when the airplane returns to wings level position.

Continued on next page ...

Flux Valve System Schematic (cont'd.)

The fact that the interconnections between these three separate circuits is mechanical results in the need for very careful positioning of the mechanical components in these items. For example, the synchro rotors in the master indicator must be care-fully positioned when the indicator is assembled so that their positions have the correct significance. The position of the flux valve in the airplane, the position of the gyro transmit synchro, and the position of indicator synchros all must be carefully shop set.

The operation of the flux valve circuit can be described as follows. As mentioned previously, the signal from the flux valve is used as a transmit synchro signal. In the master indicator the resultant field in the stator of the comparator synchro becomes the airplane heading as seen electrically by the flux valve. If the comparator synchro rotor is perpendicular to the field in the stator, there is no signal out to the slaving amplifier, no signal from

the slaving amplifier through the de-modulator to the annunciator, and therefore a null indication on the annunciator. In this case, the slaving torque in the directional gyro does not have a signal, and the position of the gyro rotor would not be affected.

A null signal from the computer synchro rotor is present if the heading indicated on the card of the master indicator is correct as seen by the flux valve (or if it is 180° away from the correct heading). If the synchro rotor signal is not a null, it is of one phase or the other, causing the slaving amplifier to give the slaving torque a signal which slowly moves the directional gyro rotor in one direction or the other. The direction in which the gyro rotor is moved causes the compass card to move toward the correct indication. During this time the de-modulator has an output, and the annunciator shows either a cross or a dot.

Continued on next page ...

Figure 6-16

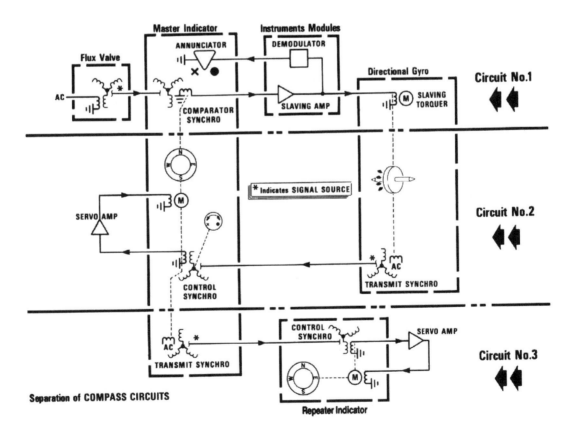

Flux Valve System Schematic (cont'd.)

Suppose that the compass card indication is 90° away from the correct indication. With the flux valve circuit torquing the directional gyro at the rate of 3° per minute, 30 minutes later the compass card indication would have moved through 90°, and the rotor of the comparator synchro would be at the correct null. Since that is much too long to wait for a correct indication, a synchronizing operation can be performed with the knob on the face of the master indicator.

When this knob is pulled out, it is connected through gears to the special movable stator of the control synchro in the master indicator servo loop. On the face of the knob are two arrows, a cross and a dot. These arrows tell you which way to turn the knob if the annunciator indication is a cross or a dot. Turning the knob mechanically rotates the stator of the control synchro, causing the servo motor loop to drive the compass card and, at the same time, the rotor of the flux valve comparator synchro to the correct null and the compass card to the correct heading.

The compass card indication now shows the aircraft heading as seen by the flux valve, so it becomes unnecessary to torque the directional gyro. The actual position of the directional gyro spin axis is of no consequence to the time we synchronize the compass indication. We will use the position of the directional gyro, whatever it may happen to be at the time we synchronize the compass.

Think of the comparator synchro as comparing aircraft heading as seen by the flux valve (field in the synchro stator) with the airplane heading as shown by the master heading indicator (synchro rotor position).

Once the compass system has been synchronized in this manner, further synchronizing is unnecessary. The directional gyro drift is considerably slower than the 3° per minute correction capability of the slaving system, probably less than 10° per hour.

Change in the position of magnetic north with respect to true north (Figure 6-2), caused by the airplane moving across the earth's surface, will also be considerably less than 3° per minute. The slaving action of circuit No. 1 is always adequate to correct both types of error.

By now it should be clear that this is a gyro stabilized, flux valve controlled system.

If the slaving system should become inoperative, as for example by failure of the slaving amplifier or loss of excitation voltage to the flux valve, the system would continue to operate, but without the correcting influence of the flux valve. The indications in the master indicator would continue to be reasonably correct for approximately 20 minutes, depending upon the rate of gyro drift and the rate of change of magnetic north with respect to true north. Since all airplanes have two of these systems, the one with the inoperative slaving could be periodically corrected to agree with the operative compass system.

Circuit No. 3 is the schematic for a repeater indicator, usually a radio magnetic indicator (RMI, Figure 9-34).

Only the compass portion of that indicator is shown here. The transmit synchro rotor in circuit No. 3 is positioned in the master indicator at the time of assembly so that the signal generated can be interpreted by the receiving control synchro in a repeater indicator as the heading shown on the master indicator compass card. The servo motor loop in the repeater indicator, therefore, keeps its compass card rotated to always indicate the same as the master heading indicator.

The actual location of the servo amps is a matter of convenience for the aircraft manufacturer. In late model RMIs the servo amp is contained within the case of the indicator itself.

Continued on next page ...

Flux Valve System Schematic (cont'd.)

A good bit can be learned about which item is faulty in the compass system by observing cockpit operations. For example, if, when you turn the synchronizing knob, the master indicator compass card follows correctly the motion of the knob, then you know that the master indicator servo motor and its servo amplifier are operating correctly, and that the signal source from the directional gyro is present. If, when the synchronizing knob is turned, the annunciator indication does not change as expected, then something is wrong with the flux valve, its signal, or the slaving amplifier.

Since the annunciator itself is a simple DC meter movement, it is probably not at fault. If a null indication in the annunciator is always present, regardless of where the synchronizing knob is turned, then either the flux valve is not excited; or there is open wiring between the flux valve and the indicator; or the slaving amplifier is dead; or there is a wiring problem between the slaving amp and the annunciator. The most likely would be flux valve excitation, or a dead slaving amp.

If the movement of the annunciator is sluggish with respect to movement of the synchronizing knob, the slaving amplifier is weak. If the repeater indicator follows correctly the heading shown on the master indicator as the synchronizing knob is moved, then the transmit synchro in the master indicator and the repeater servo loop are operating correctly.

If you move the synchronizing knob so that the heading on the master indicator is about 20° away from where you see a normal null, the slaving circuit should be torquing the gyro.

In order to see this torquing action, you will have to be a little patient, because only about 3° of change will result in one minute. If the annunciator indications are normal, the slaving amplifier is good. If slaving does not result, then the fault is in the directional gyro or wiring. If the repeater indicator does not correctly follow the master indicator as you move the synchronizing knob, then the repeater indicator servo motor or servo amplifier is probably inoperative.

If you understand these three circuits and their mechanical interconnections, you should have no difficulty with other operations of the master indicator or compass system.

Figure 6-16

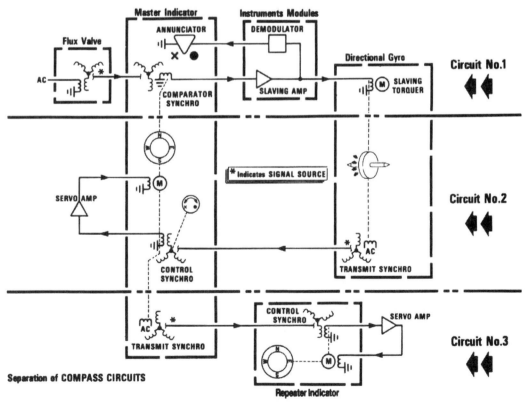

Separation of COMPASS CIRCUITS

Heading And Radio Pictorial Indicator

Figure 6-17 is an illustration of a Heading and Radio Pictorial Indicator. This instrument performs the functions of the master indicator represented in Figure 6-16.

The knob marked "heading select" knob is a two position knob; you can pull it out or push it in. If it is pulled out, it performs the synchronizing function described in Figure 6-16.

Just above this knob is the synchronizing annunciator, also shown in Figure 6-16. If the heading select knob is pushed in, then it has an altogether different function. Turning the knob changes the position of the "heading select cursor", which is shown at about 11 o'clock with respect to the compass card. With this bug we can pre-select a desired heading, and the indicator will function to provide a signal to a flight director or autopilot for heading control.

There are two lubber lines. Aircraft heading is read under the upper line. Around the fixed portion of the instrument face are other reference indexes every 450.

The "course select" knob performs functions in the compass system and in the VOR navigation system. Turning the course select knob rotates the "course select counter", the "course select cursor" deviation bar with its associated dots and the two arrow heads.

The course select cursor always points to the same number on the compass card that shows in the "course counter" digital readout. With this knob we preselect a desired heading for a VOR or localizer course. If it is a VOR selection, then with the same knob we will also have selected a desired pair of radials in the VOR system (reference Chapter 10).

The "to" arrow head is visible, obscuring most of an arrow "tail". The other arrow "tail" is fully visible because the "from" arrow is hidden under the inner mask. Both arrows are mounted on DC meter movements and spring-loaded out of view. The deviation bar is the middle section of the arrow.

The glideslope deviation bar and its associated scale to the left do not rotate.

A distance measuring equipment "distance to go" readout, similar to the course counter readout, might also be provided in the upper right-hand corner.

For the present, only the functions and operations associated with the compass system will be taken up. These are the "heading select error" signal, "course select error" signal, and heading signals.

Figure 10-25 is a photograph of an instrument similar to this one, which gives you a general idea of the appearance. This instrument is also referred to as an MHR-4 (Master Heading Reference Indicator).

Figure 6-17

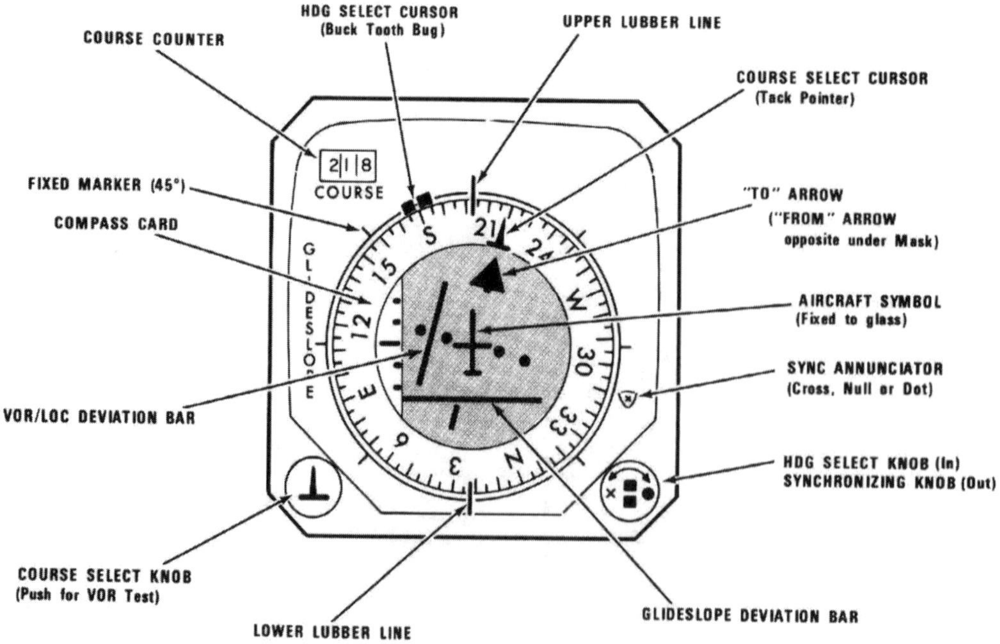

114

Differential Gear Box

Understanding some of the other functions of the master heading indicator requires understanding a differential gear box. Figure 6-18 schematically represents a differential gear arrangement. This one has been labeled so that you can relate it to the differential gear in the rear axle of a conventional automobile with a front mounted engine.

In the automobile, the pinion gear labeled "output" would be the input from the transmission. The left axle is not attached to the big ring gear, but is free to turn in it. The differential gears are pivoted and will turn only when the left axle and the right axle are not moving at the same rate or are moving in different directions.

Suppose we consider the left wheel of the automobile to be on the ground, the right wheel off the ground, and that we are turning the right wheel only. The transmission is in neutral, so the drive shaft turns as a result of our turning the right wheel. The differential gears are turning around the two axle gears, rotating the differential gear carrier and the ring gear to which it is attached. The ring gear moves the pinion gear and the drive shaft. Leaving the right wheel on the ground and turning the left wheel would also cause the drive shaft to move.

If we jack up both of the wheels and move both wheels in the same direction at the same rate, the differential gears do not turn on their pivots, and the drive shaft turns twice as fast as if we turned only one wheel at that rate.

If we turn both wheels at the same rate but in opposite directions, the differential gear carrier will not turn, but the differential gears will spin around their pivots.

The schematic symbol for a differential gear box is in the top right corner of the illustration. Inputs are represented as arrows into the symbol, and out-puts are represented as arrows out of the symbol.

Figure 6-18

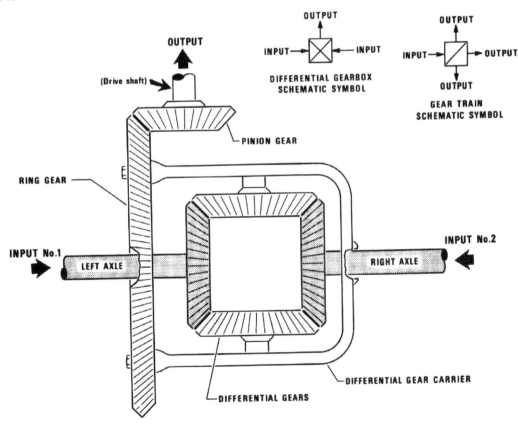

Heading And Radio Pictorial Indicator — Internal

Figure 6-19 is a schematic presentation of some of the internals of the master heading indicator we have been discussing. Mechanical connections are represented with dashed lines and electrical connections are shown as solid lines. The knob marked "heading select" at the upper left is the same knob as the one at the lower right marked "synchronize". If the knob is pushed in, it is mechanically connected as shown in the upper left. If it is pulled out, it is mechanically connected as shown in the lower right.

The two squares in the upper left heading select knob are representative of the heading select cursor. On the lower right synchronize knob, the cross and dot and their "turn instruction" arrows are shown. They tell you which way to turn the knob for the synchronizing operation when a cross or dot shows in the annunciator.

The directional gyro (D/G), the slaving amplifier, and the bottom amplifier in the middle of the drawing are made with dashed lines to indicate that they are not part of the indicator.

The movable inner mask is shown just below the compass card. This assembly moves as a unit. The VOR or localizer deviation indicator needle is shown with a "to" arrow head out, and the dots are associated with VOR/LOC deviation needle. The course select cursor (which resembles an upside down carpet tack), and the navigation system "off" warning flag are also shown because all of these items move with the inner mask.

The course select knob is shown mechanically attached to the digital readout, the VOR course resolver and the course select error synchro rotor.

The VOR course resolver is included in this series of illustrations to bring to your attention that two separate and distinct functions are associated with this knob. The other function, in the VOR system, is covered in Chapter 10.

The servo motor is shown driving a gear train, through which it drives the compass card, the four synchro rotors on the right, and the two differential gear box inputs on the left.

The two mechanical differential gear boxes each have considerable drag on their inputs on both sides. The drag results from the gear ratios and internal friction. Consequently, if one input to a differential gear box moves, the other input is not affected. An input to a differential gear box from the gear train does not affect the other inputs shown directly above or below it. Therefore, when the servo motor operates the gear train and the two inputs to the differential gear boxes, only their outputs are affected.

To recap, when the servo motor drives, it drives the four synchro rotors on the right, and through the lower differential gear box, it drives the inner mask and tack pointer. The servo motor directly drives the compass card and, through the upper differential gear box, the heading select cursor.

All of these motions exactly correspond; that is, when the gear train drives the synchro rotors 2°, all of the driven outputs on the left are also moved 2°. So, if the airplane heading changes, the compass card moves by the amount that the heading is changed, and the heading select cursor follows it with the same amount of movement, staying at the same place on the compass card. The inner mask and tack pointer also move the same amount so that the tack pointer also points to its same place

Continued on next page ...

Heading And Radio Pictorial Indicator — Internal (cont'd.)

on the compass card. The four synchro rotors are driven the same number of degrees.

You can see that all of these devices must be carefully and correctly positioned at the time of assembly so that the indications and functions will be correct.

Near the heading select error synchro and course select error synchro are drawn two reversing gear boxes (REV). These are added to the schematic so that we can significantly arrange resultant magnetic field arrows in later drawings and have them turned in a manner corresponding to the motion of the synchro rotors and control knobs. The synchros marked TX "A" and TX "B" are transmit synchros used for repeated heading to another user, such as an RMI.

Since this type of master heading indicator is sometimes used with a compass coupler, the repeated heading information to the heading select error synchro and the course select error synchro is shown coming from transmitter "A" or "B," or a compass coupler.

Regardless of whether the heading information to the heading select error and course select error synchros comes from the master indicator or from the compass coupler, it will still be aircraft heading.

If the system is not using a compass coupler, then it uses a slaved directional gyro. If the directional gyro is slaved, then the comparator synchro is used, and the circuit from flux valve to D/G is circuit No. 1 in Figure 6-16. If the compass coupler is used, it furnishes the signal to the "cross/dot" annunciator.

Figure 6-19

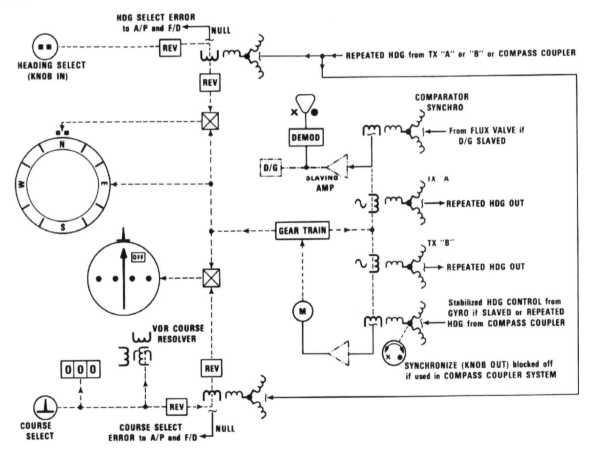

RMI Indicator Schematic

Figure 6-25 shows the principal types of radio magnetic indicators (RMI). Both present the same indications. The big difference is that the left one is an older type which has needle switching, servo amp, and monitor functions outside the indicator. The compass indications are repeated from a master compass.

The repeated heading information is given to the stator of the control synchro in the servo motor loop. The servo motor drives the compass card to the correct heading and drives the rotor of the transmit synchro ("repeated heading out").

A compass warning flag is provided in these instruments. The flag is operated by a meter movement and spring-loaded into view. Failure of the master compass system or failure of the RMI servo loop to maintain a null signal causes the monitor not to have an output.

RMIs have two indicator needles, one called a single bar needle, and the other a double bar needle. These needles are attached to the free swinging rotors of receiver synchros (Figure 3-32). The single bar needle presents information from a No. 1 system, either ADF or VOR. The double bar needle shows information from No. 2 system. They point geographically at the received station.

Figure 6-25

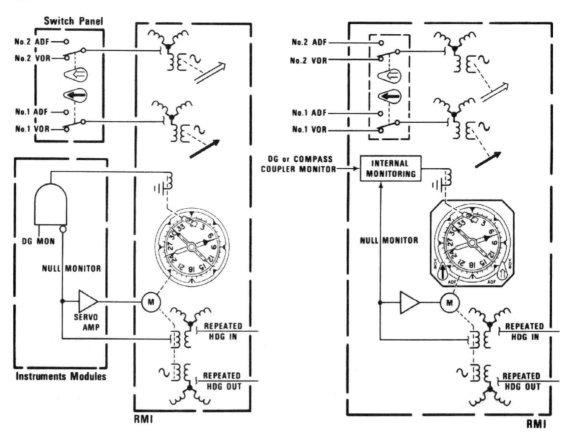

Flux Valve With Compass Coupler

Figure 6-26 shows a "compass coupler" type of compass system. The words "compass coupler" refer to a radio rack mounted box. It is the master device, and all cockpit indications or other users receive merely repeated heading information from one of the many synchro outputs illustrated at the bottom of the compass coupler. The input to this compass coupler is shown coming from a heading source; however, if the airplane is equipped with an inertial navigation system (INS), it would be using inertial navigation azimuth information instead of heading signals.

The heading source and INS systems have internal monitoring. The compass coupler also has sophisticated monitoring. If both monitoring systems in use are satisfied, a compass system "valid" voltage supplied to the users of the repeated headings will appear.

The black and white circle in the upper right corner of the compass coupler represents an external indicator on the compass coupler. This will be latched into a failure indicating position if the compass coupler monitor has seen a compass coupler failure. If the line mechanic sees this warning indication on the compass coupler unit, he knows that the box has probably failed and will need to be changed.

The signal marked "uncorrected stabilization" from the heading source indicates that this heading is not slaved. This signal passes through the slaving synchro to the servo synchro, whose rotor is kept at a null by the servo motor loop which follows up on airplane heading changes. A heading change of the airplane results in a field movement in the stator of the differential synchro, causing the servo motor to drive until the servo synchro rotor is at a null.

The servo motor also drives all the transmit synchro rotors lined up at the bottom of the box, changing their repeated heading outputs.

When the airplane heading changes, not only the uncorrected stabilization signal changes, but also the flux valve signal. Assuming that the comparator synchro was at a null to begin with, changing the airplane heading should find the comparator synchro rotor at a null after the change is completed. This is because the servo motor, which also drives all of the repeated heading outputs, drives the comparator synchro rotor by the same amount that the flux valve moves the field on the comparator synchro stator. (Remember that the position of the exciter winding in the flux valve, schematic Figure 6-6, has no significance. The direction of the earth's field with respect to the flux valve determines the orientation of the field in the comparator synchro).

Continued on next page ...

Figure 6-26

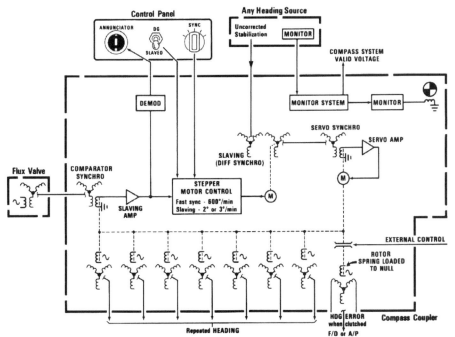

Flux Valve With Compass Coupler (cont'd.)

During heading changes as described, the comparator synchro develops signals when the flux valve is not horizontal (vertical component error).

The stepper motor control box provides a fast synchronizing speed of 600° per minute, and a slaving speed of 2° or 3° per minute. When power is first applied to the compass system, if the heading information out of the coupler disagrees with the flux valve by more than a very small amount, the step-per motor is in the fast synchronizing mode, and the signal from the comparator synchro (the difference between repeated heading and the flux valve heading) causes the stepper motor to drive the rotor of the differential synchro at a very high rate. Driving this rotor rotates the field in the servo synchro, causing the servo amplifier to drive the heading synchro rotors and the comparator synchro rotor.

This continues until the comparator synchro rotor is nulled. At that time the repeated heading agrees with the flux valve information. Once the original alignment is made, the stepper motor control drops to the slow operation and performs a "slaving" function by moving the differential synchro rotor. Whenever the comparator synchro rotor sees a discrepancy, the stepper motor provides slow

correction, keeping the repeated heading information accurate.

The annunciator in the control panel is a meter movement which, when centered, indicates no output from the slaving amplifier. If not centered it indicates that the repeated heading does not agree with flux valve information.

By turning the knob marked "SYNC" on the control panel to the right or to the left, the pilot can change his repeated heading on his flight control panel.

The knob is spring-loaded to the center "off" position. The first position to the right or to the left changes repeated heading slowly, and the second position to the right or left changes repeated heading rapidly. Moving this knob accomplishes a synchronizing operation similar to that in the previous master indicator when its synchronizing knob is moved.

If the switch marked "DG / SLAVED" is moved to the "DG" position, it disables the stepper motor control. The compass system then functions as an uncorrected directional gyro, requiring periodic correction by the pilot.

Continued on next page ...

Flux Valve With Compass Coupler (cont'd.)

Airplanes which fly polar routes without inertial navigation systems are typically fitted with high priced, low drift directional gyros. This makes it possible for them to fly in the area where the magnetic pole indications are changing too rapidly to be useful for compass information. In such cases, the switch can be moved to the "DG" position, the heading indication manually referenced to true north with the sync knob, and the information will remain reliable for considerably longer than if a less expensive gyro were used.

Moving the switch back to the "slaved" position causes the stepper motor control to revert to the fast synchronizing mode until the compass system is once again in agreement with the flux valve.

At the lower right is shown a transmit synchro whose rotor is spring-loaded to a particular position.

As drawn, the voltage on the two upper legs would be equal. Therefore, if we look at only the two leads from the two upper legs of the stator, we will see a null voltage between them. This signal can be used by an autopilot or flight director (flight guidance system) as a heading error signal for heading hold mode.

At the time that heading hold mode is initiated in the autopilot or flight director, external control engages the clutch. Any heading change after the clutch is engaged results in an error signal from the two upper legs. This is of one phase or the other, depending upon the direction of heading change. This signal is a flight guidance command to return to the heading of the airplane at the time the clutch was engaged.

Figure 6-26

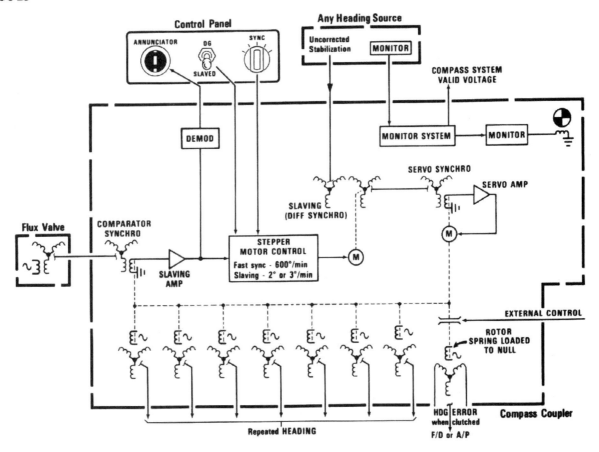

Components For "Remote" Course/Heading Select Error

Figure 6-28 is an illustration of a horizontal situation indicator such as is used in late model aircraft. A photograph of an instrument like this is shown in Figure 10-25. Since late model aircraft use compass coupler systems, this instrument is not a repeater.

There is no course select knob or heading select knob on this indicator. These functions have been moved to the autopilot control panel (Figure 6-27). In the middle of the A/P control panel is one knob marked "heading"; it controls the heading select cursor in both the captain's and the first officer's indicators. The digital readout directly above the knob should show the position of the heading select cursors on the compass cards in both indicators.

These indicators are versatile and can be used to display inertial navigation system information when

the functions are switched to that system. INS functions are to be taken up in the inertial navigation system section.

There are two knobs marked "course". One is for the captain's indicator; the other for the first officer's. These knobs control the position of the inner mask and the course select cursor. The reading of the course select cursor on the compass card should agree with the digital callout above the knob. Each knob controls the position of two synchro rotors, one in the compass system which provides the course select error, and the other which provides to the VHF navigation system the radials selected for a VOR path. These synchros perform the same functions as the corresponding synchros in the master heading indicator previously discussed.

Figure 6-27

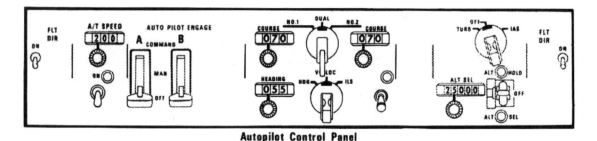

Autopilot Control Panel

Figure 6-28

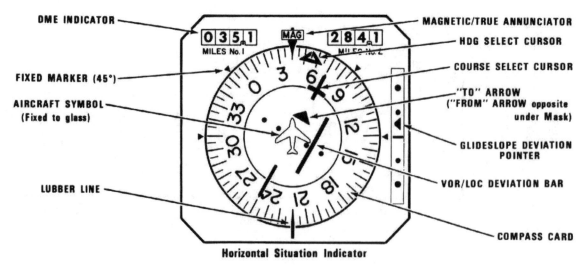

Horizontal Situation Indicator

Schematic For "Remote" Course/Heading Select Error

Figure 6-29 is a simplified schematic of the remote control of course and heading select error signals. Also included is a servo loop indicating operation of the compass card. The synchros, servo amplifiers, and servo motors indicated on the left illustration are within the horizontal situation indicator.

The top circuit shows an aircraft heading signal taken from the output of the compass coupler and delivered to the horizontal situation indicator control synchro. The servo loop brings the compass card on the indicator to agree with the output of the compass coupler.

The heading select knob on the autopilot selector panel is shown connected to the rotors of two synchros. The top synchro is a differential synchro on whose stator is aircraft heading. When the aircraft heading changes, the control synchro to the left (in the indicator) causes the heading select cursor to drive the same amount as the compass card, consequently remaining at the same point on the compass card.

When the heading select knob is moved, the rotor of the differential synchro moves. This changes the position of the field in the control synchro of the heading select cursor, driving it to a new desired position. At the same time, the knob moves the rotor of the other synchro, from which the heading select error signal is derived.

As in the master indicator of Figure 6-19, the whole compass system must be carefully assembled so that when the heading select cursor is at the lubber line, the output of the heading select error synchro is a null. The result, a desired null, is achieved when either the compass coupler or the heading select knob drives the heading select cursor to the upper lubber line.

Another system operating in the same manner controls the inner mask and the course select cursor. This is represented at the bottom of the diagram. The course select error signal has the desired null whenever the course select cursor is at the upper lubber line. Changing airplane heading drives the compass card, the inner mask, the course select cursor, and the heading select cursor all by the same amount to a new position.

In a normal operating system, the positions of the heading select cursor and the course select cursor give us a visual indication of the signals which are being developed by their respective error synchros. A cursor to the left of the lubber line develops a signal of the phase which the flight guidance system would use to turn the airplane left. A cursor to the right of the upper lubber line would be used to turn the airplane to the right.

Figure 6-29

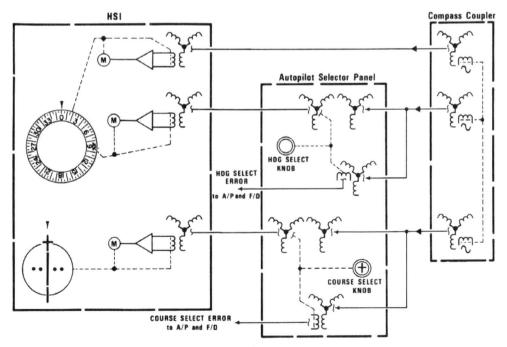

Compass System – IRS Equipped Aircraft

On modern aircraft equipped with an Inertial Reference System (IRS), aircraft heading is determined during IRS alignment.

IRS alignment consists of determining local vertical and initial heading. Both accelerometer and laser ring gyro inputs are used for alignment. The alignment computations use the basic premise that the only acceleration forces during alignment are due to the earth's gravity: the only motion during alignment is due to the earth's rotation. Accelerations due to gravity are always perpendicular to the earth's surface and define local vertical. This local vertical is used to erect the attitude data so that it is accurately referenced to vertical. Initially, only a coarse vertical is established. Once vertical is established, the laser ring gyro sensed earth rate components are used to establish true heading of the aircraft. Once true heading has been established, magnetic heading is computed based on magnetic variation data stored in the flight management computer.

As the alignment continues, both the vertical reference and heading determinations are fine tuned for maximum accuracy. During the ten minute alignment period, the initial present position entry can be entered. Earth rate sensing by the laser ring gyros allows the IRU to determine initial latitude (Figure 6-29a). This gyro determined latitude is compared to the manually entered latitude. The manually entered longitude is compared to the last stored longitude. These comparisons must be favorable to complete the alignment procedure. During the alignment period all outputs from the IRU, except for present position, are set to No Computed Data (NCD).

Figure 6-29a

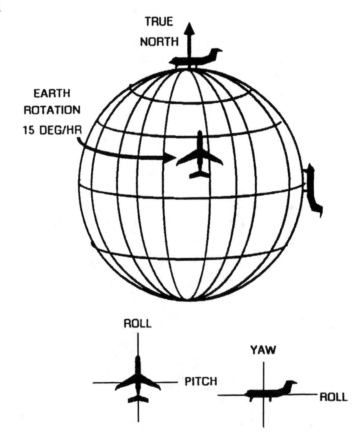

CHAPTER 7 - INERTIAL NAVIGATION / REFERENCE SYSTEMS

INS System

Inertial Navigation System (INS), is an advanced navigation concept designed as an integral part of avionics systems. This system will not only assist you in great circle navigation, but will also provide steering commands to the autopilot to steer the airplane through predetermined way-points to your destination.

In addition to steering commands, the INS navigation unit contains a gyro gimbal assembly which senses airplane attitude displacements in pitch, roll and yaw axes to maintain the airplane and weather radar display at a level attitude; to stabilize the magnetic reference heading signals; and to display the horizontal navigation data and airplane attitude on the flight instruments. Accelerometers, in the gimbal assembly, sense all vertical and horizontal accelerations (velocity changes) to solve navigational equation for steering along the desired tracks.

The INS is characterized by the following features:

Automatic self alignment and calibration are accomplished each time the INS is turned on.

The INS does not require any input from navigation aids external to the airplane.

The INS continuously monitors its own performance and furnishes warning indication and/or signals for warning indicators when output signals and displayed data are unreliable.

Insertion of present position, waypoint, and destination data is easily accomplished by using a keyboard. Each INS can be used separately to insert the waypoint latitude and longitude data; or, insertion of the waypoint data can be accomplished using one keyboard.

INS performance can be improved during flight by making a position fix when an accurate present position reference is available.

Figure 7-1

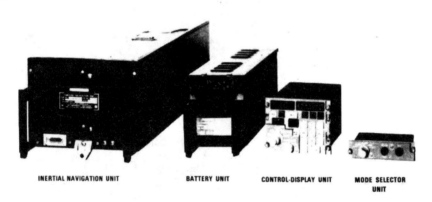

INERTIAL NAVIGATION UNIT   BATTERY UNIT   CONTROL-DISPLAY UNIT   MODE SELECTOR
                                                                               UNIT

Basic INS Principles: Accelerometers

An accelerometer is an essential component of any inertial navigation system which senses changes in airplane velocity. In its simplest form, an accelerometer consists of a small weight suspended between two springs. Acceleration sensed in a horizontal direction causes the mass to compress one of the supporting springs, and to stretch the other.

The compressive force, and the equal but opposite tensile force, are proportional to the degree of airplane acceleration sense axis. Spring displacement is directly proportional to the accelerating or decelerating forces. As long as the accelerometer remains with its sense axis perpendicular to the local vertical axis of the earth's gravitational field, only accelerations due to airplane horizontal velocity changes are sensed. Therefore, an accelerometer must be kept level at all times so that it will not misinterpret the force of gravity as an acceleration.

Mathematically, it is then possible to derive functions of velocity and distance from an original acceleration function through a process of successive integration to obtain distance traveled.

Let us take a simple example: You are taking a trip in your automobile. As you leave your home, suppose you travel a straight continuous route through the city and then out through the country.

If you kept an accurate record of your speed and time spent at each speed, you might come up with a chart like this:

Speed	Time	Distance Per Min.	Distance Covered
35	15 Mins.	.5833	8.75 Miles
40	20 Mins.	.6666	13.33 Miles
45	10 Mins.	.7500	7.50 Miles
55	15 Mins.	.9160	13.75 Miles

Distance covered at end of first hour — 43.33 miles

This was the distance traveled without regard to direction of travel; or no heading input.

Figure 7-2

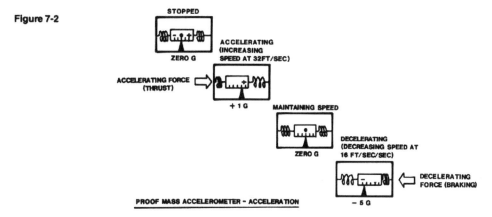

PROOF MASS ACCELEROMETER - ACCELERATION

Figure 7-3

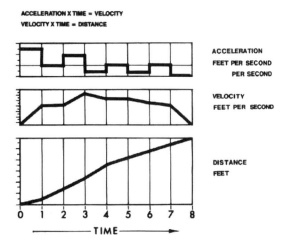

ACCELERATION X TIME = VELOCITY

VELOCITY X TIME = DISTANCE

INS Gyro

Another essential component is the gyro. The gyro operates on the principle of gyroscopic inertia, which is the characteristic of a rotating mass to resist any forces which tend to change the direction of its spin axis.

Because the earth rotates in space, the space-orientated gyro appears to rotate with respect to an earth-bound observer. This makes the gyro unsuitable for use as an earth-fixed reference unless the gyro is deliberately torqued to rotate at a rate proportional to the earth's rotation rate (earth rate). When torqued in this manner, the spin axis appears stationary, and the gyro is effectively slaved to the earth's coordinate system.

A second form of apparent precession is due to vehicle motion over the curved surface of the earth. Consider the gyro to be in an airplane flying north along a meridian from the equator to the pole.

If the gyro spin axis were orientated horizontally and parallel with the meridian when at the equator, the gyro spin axis (being space stabilized) would appear to rotate to a vertical orientation relative to earth as the airplane flies north. Consequently, in order to use the gyro as an earth-reference device in a moving airplane, it must be torqued to compensate for north-south airplane motion over the earth's surface in addition to earth rate.

Similarly, when flying east or west and starting with the gyro spin axis oriented in the horizontal and aligned east-west, the spin axis will appear to tilt due to travel east or west. This tilt is in addition to that caused by earth rate. Therefore, a gyro precessing torque is also necessary to maintain the earth reference orientation during east-west travel.

There are three of these gyros orthogonally on the platform.

Figure 7-4

Figure 7-5

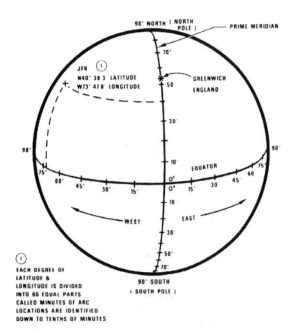

Inertial Platform

An airplane moving in three dimensions over the earth's surface has six degrees of freedom — three translational (north-south, east-west and up-down) and three rotational (roll, pitch and azimuth). Actual airplane motion is a changing combination of all six degrees of motion. Sensing of this motion is accomplished by a gyro-stabilized frame of reference known as an inertial platform.

The platform consists of a tilt table suspended in a servo-driven gimbal system. The gimbal system isolates the acceleration sensing devices from rotational motion of the airplane, and maintains the platform parallel to the earth's surface regardless of airplane motion or earth rotation. The gimbal system effectively decouples the inertial platform from the airframe and provides all-attitude freedom of motion.

Platform stabilization is achieved by use of three single-degree-of-freedom gyros mounted on the platform in the all-attitude gimbal system. The gyros are mounted with their sense (input) axes mutually perpendicular.

Any minute tilt or rotation of the platform produces an input torque about the sense axis of one or more of the gyros. This signal is amplified and used to drive the platform gimbal torque through a corresponding angle of rotation to maintain the gyro output at a null, and thus maintain correct plat-form orientation.

Stabilization of the inertial platform allows a fixed coordinate system to be established within which airplane accelerations can be measured. This is accomplished by two accelerometers, mounted on the platform with their sense axes mutually perpendicular, and aligned with reference axes on the platform.

The accelerometers sense airplane translational motion and are maintained parallel to the earth's surface. A third accelerometer is mounted on the platform to sense vertical acceleration.

Figure 7-6

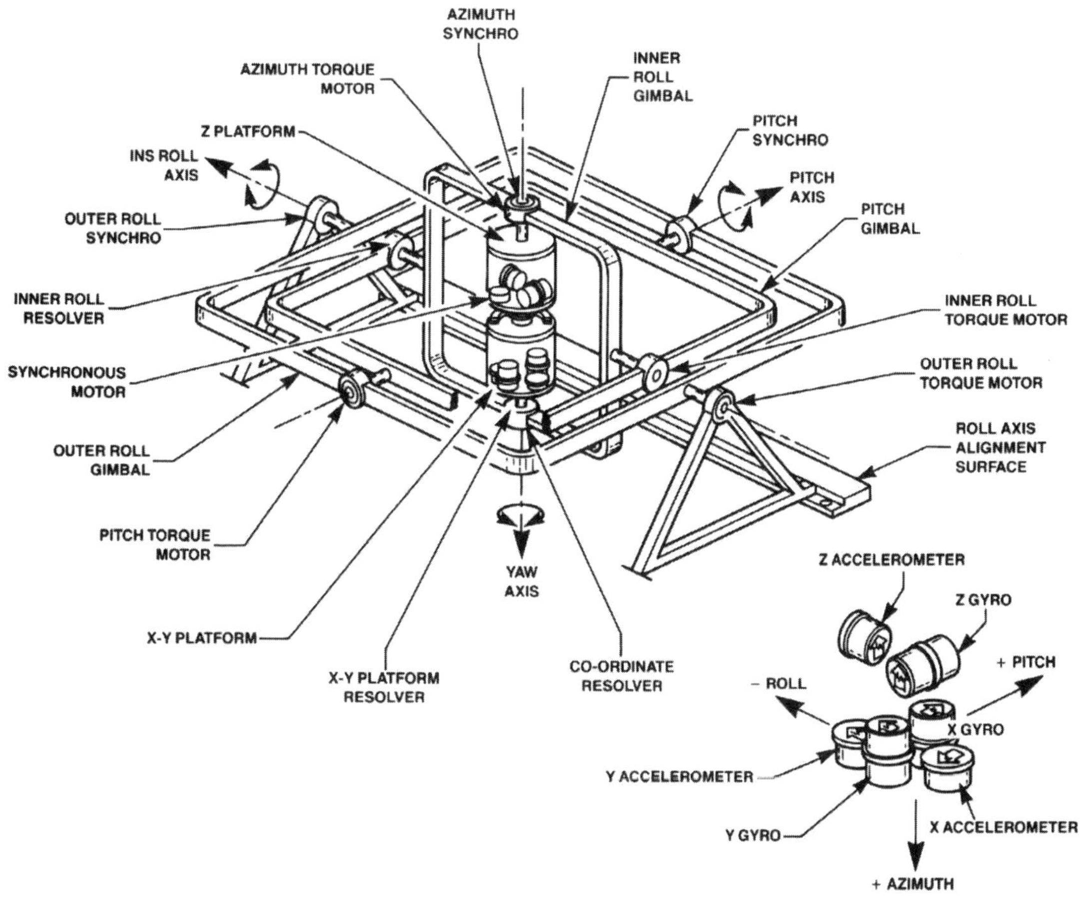

INS Computer

The inertial navigation computer uses continuously changing navigation parameters and system error correction factors to obtain a high degree of accuracy in attitude and heading computations. As the flight progresses from waypoint to waypoint, the INS will inform you of the flight leg presently being flown, and alert you to an impending course change.

If desired, the INS may be used to steer the aircraft by engaging the applicable flight director and autopilot systems. The HSI will display INS computed data.

The navigation information is displayed in digital form on the control/display unit, and in analog form on the pilot's flight instruments.

True heading (HDG) is the angle between the airplane centerline and true north.

True airspeed (TAS) is the airspeed of the airplane with respect to the surrounding air.

Wind speed (WS) is the magnitude of the wind velocity vector in knots.

Wind direction angle (WD) is the angle between true north and the wind velocity vector.

Ground track angle (TK) is the angle between true north and an imaginary line of the earth's surface connecting successive position points over which the airplane has flown (ground track).

Ground speed (GS) is the velocity with which the airplane is moving over the earth's surface.

Drift angle (DA) is the angle between the airplane's true heading and ground track.

Desired track angle (DSRTK) is the angle between true north and an imaginary line on the ground connecting successive position points desired to overfly; this line being the great circle segment that lies between two successive waypoints.

Present position (POS) is the actual latitude and longitude position of the airplane.

Cross track distance (XTK) is the shortest distance between the airplane's present position and the desired track.

Track angle error (TKE) is the angle between the airplane's actual ground track and the desired ground track.

Distance (DIST) is the great circle distance between the present position of the aircraft and the next waypoint or destination.

Figure 7-7

Figure 7-8

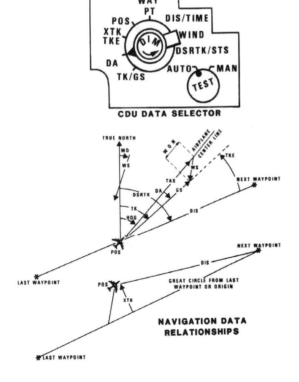

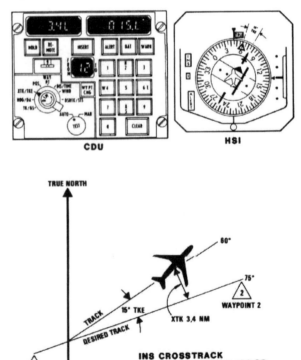

Inertial Reference System

The Inertial Reference System (Figure 7-9) provides inertial navigation data to user systems. It uses laser ring gyros, instead of the conventional spinning gyros to sense angular rate about the pitch, roll, and yaw axes. The IRS provides the inertial navigation data and inertial flight control data to user systems.

A typical IRS includes; two or three Inertial Reference Units (IRUs), an Inertial System Display Unit (ISDU), a Mode Select Unit (MSU), a Master Caution Unit, Digital/Analog Adapters (DAA), IRU Transfer Switch, IRS transfer relays, instrument transfer annunciator lights, Vertical Speed Indicators (VSIs), and Radio Digital Distance Magnetic Indicators (RDDMI).

The IRS provides basic heading and attitude reference accomplished through computations based on accelerometer and laser ring gyro sensed signals. Three accelerometers and three laser ring gyros are used in each IRU. The accelerometers and laser ring gyros are of the strap-down type and are oriented along each of the three axes of the aircraft. This orientation allows the IRU to sense acceleration along and rotation about each of the three axes.

Computer manipulation of the signals from all six sensors provide the basic heading and attitude reference signals along with present position, accelerations, ground speed, drift angle, and attitude rate information. The first requirement which must be met for proper IRS operation is alignment. The IRS alignment basically consists of determination of local vertical and initial heading.

The main function of the Inertial Reference Units (IRU) is to sense and compute linear accelerations and angular turning rates about the aircraft's pitch, roll, and yaw axes. This data is used for pitch and roll displays and navigational computations.

Each IRU contains three laser ring gyros and three accelerometers. These sense angular rates and linear accelerations, respectively. The sensed data is resolved to local vertical coordinates and combined with air data inputs to compute position (latitude/longitude), attitude (pitch, roll, yaw), true and magnetic heading, wind speed and direction, velocity, accelerations, angular rate data, and altitude.

An accelerometer is arranged along each axis to sense longitudinal, lateral and vertical G forces. Capacitive pick-off of the proof mass converts the acceleration position changes to error signals which is used to position the proof mass to its neutral position. The analog output signal is integrated once to give velocity and integrated a second time to give distance. A temperature sensor for each accelerometer is used to improve accuracy.

Figure 7-9

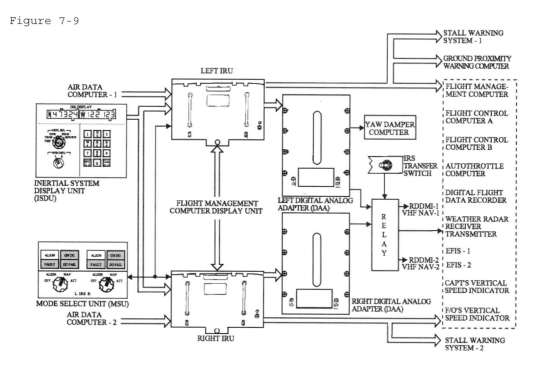

CHAPTER 8 - AUTOMATIC DIRECTION FINDER (ADF)

Propagated Wave At One Instant

Figure 8-1 introduces the principle of a loop antenna, used for Automatic Direction Finding (ADF) receivers. We are considering a radio signal whose wave length is 100 feet. The sine wave represents the different amplitudes of the voltages of the wave at a particular instant over a particular100 foot length. At the bottom of Figure 8-1 are a series of radio antennas spaced along the100 foot length.

At any given instant, each antenna sees a different voltage, indicated below each antenna.

Figure 8-1

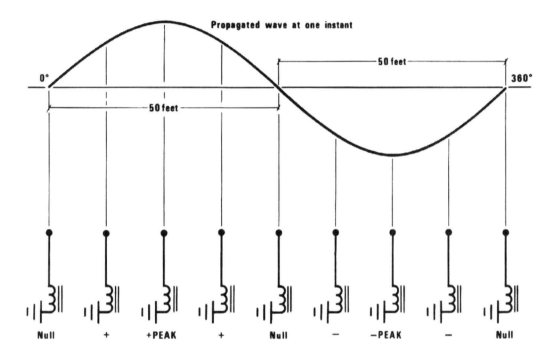

Antennas spaced along the wave at different places see different conditions at the same time.

Loop Antenna Development

In Figure 8-3 at the lower left is a pair of vertical antennas connected at the bottom to the primary of a transformer.

In Figure 8-2 you are looking down on top of a radio station. Grouped in several pairs around the radio station are the tops of these same antennas (each represented by a block dot). Each pair has a particular antenna labeled "A". If a given pair of these antennas is on a line with the radio station, the signal through the transformer primary will be a maximum of one phase. If that particular pair is rotated in azimuth 180°, the phase of its maximum signal would be reversed.

If both antennas of a given pair are at the same distance from the radio station (null indication), the voltages generated in the antennas are equal, and so no current flows through the primary of the transformer. The maximum possible signal in a pair of antennas such as these would be developed if the legs of the pair were one-half wave length long, or separated by one-half wave length (neither one of which can apply to aircraft antennas). Therefore, signals developed would be very small. The maximum for a given antenna signal is developed when the elements are in line.

In Figure 8-3, you can see that the signal from the pair of antennas on the left (two simple vertical whips) will not be much affected if we connect the tops of the whips together, as in the next illustration to the right. The one in the center is formed of wire,

and functions the same as the two whips with their tops joined.

Any of the first three illustrations on the left in Figure 8-3 cannot develop a very large signal, because only the difference between the two short whips is fed to the primary of the transformer. This difference, due to only a short distance between whips, is small. The double square antenna has an advantage because the signal is doubled, since the two squares are effectively in series.

For space or structural convenience, loop antennas are usually circular or oval. The greater the number of turns, the greater will be the signal, within limits imposed by impedance.

Be sure to notice this fact. It is not the basic radio signal itself which is being received from a loop antenna; it is the difference between two signals coming from the two edges of the loop.

Using the illustrations in Figure 8-2, if we had connected a sensitive indicator to the transformer secondary, we could discover the direction of the received radio signal by turning our pair of antennas until the signal is a null. The direction of the station from the receiving antennas is ambiguous, because there is no way of telling whether it is to one side or the other of the pair of antennas.

An ADF receiver has to resolve that problem so that it can know for sure that the transmitter received is directly in front of the loop and not in back of it.

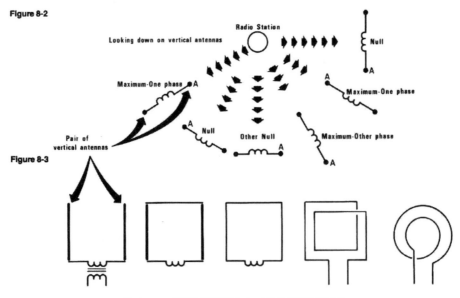

Figure 8-2

Figure 8-3

Development of LOOP ANTENNA from a pair of vertical antennas.

Loop Antenna Signals

Figure 8-4 illustrates the kind of signal developed in a loop antenna. Two different situations are shown — one in which the antenna pair is oriented in one direction, and the other in which the antenna pair has been rotated through 180° so that the sense of the signal is reversed.

The two types of broken lines represent the individual signals being received by the two vertical antennas. The heavy line is the resultant of the difference between the two leg signals, A-B (the phase difference is exaggerated so that we can clearly show the resultant of their addition). The heavy resultant line in each case peaks out where the vertical difference between the two light dashed lines is the greatest, and it is nulled where there is no difference between them.

Two main points should be understood here. First, as the antennas are rotated horizontally and the transformer signal passes through a null (antenna legs become equidistant from the station), the phase of the signal through the transformer primary reverses. This is another instance of the phase reversal of a signal passing through a null.

Second, the resultant signal (through the transformer primary) is very nearly 90° out of phase with the signal in either vertical antenna.

In aircraft ADF loop antennas, the phase difference between the loop signal and the transmitted signal received on either side of the loop is very close to 90° because the phase difference between the two sides of the loop is quite small, much smaller than illustrated here.

The cockpit control of the typical aircraft ADF receiver installation has an arrangement by which the loop antenna can be manually rotated from the cockpit. An audio control switch permits listening to the loop signals only. If you listen to loop signals only and rotate the antenna through 360°, you will notice two nulls and two peaks. You will also notice that the two nulls are fairly sharp and easy to define, whereas the two peaks are broad and hard to define accurately.

This is typical of null versus peak signals in general. It is also the reason why the direction of the transmitting station can be discovered accurately by finding a null in the loop antenna.

Figure 8-4

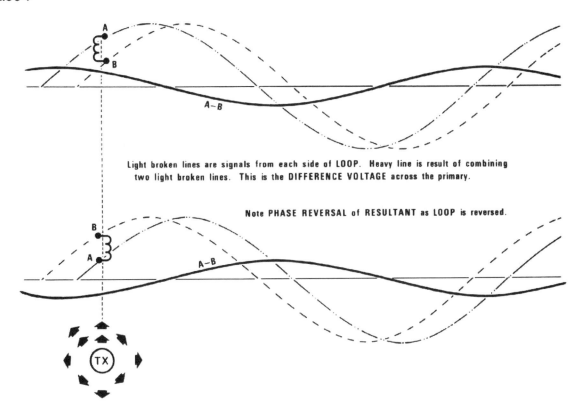

Light broken lines are signals from each side of LOOP. Heavy line is result of combining two light broken lines. This is the DIFFERENCE VOLTAGE across the primary.

Note PHASE REVERSAL of RESULTANT as LOOP is reversed.

ADF With Rotating Loop

Figure 8-5 shows only a simplified operation system of an ADF (rotating loop) in the ADF mode. Other optional modes and switching operations are omitted. This illustration shows the basic principle of automatic direction finding.

The system consists of five main components: the ADF receiver; the ADF loop antenna; the RMI, whose needles geographically point to the station; the sense antenna; and a control panel.

The sense antenna takes many different forms in different aircraft, but it functions the same as if it were a straight piece of wire hanging out of the airplane. The function of the sense antenna, so far as automatic direction finding is concerned, is to eliminate the inherent ambiguity of a loop by distinguishing between signals received from one side of the loop and signals received from the other side of the loop. This must be done to be sure the loop antenna points toward the received station.

The signal from the loop antenna is fed into the receiver and phase shifted by 90°. The purpose of the phase shift is to have a loop signal which will be either in phase, or 180° out of phase, with the signal from the sense antenna. Remember that the loop

signal is 90° out of phase in one direction or the other with the signal received on a whip antenna at the same location. Since the loop antenna signal, after the 90° phase shift, is always in phase or 180° out of phase with the sense antenna signal, after the 90° phase shift, is always in phase it is possible to know which way to turn the loop to point a particular face toward the transmitting station.

Just as in the heading select and course select synchro signals of Figure 6-19, if the 90° shifted loop antenna signal is in phase with the sense antenna signal, we will turn the loop in one direction toward the desired null. If the signal from the loop is 180° out of phase with the sense antenna signal, we will turn the loop in the other direction toward the desired null.

Discriminating phases of radio frequency signals is more difficult than discriminating the phase of audio frequency signals. Therefore, the phase comparison job is given to an audio section.

A 47 Hz oscillator is indicated. This actual frequency will vary from one model to another, but will be in the low audio range.

Continued on next page ...

Figure 8-5

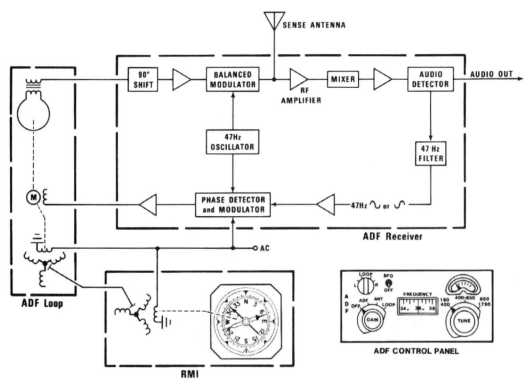

ADF With Rotating Loop (cont'd.)

Without trying to explain the operation of the balanced modulator, let's say the result is that, when its output is mixed with the sense antenna input, the mixture becomes the radio frequency, modulated by the audio frequency, but in a very special way. It is special because the phase of the audio modulation, referenced to the oscillator, will be the same phase as the loop antenna signal referenced to the sense antenna signal.

The phase of the audio modulation into the RF amplifier is, therefore, determined by the orientation of the loop antenna with respect to the radio transmitter. The conventional radio receiver is briefed down to two amplifiers, a mixer, and an audio detector. The detected audio goes to an audio control panel in the cockpit. It is also passed through a filter where only the 47 Hz signal emerges.

From there it is phase detected, using the oscillator for reference, and modulated by changing it to a 400 Hz signal referenced to aircraft power. The phase of the ultimate 400 Hz signal therefore determines the desired direction of rotation of the loop.

The loop is driven toward a particular null, as we have seen other servo systems driven, toward a particular null. When the loop has been driven to this particular null, a particular face of the loop is oriented toward the received station. The loop in the transmit synchro and the needle receiver synchro are so assembled that, when the received station is directly ahead of the airplane, the needle is pointing toward the upper lubber line on the RMI case, indicating straight ahead. Notice that the needle is oriented only to the case of the RMI — not to the compass card. (Refer to Figures 3-32 and 6-25 for other information on the RMI). The compass card is a separate function.

For every degree that the loop antenna turns, the transmit synchro rotor also turns one degree. If the received station is directly to the right of the airplane, the needle will point straight off to the right.

The RMI needle information is such that, if we lay the instrument panel down flat with the top end forward, the ADF needle points directly at the received station. If the compass card of the RMI is operative, we are able to read also the magnetic bearing of the received station on the compass card directly below the needle point. But, whether the compass card is working or not, the ADF needle still points at the station.

Figure 8-5

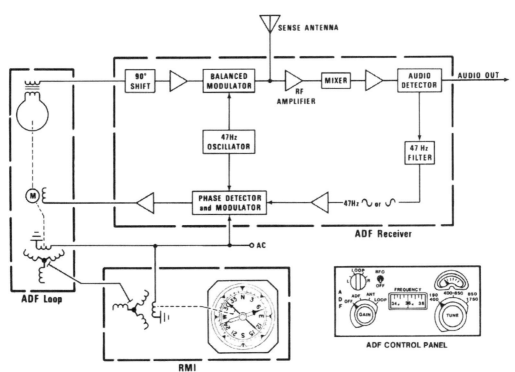

ADF With Non-Rotating Loop

Figure 8-6 shows an ADF system using a fixed loop. The fixed loop is preferred because it is more nearly trouble-free, due to fewer moving parts. The fixed loop actually consists of two loops oriented 90° to each other. These are represented with separate loops in the illustration. Each loop is connected to an individual stator of a receiving resolver in the ADF receiver.

If the received station is directly ahead of the airplane, the "A" loop will have a maximum signal and the "B" loop will have a null signal. The resolver rotor in the receiver would see a null signal because it is perpendicular to the stator which has a signal, and parallel to the stator which has a null.

In other words, the resultant field in the stator of the resolver is perpendicular to the rotor. If the received station is to the right of the airplane, then the "B" loop has a maximum signal and the "A" loop has null signal. The rotor of the resolver in the

receiver would see a maximum signal, because it is parallel to the resultant field in the stator.

Intermediate positions of the received station would result in intermediate positions of the resultant field in the stator of the resolver.

The fixed loop functions as a transmit resolver. (Refer to Figures 3-85 and 86 for transmit and receiver resolvers.) The ADF receiver itself operates the same as the receiver in the previous illustration with one exception: The servo motor drives a resolver rotor instead of a rotatable loop.

The resolver rotor and the transmit synchro rotor in the receiver are mechanically driven degree for degree. When the synchros are assembled in the receiver, the transmit synchro rotor is so oriented that, in effect, one end of the transmit synchro rotor points at the received station. That information is given to the RMI needle synchro, causing the needle to point at the station.

Figure 8-6

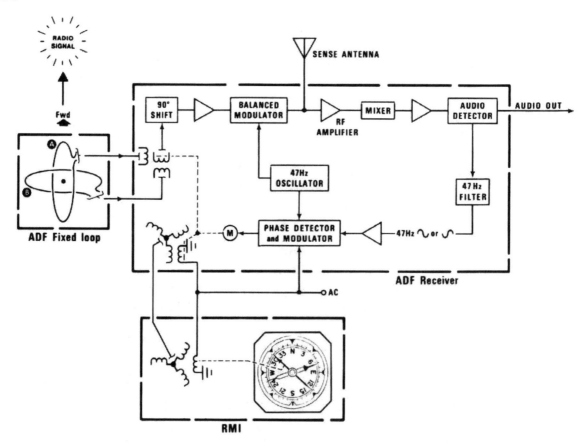

CHAPTER 9 - GLOBAL POSITIONING SYSTEM (GPS)

Global Positioning System

The global positioning system (GPS) is a satellite-based radio navigation system which uses navigation satellites to calculate accurate airplane position and time (Figure 9-1).

The first GPS satellite was launched in the summer of 1983. Today there are 24 GPS satellites in orbit, 21 primary and 3 spares. The life span of a GPS satellite is approximately 12 years. Currently, 2 replacement satellites are launched into orbit each year.

The 24 satellites orbit the earth at approximately 10,900 nm, effectively forming a constellation of satellites. There are at least eight satellites visible to a GPS receiver at any time, anywhere on (or above) the earth.

For accurate positioning information, a GPS receiver must have a minimum of 4 satellites locked on. Three, will provide precise latitude and longitude identification, the fourth will add altitude

information. In the event that only 3 satellites are locked on, the IRS system can act as the fourth satellite.

There are two GPS systems installed on most widebody commercial aircraft. Each system is identical. The left GPS System would be connected to the left FMC system and the left CDU.

Civilian users have access to standard positioning service (SPS). SPS has an accuracy of 15 – 25 meters in 95% of the position fixes.

The global positioning system (GPS) calculates:
• Latitude
• Longitude
• Altitude
• Accurate time
• True heading
• Ground speed.

Continued on next page ...

Figure 9-1

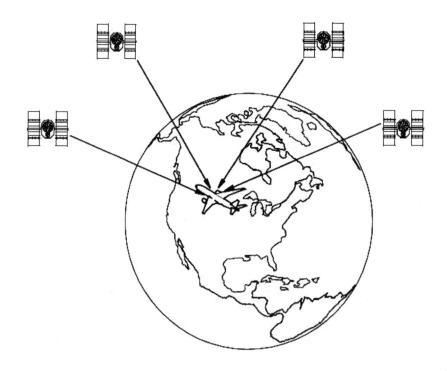

Global Positioning System (cont'd.)

The GPS antenna receives an extremely small L-band frequency signal from the satellite. Antenna cable impedance is 50 ohms, and the length of the antenna cable is limited to 15 feet due to signal loss.

The Flight Management Computer uses the GPSSU input and other navigation sensors to calculate airplane position and uses this information for flight plan management (Figure 9-2). Some airplanes use a Multi-mode Receiver (MMR) in place of the GPSSU. The same functions are performed; however the GPS receiver is contained in the same Line Replaceable Unit as the ILS receiver.

One of the Control Display Unit functions is to display the airplane's latitudinal and longitudinal position, and the source of this information. For example, if the GPS system was not operational, then the radio update mode might be DME-DME, or VOR-DME.

The Data Management Unit monitors and stores airplane data, in this case, GPSSU data. Reports can be produced from the DMU for troubleshooting and performance monitoring.

The Inertial Reference Units send latitude and longitude to the GPSSUs for initialization. This allows the first satellite position fix to take place within 10 minutes from power-up. Short periods of adverse satellite coverage can occur. The GPSSUs use IRU data to aid in continued calculation of airplane position when not enough satellites are in view. This input also lets the GPSSUs require the satellites to re-enter the navigate mode quickly.

On some aircraft types, GPSSU sends GPS time to the clocks in the flight deck. The clocks can show GPS time. GPS time is the same as universal time (coordinated) (UTC).

Most aircraft currently use the GPS system to monitor IRS performance to determine Actual Navigational Performance (ANP). In newer fleet types such as the Boeing 777, GPS is the Primary mode of navigation and directly provides position information to the Flight Management Computer.

Continued on next page ...

Figure 9-2

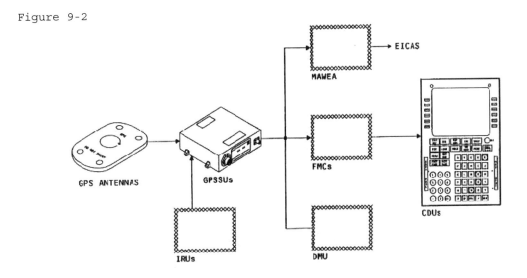

Global Positioning System (cont'd.)

WAAS

WAAS stands for "Wide Area Augmentation System" (Figure 9-3). WAAS is really the same thing as DGPS (Differential GPS) except it uses new version of satellites instead of a ground station. The FAA is launching satellites into space that transmit the correction information to your GPS using the SAME radio band as the current GPS system. This is economical because it doesn't require you to buy anything; you just upgrade your GPS (if the manufacture built in this capability).

The way it works is the WAAS satellites listen to transmitting ground stations that figure out what the error is for each satellite orbit. This error is typically induced by the atmosphere but can be due to other causes such as satellite clock error. The ground station then transmits this calculated error to the WAAS satellite system and in turn transmits the calculated error to your GPS receiver. This correction typically reduces GPS error to less than 2 meters vertically and 3 meters horizontally. The WAAS satellites are geo-stationary and the plan is to only position them over the US, therefore, if you fly to another country, this improved capability will not be available. Several countries are planning their own signal correction systems including Japan, and Europe.

In addition to accuracy improvements WAAS also provides information on the status of the GPS system. Should there be any problem, it will send a signal to your receiver that the GPS system should not be trusted (this is aimed mainly at IFR approach approved receivers.

WAAS satellites also act as standard GPS satellites, so the GPS receiver uses them if they are available. The FAA will be able to quickly certify airports to CAT I for WAAS enabled IFR GPS systems for many airports.

WAAS is currently operational in the US and has been instituted to provide a higher level of accuracy and data integrity for aircraft operating in US airspace. Although GPS has an excellent track record, it was never designed to perform aircraft applications and lacks real time monitoring of signal reliability and accuracy.

Although not yet certified by the FAA for aircraft use, WAAS is in widespread use today by agricultural and other applications.

In the first level of service, WAAS enabled LNAV/VNAV will be offered to aircraft operators for approach use. In later phases, better availability and lower decision heights will become possible.

Continued on next page ...

Figure 9-3

Global Positioning System (cont'd.)

LAAS

LAAS stands for "Local Area Augmentation System" (Figure 9-4). LAAS is basically the same as DGPS. It uses a ground station like DGPS and improves the accuracy even more than WAAS for a small area. The system will be so accurate that CAT II/III approved approaches (with autoland) will be possible with the required equipment installed in the aircraft and on airfield being used. It will require additional hardware to receive the special signal. The reason that LAAS is on the drawing board is because WAAS will only be accurate enough for CAT I approached. In addition, LAAS will also be used where WAAS doesn't reach An example might be where there are mountains blocking WAAS signals.

Figure 9-4

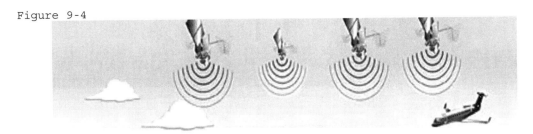

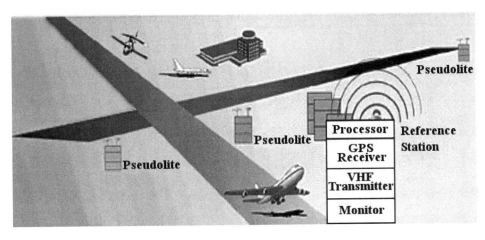

CHAPTER 10 - VERY HIGH FREQUENCY OMNI-RANGE (VOR)

Radial/Bearing/Heading/Course

Figures 10-1 through 10-4 introduce terms associated with the very high frequency omni-range system (VOR). There are four terms which must be defined and clearly understood. These are radial, bearing, heading, and course.

All directions associated with a VOR station are related to magnetic north. All airway and navigation maps refer to magnetic directions. (This is true unless the maps are especially intended for use with inertial navigation systems, or area navigation systems using true headings.)

Figure 10-1 illustrates several radials drawn from a VOR station. These radials are assigned numbers which pertain to their situation around the magnetic compass card. They are considered as being drawn away from the VOR station to a particular magnetic direction.

Figure 10-2 indicates that bearing refers to the direction from a certain place towards another. The only bearings in which we are interested are the bearings from an airplane to a VOR station. If the airplane is north of the station, the bearing of the VOR station is 180°; if the airplane is east of the station, the bearing is 270°. Radials are shown to be reciprocals of bearings. If you know the radial on which the airplane is located, you know the bearing of the station (its reciprocal). This has nothing to do with the heading of the airplane, only its position.

Figure 10-3 indicates that heading refers only to the direction in which an airplane is pointed. It has nothing to do with anything else other than the compass heading which the airplane has as a result of being pointed in that direction. It has nothing to do with the direction of travel. The airplane could be moving sideways or backwards while pointed north and its heading would still be north.

The term course is often misunderstood. It causes no particular confusion among the flight crews because everything in the cockpit is set up so the use of this word "course" and related instruments and systems is clear and usable from their point of view. The confusion arises among maintenance people because the course select knob in an airplane functions in two completely separate aircraft systems, the compass system and the VOR system (refer to Figures 6-17, 6-19, and 6-28.)

A course is a path. But a path goes in two directions, so the term course identifies a particular path and the direction which is intended to be traveled on that path. In this section we use this term with respect to VOR operations, so the term course identifies a path consisting of two reciprocal VOR radials and the direction intended to be traveled on that path.

Since the VOR system has no way of knowing whether an airplane is moving along a particular radial, or merely parked on that radial, movement of an airplane along the radial is not significant. The only significant thing is whether the airplane is on a particular selected radial or not.

Continued on next page ...

Radial/Bearing/Heading/Course (cont'd.)

For example, the upper illustration in Figure 10-4 is for a 90° course. This 90° course consists of a 90° radial and/or the 270° radial, with an intended direction of travel of 90°. The lower illustration is for a 270° course consisting of a 90° radial and/or the 270° degree radial, with an intended direction of travel of 270°.

The term course includes two ideas — a path consisting of two radials, and a direction or heading. These two ideas are often confused. Sometimes the heading concept is mixed up with the position concept, or one is ignored entirely. Most manual publications do not help much in clearing up this confusion. Sometimes you will have to work with what is meant instead of what is said. The important thing to remember in reference to the term course is that it involves the compass system and the VOR system.

Later this VOR section shows that intended direction of travel affects the sense of a displayed deviation signal, but position information is basic to VOR; and airplane heading has no function in the basic operation of a VOR system. Intended direction of travel is not airplane heading.

When we select a course with the course select knob, we are actually performing two operations independent of each other. In one operation, we set up a section in the compass system giving a heading reference to that particular course. The signal that we get from the compass system is a heading error signal.

In the other operation, we set up the VOR system to determine our position with reference to that particular course.

This is delightful from the flight crew's point of view because they do not particularly care how their objectives are reached with the navigational equipment. But it can and does cause confusion on the part of the maintenance crews.

Figure 10-1

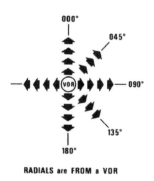

RADIALS are FROM a VOR

Figure 10-2

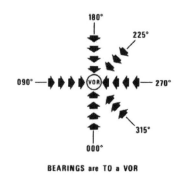

BEARINGS are TO a VOR

Figure 10-3

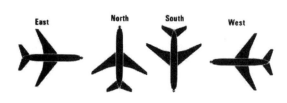

HEADING: A COMPASS FUNCTION only — Nothing to do with radio.

Figure 10-4

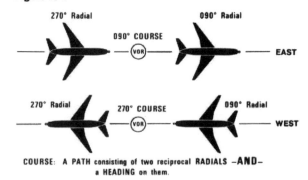

COURSE: A PATH consisting of two reciprocal RADIALS —AND— a HEADING on them.

VOR Receiver Location

The basic VOR system by itself can only tell which radial is occupied by an airplane. How it does this is shown later. Now imagine yourself in any one of the airplanes in Figure 10-5. You know that you are on a particular radial, but also imagine that you do not know the heading of the airplane.

Even though you are able to point out on the map the radial on which you are located, you are not able to point toward the VOR station. You would need to know also the heading of the airplane to be able to point toward the station.

This is exactly the situation with the VOR/RMI needles. The needles can point to a VOR station providing that both types of information are available. Therefore, the compass system must be furnishing correct heading information, and the VOR system must be furnishing correct position information if the VOR/RMI needle is to point to the station.

The position information and the heading information are added electrically in a differential synchro located in the VOR system. The result is fed to the stator of the free-swinging rotor synchro in the RMI.

Now you have an overall preview of what is covered in this chapter.

Figure 10-5

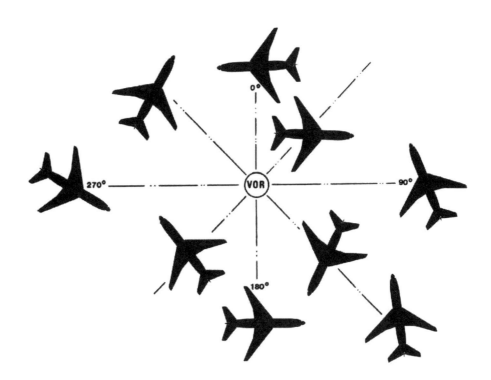

VOR SYSTEM by itself can only tell what RADIAL we are on. Without knowing the PLANE'S HEADING, we could not POINT to the station.

VOR Transmitter Block Diagram

Figure 10-6 is a block diagram of a VOR transmitter. It indicates the basic sections and the type of signal transmitted. The RF carrier frequency (100 MHz range) is different at different stations.

The modulating signal which contains the VOR information is the same for all stations. It is the 9960 Hz subcarrier. The final amplifier is modulated by this constant amplitude, frequency modulated 9960 Hz signal.

A 30 Hz oscillator frequency modulates the output of the 9960 power amplifier. This power amplifier output varies in frequency 30 times per second from 10,440 to 9480 Hz while the amplitude is held constant.

Therefore, the output of the amplitude modulator is changing through its range of 10,440 Hz to 9480 Hz 30 times per second while maintaining a constant amplitude.

The output to the antenna is a radio carrier, amplitude modulated with a constant amplitude 9960 Hz subcarrier which is frequency modulated 30 times per second. Bear in mind that there is no amplitude modulation of the subcarrier by the transmitter.

Figure 10-6

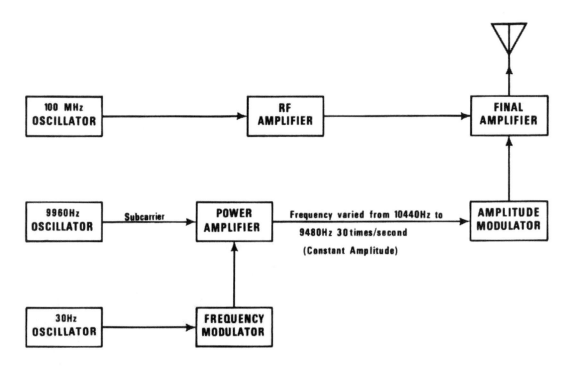

VOR Transmitted Signal

Figure 10-7 illustrates the functioning of the VOR station transmit antenna. It is a directional antenna which rotates its signal at 1800 RPM, or 30 revolutions per second. The effect is exactly the same as if we had a single directional antenna which we mechanically rotate at 30 revolutions per second, even though in fact the antenna is not physically rotated.

The direction which the antenna points at any instant is very carefully coordinated with the phase of the 30 Hz FM signal at that instant. When the antenna is pointed north, the phase of the 30 Hz FM signal is peaked positive. When the antenna is pointed south, the phase of the 30 Hz FM signal is peaked negative. It is this careful coordination between the 30 Hz FM signal and the direction which the antenna is pointing that makes it possible

for the receiver to establish its position with respect to the transmitter.

That is all it can learn from the VOR station directly. The receiver cannot tell from the VOR signal, by itself, where the station is located with respect to the airplane. In this regard it differs basically from the ADF information. The ADF receiver points the RMI needle directly at the station, independent of compass information, but, by itself, gives no information as to the compass direction of the station.

In contrast, the VOR signal and the VOR receiver by themselves tell nothing about where the VOR station is located with respect to the airplane, but they do tell where the receiver is located with respect to magnetic north and the VOR station; that is, which radial the receiver is on.

Figure 10-7

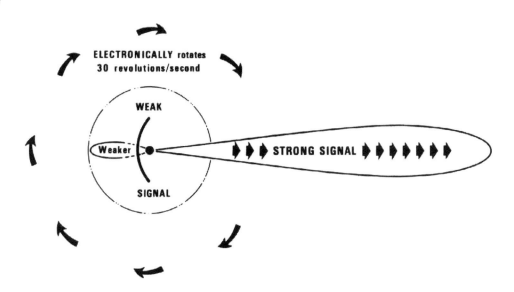

ANTENNA is electronically rotated, not physically rotated. 100 MHz CARRIER is AMPLITUDE MODULATED with 9960Hz SUBCARRIER. 9960 SUBCARRIER is FREQUENCY MODULATED with 30Hz SIGNAL called REFERENCE SIGNAL.

Subcarrier Transmitted

Figure 10-8 illustrates the coordination between the phase of the FM signal and the direction the antenna is pointing. On the left, the antenna is shown pointing north and the frequency of the 9960 Hz signal is peaked positive at 10,440 Hz.

While the antenna is rotating toward east, the frequency of the subcarrier is diminishing from 10,440 Hz to 9960 Hz (the null point of the FM signal). This requires one 120th of a second. The waveform on the bottom indicates the FM phase amplitude and time of the FM signal corresponding to the upper portions of the diagram.

During the time that the antenna is rotating from east to south, the frequency of the subcarrier drops from 9960 Hz to 9480 Hz, the negative peak of the FM signal.

Another quarter of revolution and the antenna is pointing west. The frequency of the subcarrier is back to its null of 9960 Hz.

While the antenna is rotating from west to north, the frequency of the subcarrier is increasing from 9960 Hz to 10,440 Hz. At 10,440 Hz, the signal has returned to its positive peak and the antenna has returned to the north pointing direction.

The accuracy of coordinating antenna direction with the FM phase of the subcarrier by the transmit station is the ultimate limiting factor in the accuracy of information from the VOR station. This is the signal as transmitted.

As received, however, the signal has an additional modulation. This additional modulation is a 30 Hz amplitude modulated signal which results from directional antenna rotation. For each rotation of the antenna, the receiver sees a very strong signal when the antenna is pointing at the receiver, which diminishes as the antenna moves away from that position.

The amplitude of the signal as seen by the receiver is at a minimum when the antenna has moved half a turn and is pointing away from the receiver. The amplitude of the received signal increases to a peak during the next half turn, when the antenna is once again pointing at the receiver.

So although the transmitter does not amplitude modulate its transmitted 9960 Hz subcarrier, the receiver receives the total signal with 30 Hz amplitude modulation, which includes amplitude modulation of the subcarrier.

Figure 10-8

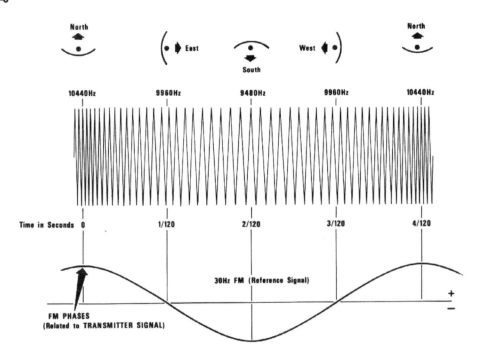

VOR 9960Hz SUBCARRIER as TRANSMITTED - CONSTANT AMPLITUDE. 30Hz FM PEAKED POSITIVE when antenna pointed NORTH. Rotating directional antenna causes RECEIVER to see AMPLITUDE MODULATION. AM FREQUENCY is antenna rotating frequency–30Hz.

Subcarrier Received

Figure 10-9 illustrates the 30 Hz AM signal and the 30 Hz FM signal as they are seen by a receiver located directly north of the VOR station. They are shown in phase with each other.

The upper waveform is one-half of the complete 30 Hz AM envelope of the 9960 Hz subcarrier. The amplitude modulation in the received signal that results from antenna rotation causes amplitude modulation, not only of the RF carrier, but also of the subcarrier. Here we are looking only at the subcarrier.

In this example, the amplitude of the received signal is maximum when the directional antenna is pointed north toward the receiver. This is the same time at which the transmitter is peaking the FM signal positive. The amplitude of the AM signal falls off as the antenna turns until, when the antenna is pointed south, the received signal is minimum in amplitude. At this time, the FM signal is peaked negative. As the antenna continues to rotate, it builds up its amplitude toward another peak which is reached when the antenna is pointed north, when the FM signal is also peaked positive.

The detected AM and detected FM signals are exactly in phase if the receiver is located exactly on the north radial. If the receiver should be off to one side of the north radial, these signals could not be exactly in phase.

Figure 10-9

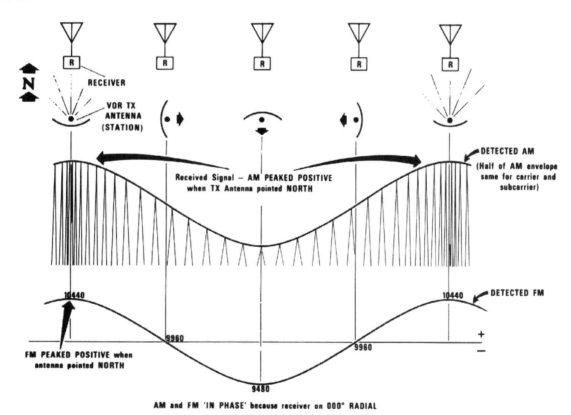

Receiver East Of Station

Figure 10-10 illustrates signal waveforms as received by a receiver located directly east of the VOR station. The FM signal is peaked positive when the antenna is pointed north. The AM signal is peaked positive when the antenna is pointed east.

Therefore, the AM signal lags the FM signal by 90° when the receiver is on the 90° radial. If the receiver is on the 45° radial, the AM lags the FM by 45°.

Figure 10-10

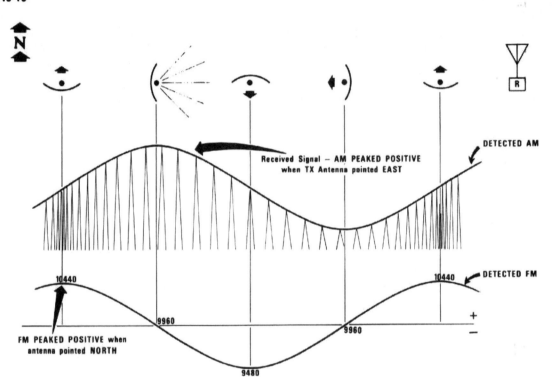

Receiver South Of Station

Figure 10-11 shows signal waveforms as received by a receiver located directly south of the VOR station. Since the AM signal is not at a positive peak until the antenna points at the receiver, it lags the FM signals by 180°.

The numerical designation in degrees of the AM lag is always the same as the numerical designation in degrees of the radial on which the receiver is located.

Figure 10-11

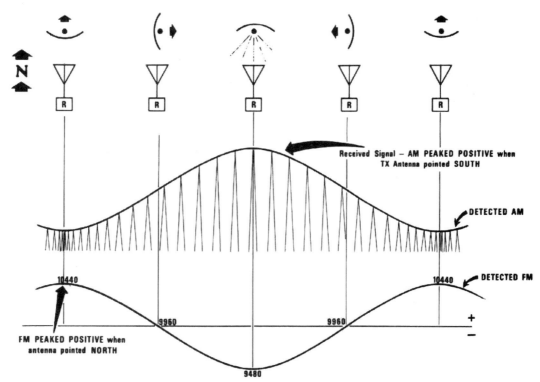

Receiver West Of Station

Figure 10-12 illustrates signal waveforms as received by a receiver located directly west of the VOR station.

Since the AM signal cannot peak positive until the antenna points at the receiver, the antenna will have rotated through 270° beyond its north pointing position before the AM peaked positive. The AM lags the FM by 270°. Therefore, the receiver is on the 270° radial.

Figure 10-12

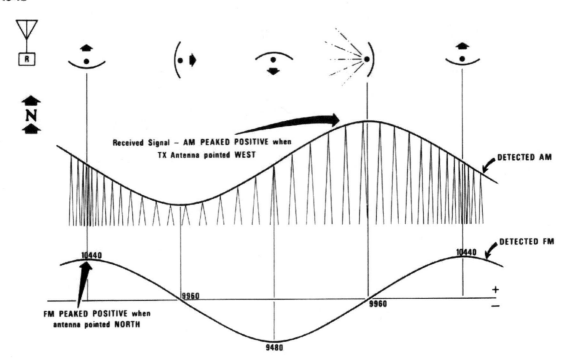

Received Signal – AM PEAKED POSITIVE when TX Antenna pointed WEST

DETECTED AM

DETECTED FM

FM PEAKED POSITIVE when antenna pointed NORTH

AM lags FM by 270° because receiver on 270° RADIAL

VOR Receiver Principle

Figure 10-13 illustrates the fundamental task of a VOR receiver. This task is to compare the phases of the detected 30 Hz AM and 30 Hz FM signals, in order to discover on which radial the receiver is located. The VOR system can give no information without having accomplished this task.

Once the AM and FM signals have been separately detected, they can be compared for phase difference by the phase detector section. When the phase detector section establishes the phase difference between these signals, it has established the radial on which the receiver is located.

Once this fundamental knowledge has been derived, it can be used in a deviation indicator by comparing it to the radials which have been selected.

It can also be used in a needle-pointing RMI by adding compass information to it.

When we select a VOR course with the course select knob, we are telling the VOR receiver the radials on which we intend to travel. The VOR receiver can compare the selected radial with the actual radial on which the receiver is located.

The deviation indicator shows the difference between the selected and the actual. One dot is 5° and two dots is 10°. The maximum indication of the deviation indicator is limited to about 12° or 15° difference between selected and actual.

When the compass system tells the VOR receiver the heading of the airplane, the VOR/RMI needle can point at the VOR station.

Figure 10-13

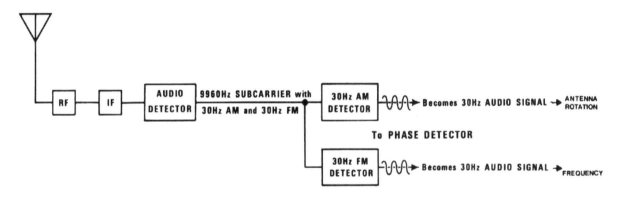

IF 'IN PHASE' — RECEIVER is NORTH of station (on 000° RADIAL)
IF AM is 090° behind FM — RECEIVER is EAST of station (on 090° RADIAL)
IF AM is 180° behind FM — RECEIVER is SOUTH of station (on 180° RADIAL)
IF AM is 270° behind FM — RECEIVER is WEST of station (on 270° RADIAL)

VOR RECEIVER PRINCIPLE — Discovers LOCATION of RECEIVER by comparing phases of 30Hz AM and 30Hz FM.

VOR Test Station (VOT)

A VOR test station (Figure 10-14) is a very low power station, usually located near the larger airports. Its use is mainly for the convenience of maintenance people. The transmitted power is too low to be usefully received by most aircraft in flight. It is an interesting little station because, no matter where the receiver is, it always thinks it is on the north radial.

The VOR test station deceives the receivers very easily. It does this by using a non-directional, non-rotating antenna and transmitting a subcarrier which it modulates not only with the 30 Hz FM signal, but also with the 30 Hz AM signal; and it transmits these two signals in phase.

No matter where the receiver may be located, the detected AM and FM signals are always in phase, so the receiver comes to the conclusion that it is located north of the VOR station. For example, if a course of 180° or 0° is selected, the deviation needle centers, indicating that the receiver is on one of the selected radials. The VOR/RMI needle points south.

Figure 10-14

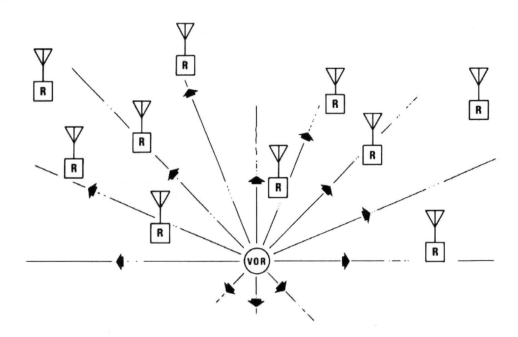

OMNI-DIRECTIONAL ANTENNA transmits signal in 100MHz BAND modulated by 9960Hz SUBCARRIER modulated by 30Hz AM and 30Hz FM in phase. All receiver locations appear to be north of test station.

VOR Receiver Test

In flight, if the VOR systems should disagree with each other, it is nice to know which one is probably at fault.

Figure 10-16 indicates how a late model VOR receiver is self-tested from the cockpit. A test signal generator is switched into the receiver system fairly close to the antenna so that practically all of the VOR receiver components are tested.

This test signal is such that the indications given by the tested receiver are the same as if it were receiving a VOR test station.

Figure 10-16

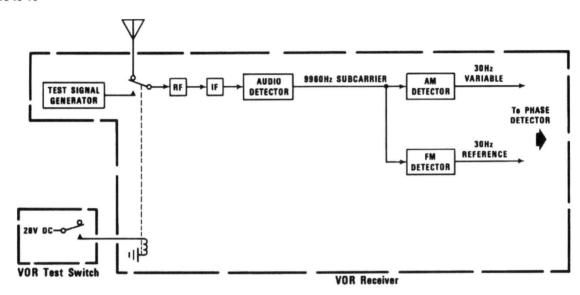

Deviation and To/From Signal Development

Figure 10-17 is a schematic of the VOR portion of a VOR/LOC receiver, usually called a "navigation" receiver. Although the VOR functions and the localizer functions are separate and distinct, the radio frequency portion of the receiver is common to both systems.

The radio frequency section (dashed lines in upper left corner) in older aircraft may be a separate unit. Later units combine the radio frequency portion with the VOR section. In addition, these later units usually contain the glideslope receiver.

In this schematic we are limiting the indicated operations to the VOR deviation and "To/From" functions. The differential synchro shown without a circuit is used for the VOR/RMI needle and will be explained later. In the indicator, only the deviation needle, the To/From arrow heads, and the course select knob are indicated. The course select knob selects the desired pair of radials for the VOR path that we intend to fly.

Whether the resolver shown at the bottom of the indicator exists in the indicator, as in the type illustrated in Figure 6-19, or whether the knob is located on an autopilot control panel as in Figure 6-27, make no difference to the operation of the system.

Previous pages show that the basic task of the VOR receiver is to discover its location with respect to the VOR station; that is, on which radial it is located. The servo motor loop in the receiver demonstrates how that is done. (Figures 3-98 through 3-114 explain how a phase shifter works.)

The servo motor loop using the phase shifter mechanically establishes the phase relationship between the 30 Hz AM signal and the 30 Hz FM signal. The detector limiters carefully limit the amplitudes of their outputs so that their amplitudes are equal at the top summing point. If these signals are exactly 180° out of phase with each other, they cancel each other and there is no input to the servo amplifier. If they are not exactly 180° out of phase, there is a difference signal from the top summing point into the servo amplifier of one phase or the other, depending upon which side of null the difference signal lies.

The servo motor then drives the phase shifter resolver rotor in a direction which depends upon the phase of the difference signal. The servo motor operates to shift the phase of the signal from the phase shifter summing point until the two signals at the top summing point cancel. The servo motor then stops driving.

If the receiver moves away from the radial on which it was located, the phase difference between the AM and FM signals changes. If the receiver moves 1° to one side, then the phase shifter resolver rotor moves 1°.

If the receiver moves 20° away, for example, from a 90° radial to a 110° radial, then the phase shifter rotor is driven 20°. If the receiver should move all the way around the VOR station through 360°, then the phase shifter resolver rotor would be driven through 360°. The resolver rotor position is a direct function the receiver position on a particular radial.

When the receiver is assembled, the position of this resolver rotor must be carefully set as well as the positions of the differential synchro rotor and the bottom resolver rotor. There, three synchro rotors move as if they were all on the same shaft; for each degree that the phase shifter resolver rotor moves, the other two synchro rotors also move one degree. The positions of the two lower synchro rotors also become direct functions of the position of the receiver on a particular radial.

The position of the resultant field in the stators of the bottom resolver represents the radial on which the receiver is located. This information is transmitted to the stators of the resolver in the indicator.

Old VOR systems furnish a direct readout of radial position information. It is available on the front of an accessory box which contains the servo motor system, and is called "omni-bearing indicator". It consists of a disk, driven by the servo motor synchronously with the synchro rotors, with a readout similar to a compass card readout.

An opening in the box shows the indication at the bottom of the disk. This indication is the bearing from the receiver to the VOR station. Since radials and bearings are reciprocals of each other, if there were an opening in the box at the top of the card, that opening would give a reading of the radial on which the receiver is located.

Continued on next page ...

Deviation and To/From Signal Development (cont'd.)

The radial on which the VOR receiver is located is represented by a resultant field in the stators of the resolver in the indicator, whereas the position of the two crossed rotors in that resolver represent the radials which the pilot has selected with the course select knob.

One of the two crossed rotors is connected to a meter movement which indicates deviation. When the resolver is assembled in the indicator or in the autopilot control panel, its position is carefully set so that the deviation rotor will be perpendicular to the field in the stators of the resolver when the airplane is located on one of the desired radials.

The field position will be of one sense (direction) if the receiver is on the 90° radial and of the opposite sense if the airplane is on the 270° radial. This reversal occurs because, as the receiver moves from 90° to the 270° radial, the servo in the receiver had to drive its synchro rotors through 180°. In either case, the deviation rotor is perpendicular to that field and sees one null or the other.

The deviation meter is a zero centering meter. If the course select resolver is perpendicular to the field, there is no current through the deviation meter and the indicator needle centers, showing that the aircraft is located on one of the selected radials.

If the airplane moves to one side of a selected radial by 5°, the deviation rotor signal strength is such as to cause the deviation needle to move to the first dot on one side. If the aircraft is located 5° to the other side of the selected radial, the deviation meter moves to the first dot on the other side of the indicator. If the airplane is 10° away from a selected radial, the deviation needle moves to the second dot away from center.

The deviation indicator meter is a DC meter, but the signal from the resolver rotor is an AC signal. How can the AC signal drive this DC meter? The secret lies in the deviation discriminator that we do not attempt to diagram. Its functions, however, can be described.

The deviation discriminator is a phase discriminator which can tell whether the phase of the signal from the indicator resolver rotor is 0° or 180° with reference to the AC source power to the rotor of the bottom resolver in the receiver.

The discriminator also functions as a rectifier. Its task is to eliminate the negative halves of the AC signal that it receives if the phase of that signal is of one sense, and to eliminate the positive halves of that AC signal if its phase is of the opposite sense.

The output from the deviation discriminator is, therefore, a half wave rectified AC signal. This would be a very rough signal to supply to a DC meter movement, so large capacitors are connected across the output of the deviation discriminator to smooth the signal and make it approximately DC.

Continued on next page ...

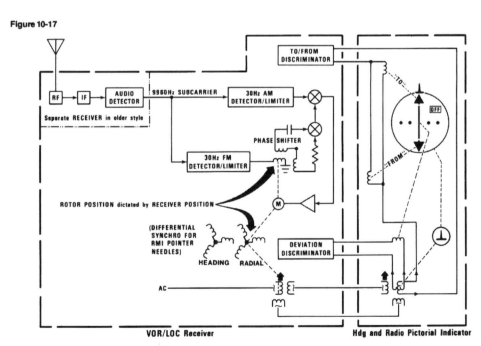

Figure 10-17

Deviation and To/From Signal Development (cont'd.)

In older systems, these large capacitors are located either in radio rack accessory boxes or in some convenient spot in an aircraft wiring panel. For example, in older DC-8s they are at the top of the radio rack on the side of the cockpit companion way. In older 727s they are in a radio rack accessory box. Later model receivers have these capacitors incorporated into the receiver itself.

In Figure 10-17, the other resolver rotor is connected to the To/From arrowhead meter movements. Both arrowheads are spring-loaded out of view. If the signal is of one polarity one arrowhead appears, and if it is of the opposite polarity the other arrowhead appears. If there is no signal, both arrowheads disappear.

The To/From discriminator function is exactly like that of the deviation discriminator function. The signals that it sees are, however, at 90° to the signals that go to the deviation discriminator be-cause its rotor is at 90° to the deviation rotor.

If the deviation signal is a null, (rotor perpendicular to the resultant field), the signal to the To/From discriminator is a maximum. Conversely, if the airplane is on a radial at right angles to those selected (for example, selected radials are 90° and 270° and the airplane is on the 180° radial), then the To/From discriminator sees a null, both arrowheads are out of view, and the deviation needle is pegged to one side.

These operations are the essence of a VOR receiver operation. They can be summed up as follows: The radial on which the VOR receiver is located determines the phase difference between the 30 Hz AM and the 30 Hz FM signals. The servo motor loop (including the phase shifter) causes the rotors of three synchros to be turned to a definite position, corresponding to the radial on which the receiver is situated. The transmit resolver in the receiver causes the radial position information to appear on the stator of the course select resolver in the deviation indicator or autopilot control panel.

The two rotors of the course select resolver are positioned in accordance with the radials selected. The difference between a selected radial and the radial occupied by the receiver shows on the deviation meter as a result of a "not null" signal in the deviation rotor. If there is no difference, then the deviation rotor is perpendicular to the resolver field and a null signal permits the DC meter movement to go to its center rest position.

The phase of the To/From signal is such that if the receiver is on the radial selected to fly toward the station, the "To" arrow is out. If it is on the radial selected to fly away from the station, the "From" arrowhead is out.

Later we will deal with the Left/Right and To/From functions pictorially, but for now you should see that if the deviation signal passes through a null, the phase of the signal is reversed. Similarly, if the To/From signal passes through a null, its phase is reversed. These reversals cause reversals in indications.

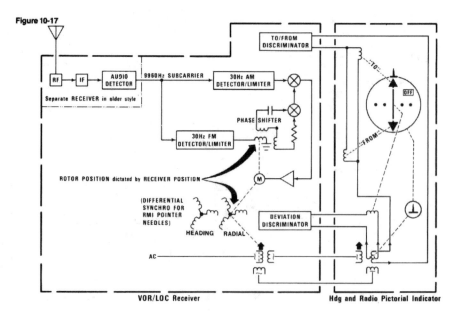

Figure 10-17

VOR/LOC Receiver   Hdg and Radio Pictorial Indicator

Alternate Deviation and To/From Signal Development

Figure 10-18 illustrates an alternate method of developing the To/From and deviation signals. Although generally the oldest and the newest receivers use the servo motor system just described, some late models use this system. In this type of receiver, the servo motor is still used to operate the VOR/RMI needles, as will be illustrated later, but the deviation and To/From indications are derived directly from the 30 Hz AM and FM signals.

It is more difficult to describe and understand without getting into the electronic component details, but if you understand Figure 10-17, you will grasp the concept of this one.

In this receiver, the phase shifter operation is performed partly with the resolver attached to the course select knob, and partly by the condenser resistor combination in the receiver. The combination of these two sections constitutes a phase shifter.

In looking at the deviation discriminator only, the 30 Hz AM signal goes directly to the discriminator. The 30 Hz FM signal (indicated by the asterisks) goes to the rotor of the course select resolver in the indicator. The rotor of this resolver is positioned in accordance with a selected pair of radials.

If the VOR receiver is located directly on one of the desired radials, the 30 Hz FM signal out of the phase shifter is such that its phase is the same phase as the 30 Hz AM signal (or 180° out of phase with it). If the deviation discriminator sees either of these

conditions, it has a null output and the deviation needle is centered.

In this case, the two signals in the To/From discriminator are 90° apart, giving a maximum output from the To/From discriminator. It is of one polarity or the other, causing one arrowhead or the other to appear.

If the 30 Hz signals to the deviation discriminator are not exactly in phase or exactly 180° out of phase, then the receiver is not located on one of the selected radials and the deviation discriminator has an output of one polarity or the other, moving the deviation needle to one side or the other.

In any VOR receiver there are two ways of changing the deviation indication. One is to move the airplane off to one side or the other, and the other is to move the course select knob in one direction or the other.

In either event, a separation of the airplane position from a selected radial by 5° causes one dot indication, and a separation of 10° causes a two dot indication. If the airplane is actually on the 90° radial, for example, and we select 100° with the course select knob, we see a two dot indication to one side of the indicator. If we select an 80° radial with our course select knob while the airplane is on 90°, we see a two dot indication on the other side. Whether we move the airplane or the course select knob, we change the phase of the signal out of the phase shifter into the deviation discriminator.

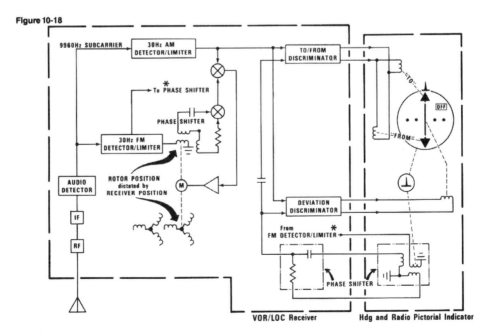

Figure 10-18

Deviation Indications When Passing VOR

Figure 10-19 illustrates the map-like presentation which a radio pictorial indicator (HSI, Figure 6-17 or 10-25) gives to the pilot. The indicator is shown as if the pilot's instrument panel were tipped forward far enough to be horizontal.

In these illustrations, the top of the panel is toward the front of the airplane; and the compass card, therefore, is always situated so that north is at the top of the illustration. Consequently, in these illustrations the compass card does not appear to move, but as the airplane changes heading, the case of the indicator moves around the compass card and the inner mask.

The miniature airplane in the center of the indicator is fixed to the glass and always appears to the pilot to be pointing up. In the indicator alongside the first position of the airplane, the pilot can see that he is headed toward one of the radials which he has selected. The To/From arrow indicates that the VOR station has not yet been passed.

In the indicator alongside the second airplane position, the pilot can see that he is still heading toward one of the selected radials and that he has not yet passed the VOR station. His intercept angle, however, has diminished.

In the third position of the airplane, the indicator shows that the VOR station has been passed; the to/from indication changes when the airplane crosses over a radial which is at 90° to the selected radials. Also, in the third position, the indicator shows the airplane still approaching the selected course, but at a still smaller intercept angle.

The indicator in the fourth position shows the deviation needle lined up with the miniature airplane in the center, indicating the airplane is on the selected radial. The To/From arrowhead shows the VOR station to be behind the airplane.

These arrowheads in an indicator of this type do not point at the station unless the airplane is directly on one of the selected radials. Therefore, these arrowheads do not function like the VOR/RMI indicator needles unless the airplane is on the selected course.

Figure 10-19

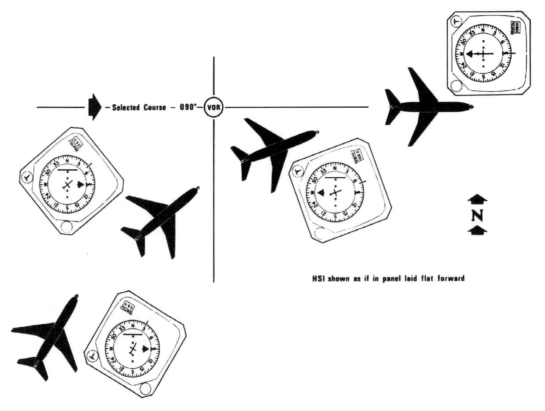

HSI shown as if in panel laid flat forward

VOR Deviation – "Left" Or "Right" Sense (Selected Course 90°)

Please refer back to Figures 10-17 and 10-18 and notice that aircraft heading has no function what so ever in determining the deviation indication. Figure 10-20 was made up to emphasize this fact since it is often a matter of confusion.

In the schematics just referred to, notice also that the sense of the deviation signal "left" or "right" does not reverse until the signal goes through a null. Putting it another way, the sense of the signal, left or right, does not change unless the receiver crosses over one of the selected radials.

In this figure, the sense of the deviation signal has been designated as calling for a right or left turn because that is the way an autopilot or flight director uses them.

There is no confusion as presented to a pilot in an integrated instrument such as the one in Figure 10-19.

The operation of the compass system in a pictorial indicator causes the inner mask containing the deviation meter to be moved around in the indicator simultaneously with the compass card. Hence, the presentation to the pilot is always correct with reference to his little airplane symbol fixed on the glass of the indicator.

Figure 10-20 illustrates the condition for a selected course of 90°. No matter where the airplane may be located within the 180° sector above the selected course in this drawing, the sense of the deviation signal is the same. The sense of the signal changes only if the airplane crosses over the selected course.

In a flight director or autopilot, not only the deviation signal, but also the course select error signal from the compass system is used in guiding the airplane along the selected course.

Figure 10-20

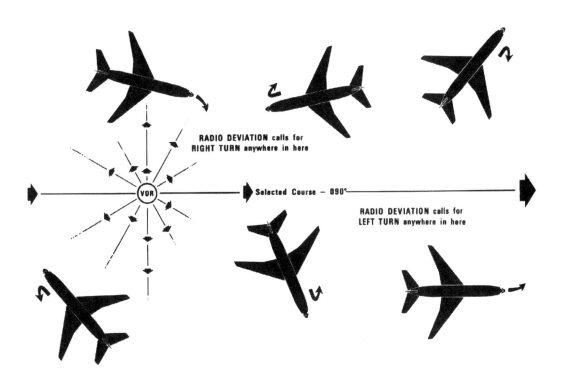

REGARDLESS of AIRPLANE HEADING

VOR Deviation – "Left" Or "Right" Sense (Selected Course 270°)

Figure 10-21 indicates the sense of the deviation signals if we have selected a 270° course. When we changed the selected course from 90° to 270°, we rotated the rotor of the resolver connected to the course select knob by 180°. This exactly reverses the sense of the deviation and to/from signals.

Figure 10-21

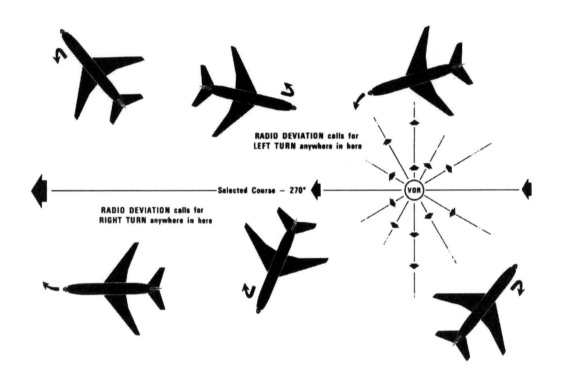

REGARDLESS of AIRPLANE HEADING

RADIO DEVIATION calls for LEFT TURN anywhere in here

Selected Course – 270°

RADIO DEVIATION calls for RIGHT TURN anywhere in here

VOR

To/From Signal "To" Or "From" Sense

Figure 10-22 illustrates a similar principle operating in the to/from signals. The sense of To/From signals does not change unless the airplane crosses over one of the two radials perpendicular to the selected course.

Remember that deviation and To/From signals are exactly alike, except that the To/From signal is associated with a pair of radials perpendicular to the selected course. If, in a parked airplane for example, we see a "to" arrowhead out when we have selected a 90° course, selecting a 270° course will bring the "From" arrowhead out.

To expand on this idea a little further, suppose that we are in a parked airplane situated on the 270° radial. If we select 270° with our course select knob, the airplane is on one of the selected radials. (Refer to Figure 6-17 or 10-25 for indicator).

The deviation indicator is centered. The "From" arrowhead is out. It makes no difference what the compass heading is, it cannot affect deviation of "To"/"From" signals.

If we turn the compass card so that a 270° heading is shown, the "from" arrow will be at the bottom of the indicator. If we now change our selected course to 260°, we have introduced a deviation signal of 10° because the airplane is 10° away from the selected course.

The deviation indicator is on the second dot to the left of the indicator showing the selected course to the left of the airplane. If we continue changing the selected course in the same direction until we have selected 180°, the meter mask will have been rotated until the deviation needle is horizontal. It has not changed its position in the meter mask. In other words, the sense of the deviation signal has not changed. The deviation needle is now at the bottom of the indicator.

In one type of indicator, since the airplane on the 270° radial is on a radial 90° away from the selected 180°, the To/From indication is nulled and both arrowheads disappear. In another type of indicator, one or the other arrowhead will be visible; no null signal is permitted.

Selecting 175° causes the "to" arrowhead to appear and rotates the mask a little farther. Continuing the change of selected course until we have selected 90° centers the deviation needle and positions the "to" arrow at the bottom of the indicator, showing the VOR station in back of the airplane.

Figure 10-22

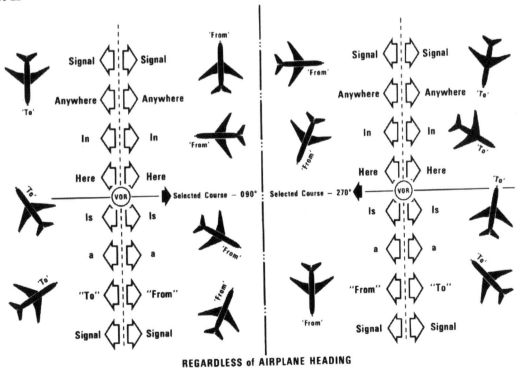

REGARDLESS of AIRPLANE HEADING

Simple Deviation Indication

The purpose of Figure 10-23 is to emphasize one of the facts that we have been talking about. It is that the polarity of the deviation signal itself is in no way involved with the aircraft heading.

The position of the needle in the mask itself is strictly a function of the position of the aircraft with respect to the selected course.

Figure 10-23

DEVIATION shows ONLY whether to RIGHT or LEFT of course - REGARDLESS of HEADING.

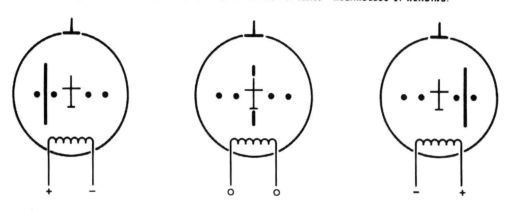

Heading is MECHANICALLY indicated by moving the whole INNER MASK around in accordance with a COMPASS SIGNAL.

Deviation Indicator Rotating Mask

Figure 10-24 shows the effect on the inner mask (which contains the deviation needle and the arrowheads) of a change in selected course. The effect is the same whether the course select knob is located on the pictorial indicator, or whether it is located on the autopilot control panel. In either case, the course select cursor (which may look like a carpet tack), and the inner mask with deviation needle, and the To/From arrows are all rotated together so that the course select cursor points to a compass card reading the same as the digital read-out associated with the course select knob.

If we are actually flying a selected course, the heading of the airplane is the same as the selected course and, therefore, the position of the mask is upright as shown in the illustration on the left.

Figure 10-24

Selecting a COURSE, positions the INNER MASK for CORRECT mechanical presentation.

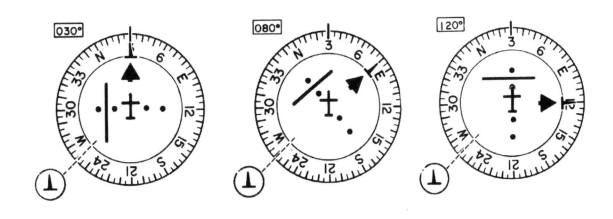

This allows the INNER MASK to be RIGHT SIDE UP when on a desired hdg (VOR or LOC). The presentation will then be map-like with respect to the compass.

Horizontal Situation Indicator - "Course Select" Function

Figure 10-25 shows an indicator as it would look with an airplane heading of 0°, a selected course of 30°, positioned on the 210° radial.

Figure 10-26 shows the indicator as it would look with an airplane heading of 0°, a selected course of 0°, and the airplane positioned on the 180° radial.

Figure 10-27 shows the airplane with the same north heading, positioned on the 150° radial, and a 330° course selected.

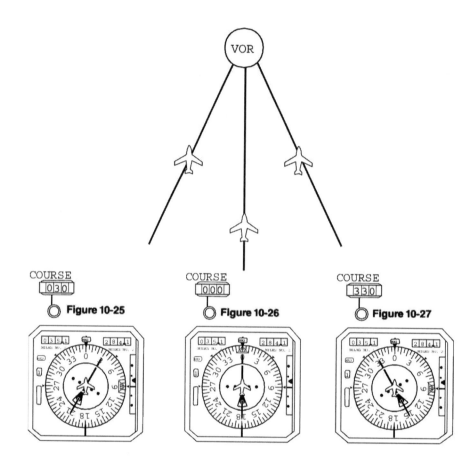

Horizontal Situation Indicator - "Radio Deviation" Function

Figure 10-28 shows the airplane 10° off to the right of the selected course, while maintaining its north heading.

Figure 10-29 illustrates the airplane on the selected course with a north heading.

Figure 10-30 shows the indicator as it would appear if the airplane moved off to the left of the selected course by 5°, while maintaining its north heading.

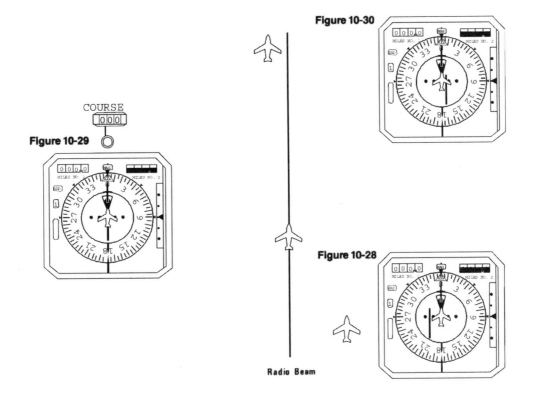

Figure 10-30

COURSE

Figure 10-29

Figure 10-28

Radio Beam

Horizontal Situation Indicator - "To/From" Function

Figure 10-31 shows the airplane approaching the VOR station with a heading of 180°. Notice that North is to the bottom in this illustration. 180° is the selected course as shown by the course select cursor, appearing over the 18 on the compass card. The "To" arrow is out because the airplane has not yet reached the VOR station.

On the other side of the VOR station, Figure 10-32, the deviation and To/From signals have been reversed by 180° in the VOR receiver. The deviation is still a null, but now The "From" arrow is out.

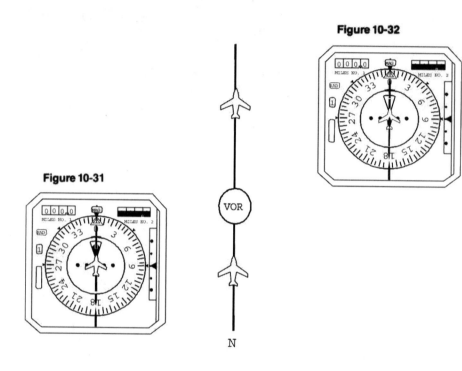

Figure 10-32

Figure 10-31

Typical Instrument Panel

Figure 10-33 shows the captain's instrument panel in a Boeing 747. At the upper left is the indicated airspeed/mach meter. Below it is a DME indicator, and below that is an RMI. The attitude director indicator has a decision height light in the upper right corner of the instrument and a test button in the lower left corner. The two blank windows at the top of the horizontal situation indicator show No. 1 and No. 2 INS distance to go.

Below the HSI is the captain's flight director computer selector. At the top right is a panel which contains flight director and autopilot mode annunciation in the center. The warning light to the

left is used by the auto throttle system, and the warning light to the right is used by the autopilot system. The circles above the warning lights are photo-electric cells used to regulate the illumination intensity on this panel.

At the middle left in the radio altitude indicator is the index bug (white triangle), set by the decision height selector knob at the bottom. The barometric altimeter is to the right of the radio altitude indicator, and to the right of that are the three marker beacon lights. To the right of the vertical speed indicator is a 24-hour clock.

Figure 10-33

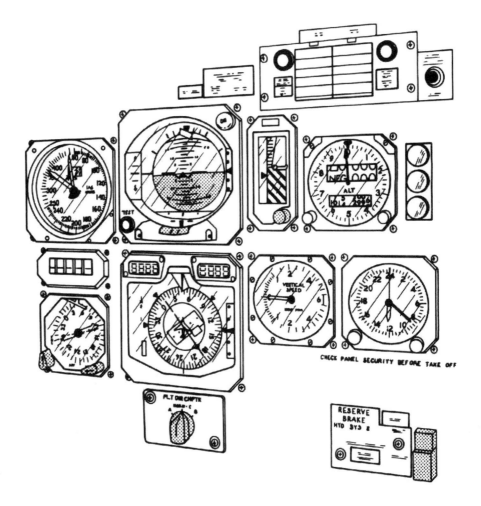

Electronic Flight Instrument System (EFIS)

Late model aircraft use an information display system referred to as the Electronic Flight Instrument System (EFIS). These systems are digital in nature and depending upon particular aircraft configuration can display primary flight information such as attitude, altitude and airspeed as well as navigational information (Figure 10-33a).

The displays will usually be configured one of two ways; Boeing 737, 757 and 767 aircraft use an Electronic Attitude Director Indicator (EADI) and an Electronic Horizontal Situation Indicator (EHSI) configured EADI over EHSI. The EADIs typically will display basic attitude information similar to the older ADI equipped aircraft with the addition of flight mode annunciation and ground speed. The EHSIs typically will display aircraft planned route, waypoints, heading, track, VOR / ILS, weather, GPWS and TCAS information. Newer such as Boeing 747, 777 and Airbus A320 aircraft will typically use a Primary Flight Display (PFD) and a Navigation Display (ND) for display of the flight data. The PFD / ND equipped aircraft will be positioned side by side. The PFD will display information such as attitude, heading, airspeed altitude, vertical speed, Marker beacon, flight mode

annunciation, TCAS and an assortment of other flight related data. The ND displays information such as aircraft planned route, waypoints, heading, track, VOR / ILS, ADF, prevailing winds, ground speed, true airspeed weather, GPWS and TCAS information.

A typical Boeing 757/ 767 electronic flight instrument system consists of 3 (left, center and right) symbol generators, and cathode ray tubes (EADIs and EHSIs) with associated control panels. The system receives attitude, heading and track information from the inertial reference systems, and flight progress and map background information from the FMCs. It presents this information to the pilots in the form of moving map and dynamic graphic displays.

The symbol generators receive inputs from various avionics systems and process the inputs to generate displays on the EADIs and EHSIs. Normally the left symbol generator supplies the left EADI and EHSI, and the right supplies the right EADI and EHSI. The center generator serves as a spare which can be selected to supply either or both sets of instruments if one or two symbol generators fail.

Continued on next page ...

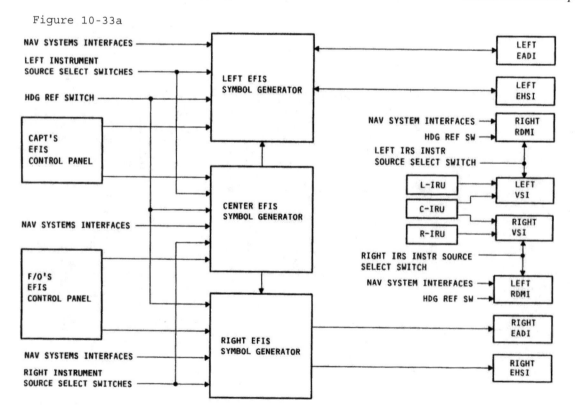

Figure 10-33a

Electronic Flight Instrument System (cont'd.)

The electronic attitude director indicator (Figure 10-33b) presents conventional airplane attitude indications, flight director commands, and deviation from localizer, glide slope and selected airspeed. A/P operating modes, ground speed, radio altitude and decision height are also displayed. Attitude data is unavailable until the associated IRU has completed alignment and entered the NAV mode. The ATT flag does not appear in this case.

The electronic horizontal situation indicator (Figure 10-33c) presents a selectable, plane view dynamic color display of flight progress. The associated control panels allow the pilots to select any of four operating modes for their respective HSI's: MAP, PLAN, VOR or ILS. The mode selection using the onside EFIS control panel indirectly controls the operation of other avionics equipment.

Continued on next page ...

Figure 10-33b

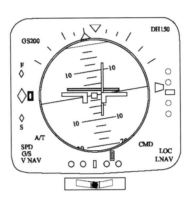

Figure 10-33c

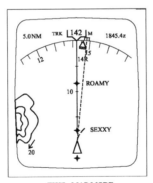

EHSI - MAP MODE

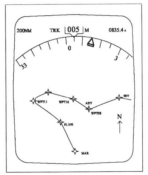

EHSI - PLAN MODE

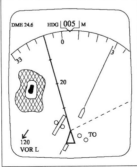

EHSI - VOR MODE

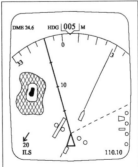

EHSI - ILS MODE

Electronic Flight Instrument System (cont'd.)

The MAP mode presents information against a moving map background, oriented with the airplane track at the top of the display. Information includes heading, trend vectors, range to altitude, relative wind, distance, ETA, and selected navigation data points as provided in the FMS data base. The map range can be varied from 10 to 320 nautical miles. Weather radar returns can be displayed.

The PLAN mode presents a static map display on the lower two thirds of the HSI, oriented to true north. The top third of the HSI retains the same dynamic presentation as in the MAP mode. Map range and radio tuning also remain the same as in MAP. In addition, the center point of the map can be shifted (stepped) to any waypoint on the route through the FMS control display units. Weather radar returns cannot be displayed in the PLAN mode.

The VOR mode provides an expanded compass rose with airplane heading at the top, along with VOR course deviation information. The selected VOR and DME must be tuned manually through the associated VOR control panel. Selected range, wind information and system source annunciation is provided. Weather radar returns can be displayed.

The ILS mode provides essentially the same heading at top display as the VOR mode plus localizer and glide slope deviation information. Selected range, wind information, and system source annunciation is provided. Weather radar returns can be displayed.

HSI heading information is provided by the associated IRU (left or right), with the center IRU serving as a selectable spare for either or both HSI's. The HSI compass rose is automatically referenced to magnetic north when between 73° north and 60° south latitude with the heading reference switch in NORM, and to true north when outside those latitudes. The compass rose may be referenced to true north at any latitude by manually selecting TRUE with the heading reference switch (B757 & B767-322 only). HSI track information is provided by the associated FMS (left or right).

Late model aircraft such as the Boeing 777 use side by side EFIS displays as illustrated in Figure 10-33d. The Primary Flight Display (PFD) shown left has consolidated airspeed, V-speeds, altitude, vertical speed, radio altitude and heading into one display. The Navigation Display (ND) shown right displays much of the same information as an EHSI including the various map modes. There are only minor inclusions such as ground speed, true airspeed, radio station tuned and DME distance.

Figure 10-33d

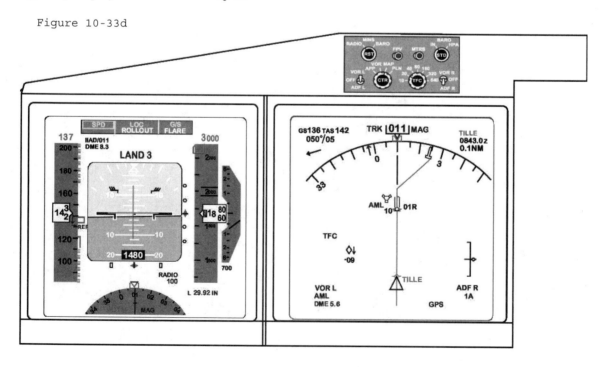

RMI Pointer Signal Development

Figure 10-34 is concerned only with the development of the signal which causes the RMI needle to point at the VOR station.

The VOR/RMI needle position is relative to the case of the instrument. Regardless of what the compass card may show (compass section of the RMI is operative or not), as long as the signal to the needle is correct, it points at the station in the same way that an ADF needle points at its received station.

Remember that in order for this needle to point at the station, two kinds of information must be combined (Figure 10-5). One is the receiver position relative to the VOR station (on which radial it is located), the other is the airplane heading. The combining of these two information bits is accomplished in the differential synchro in the VOR receiver.

Heading, from one of the repeated heading transmitters in the compass system, is given to the stator of the differential synchro. The position of the rotor of the differential synchro is a direct function of the particular radial on which the receiver is located. So the output from the differential synchro is the combination of heading and position information.

This output goes to the stator of the "free swinging rotor synchro" in the RMI indicator to which the RMI needle is attached. The rotor lines itself up with the magnetic field appearing in the stator. If we keep the airplane in the same position while changing its heading, as we do on a compass rose, the field in the stator of the differential synchro moves in accordance with heading changes. This causes the RMI needle to move by the same amount. If the RMI compass card is functioning, the card and the needle move together.

If we maintain a constant airplane heading while changing radials, as in flying by a VOR station off to one side of the airplane, the heading signal does not change, but the differential synchro rotor moves, causing the RMI needle to move around, pointing to the station all the time.

Compare this diagram with Figure 10-17. The differential synchro rotor is positioned according to the VOR radial position of the receiver. An aircraft heading signal from the compass system is present in the stator of that synchro; the signal handed over to the RMI needle synchro is therefore the algebraic sum of position and heading.

Only two things can change the position of the RMI needle — change in airplane position, or a change in airplane heading.

Figure 10-34

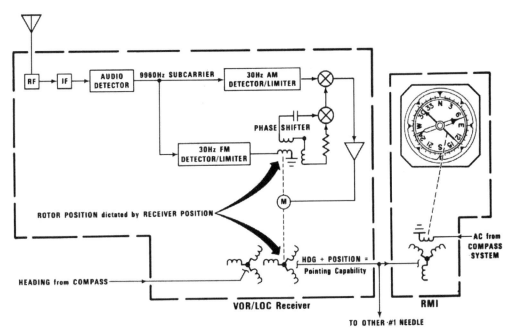

RMI Indication When Passing VOR

Figure 10-35 illustrates some RMI indications. As the airplane changes its position with respect to the VOR station, it simultaneously changes its heading.

The result of this combination of signals always results in the needle pointing at the station.

Figure 10-35

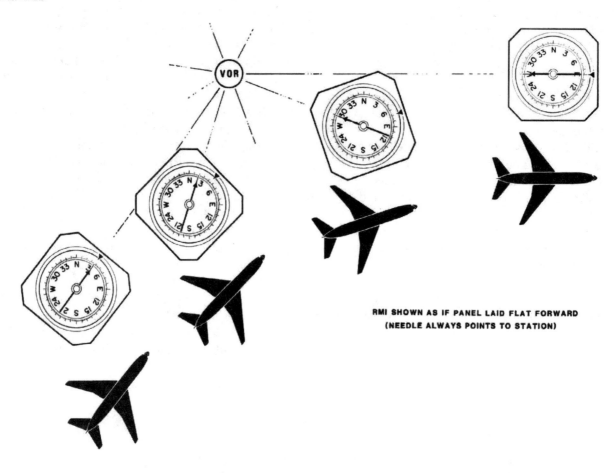

RMI SHOWN AS IF PANEL LAID FLAT FORWARD
(NEEDLE ALWAYS POINTS TO STATION)

RMI Indication Around VOR

Figure 10-36 shows a variety of airplane situations and the corresponding RMI indications. In order to see the indicator as the pilot sees it in his panel, rotate the indicator so that the lubber line appears at the top.

Figure 10-36

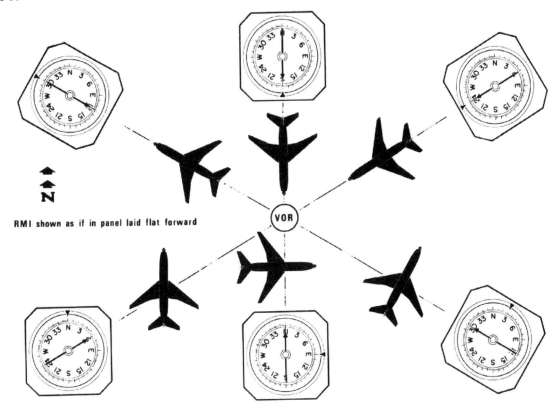

RMI shown as if in panel laid flat forward

HSI VOR Presentation Illustrations

The primary purpose of the following series of illustrations is to demonstrate the Horizontal Situation Indicator "map-like" presentation of airplane position relative to a selected VOR course. Here 0° has been selected as shown by the course select cursor at 0° (a localizer presentation would be similar). In the HSI, airplane position is represented by the airplane symbol fixed to the glass, while radio course is represented by the deviation needle.

The hypothetical "tear drop" flight path does not cross the "to-from" radials (90° and 270°), so the "To" arrow is always in view. Only when the airplane crosses the selected course (0° or 180°) does the deviation needle move to the other side of center (reference Figures 10-20 to 10-22, and 10-28 to 10-32). Validity of the HSI course indication

relative to the airplane symbol results from rotation of the inner mask in following the compass card movements as the airplane turns. For example, in this series, when the airplane is not on the radial, it is always to its left of the radial, yet the deviation needle crosses over center when the airplane crosses the 180° radial.

The flight path section of these illustrations is rotated each time as necessary to show the airplane headed toward the top of the page. This technique facilitates the viewer's correlation of airplane flight position with instrument information.

The Distance Measuring Equipment distance from the VOR station is shown in the upper left corner of the HSI. RMI needles always point at the station.

On Selected Radial — "To" Arrow Points At Station

Figure 10-37

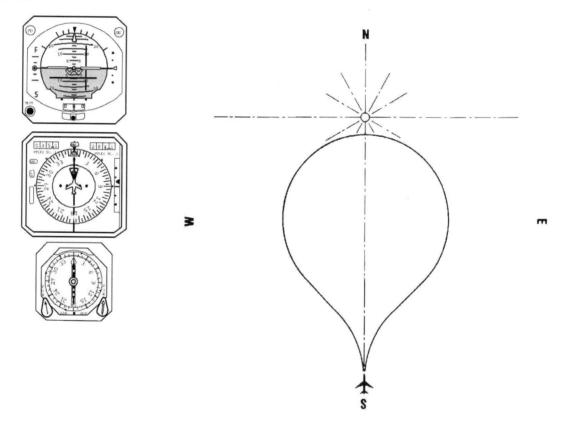

Leaving Selected Radial at 35° Angle — 3/4 Dot Deviation

Figure 10-38

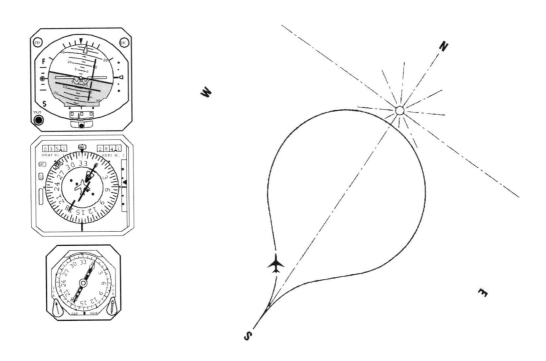

Selected Radial Behind And To The Right — Deviation Needle Pegged

Figure 10-39

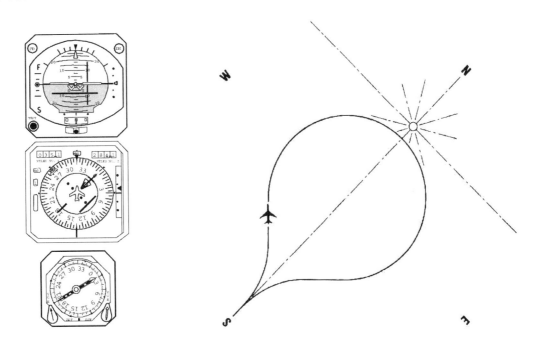

Parallel To Selected Radial On The Right

Figure 10-40

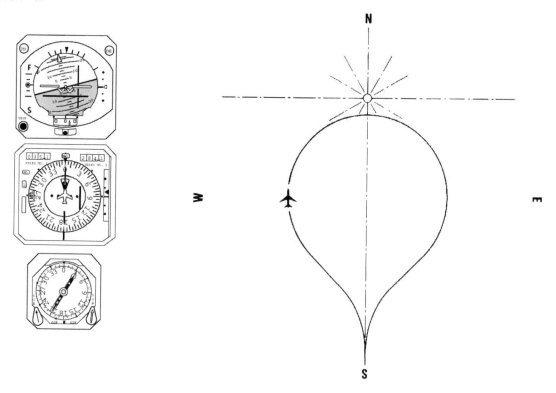

Selected Radial Ahead And To The Right — "To" Arrow Does Not Point At Station

Figure 10-41

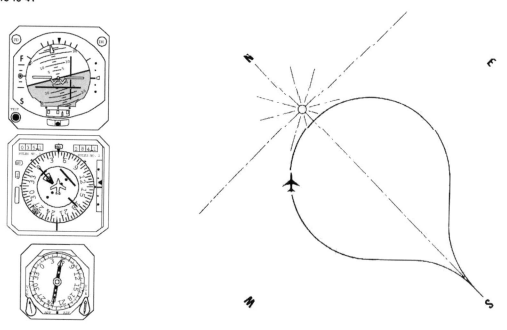

Two Dots (10°) Away From Selected Radial

Figure 10-42

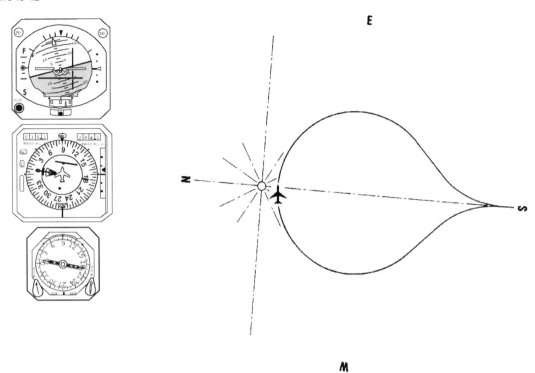

On Selected Radial — Deviation Needle Centered — "To" Arrow Points At Station

Figure 10-43

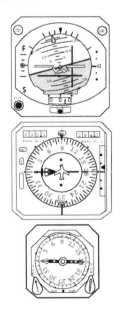
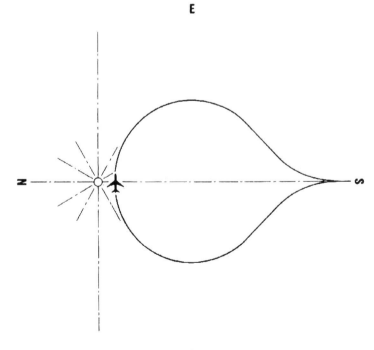

Two Dots Away From Selected Radial —
Deviation Needle On Other Side Of Center

Figure 10-44

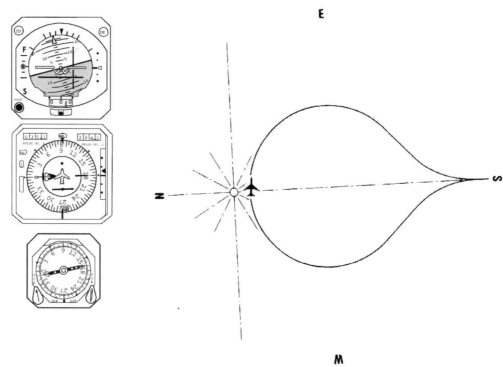

Selected Radial Behind And To The Right —
"To" Arrow Does Not Point At Station

Figure 10-45

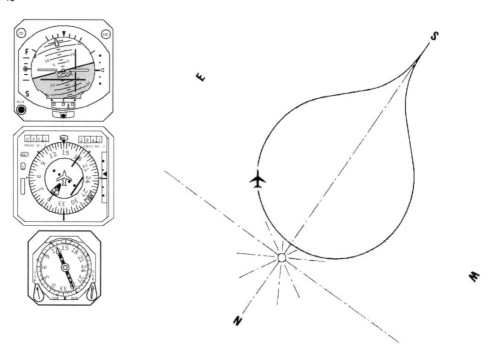

Parallel To Selected Radial On The Right
(Reference 10-40)

Figure 10-46

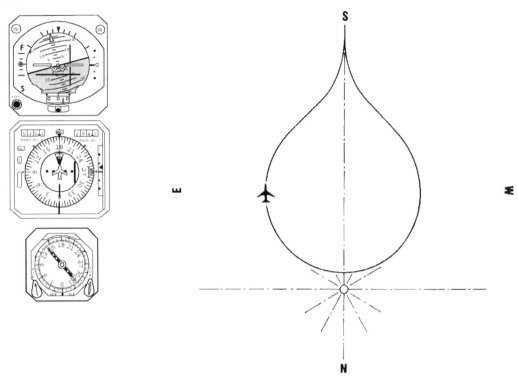

Approaching Radial At 28° Angle — 1/2 Dot Deviation

Figure 10-47

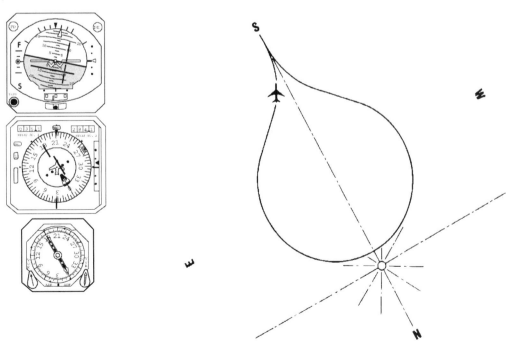

On Selected Radial — "To" Arrow Points At Station

Figure 10-48

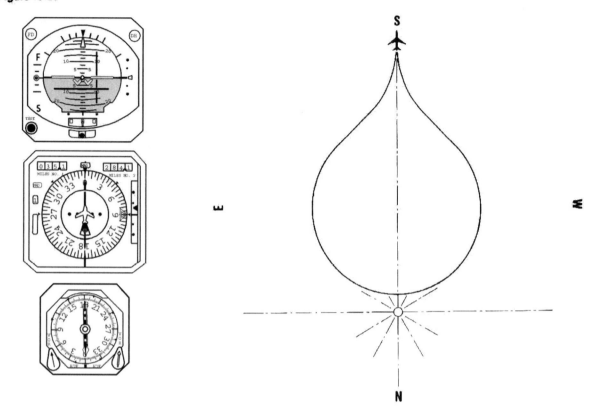

CHAPTER 11 - INSTRUMENT LANDING SYSTEMS (ILS)

Category I, II and III Capability

Autopilot performance categories are established by federal regulatory agencies based on airplane certification, airport facilities, flight crew proficiency, airline maintenance, runway visual range and decision height requirements.

Typical category minimums are as follows;

Category	DH	RVR	
I	200'	2600'	(Auto Appr)
II	100'	1200'	(Auto Appr)
IIIa Fail Pass	50'	700'	(Autoland)
IIIa Fail Op.	0'	700'	(Autoland)
IIIb	0'	150'	(Autoland)
IIIc	0'	0'	(Autoland)

Category I airplanes require one A/P and one F/D computer.

Category II airplanes require two A/P F/D computers.

Category three airplanes require triple channel redundancy.

Fail operational; the level of redundancy is such that any single failure occurring below alert height allows the landing to be continued using the remainder of the automatic system.

Fail passive; any single failure leaves the airplane substantially in trim and doesn't cause a significant deviation from the flight path

Land 3; there is triple redundancy of power sources, engaged FCC's, sensors and servos (fail operational).

Land 2; before multi-channel engage, a fail operational level will not be achieved. After multi-channel engage there is at least dual redundancy of engaged FCC's sensors and servos (fail passive).

No land 3; a fail operational level will not be achieved.

No autoland; neither a fail passive nor a fail operational level has been or will be achieved.

Category II Airport Equipment

In a Category II approach, an autopilot is permitted to bring its airplane down the approach path toward a runway, even though the pilot may not be able to see to land until he is 100 feet above the altitude of the runway (decision height). If the pilot can see to land at this point, then he may legally land his airplane. If he cannot see to land when he reaches that point, he must go around.

Only the autopilot, not the pilot, may control the airplane down to the decision height when the pilot cannot see to land.

A Category II approach requires an airplane whose Category II equipment has been certificated, in operation, by the FAA. In addition, special training programs must be certified for the flight crews and for the maintenance people.

A Category II approach can only be made at an airport which is Category II certificated. The localizer and glideslope transmissions must meet stricter standards than those for a Category I (higher minimums) certificated airport, and must be

monitored with failure indications available in the control tower.

Two RVR transmissometers (Figure 11-2) must be operating on the runway, and extensive lighting requirements must be met.

Figure 11-1 is an explanation of Category II Airport Equipment lighting. Before we discuss the illustration, familiarize yourself with the light legend centered in the drawing. On the left of Figure 11-1 are approach lights, which constitute a visible extension of the runway. These are designed so that the pilot can judge his position and altitude before touchdown. Sequenced flashing lights cause an apparent movement of the light toward the run-way. The 1,000 and 2,000 yr distances indicated are approximate.

The right illustration represents a runway for a Category II approach. The 1,000 FT distances indicated for the locations of the localizer and glide-slope antennas are approximate, and vary from one installation to another.

Figure 11-1

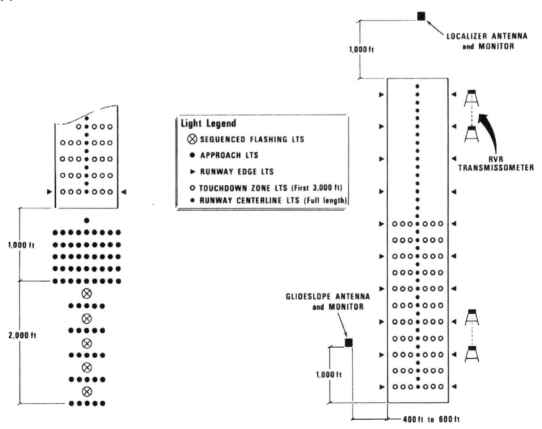

Category III Airport Equipment

As with Category II approaches, a Category III approach requires an airplane whose Category III equipment has been certificated, in operation, by the FAA. In addition, special training programs must be certificated for flight crews and for the maintenance people.

In general, a Category III equipped airfield will incorporate all equipment described in the Category II section of this chapter with few exceptions (Table 11-1). The major differences lie in the addition of airborne equipment and in aircraft system redundancy.

A Category III approach can only be made at an airport which is Category III certificated.

Table 11-1

Ground Equipment Required for Precision Approaches

Component	CAT I.	CAT II	CAT III
Localizer/Glide Slope	Yes	Yes	Yes
Outer Marker	Yes	Yes	Yes'
Middle Marker	No	No,	No
Inner Marker	No	Yes'	Yes
Approach Light System	Yes	Yes	Yes'
Sequenced Flashing Lights	Yes'	Yes'	No
High Intensity Runway Lights	No	Yes	Yes
Touchdown Zone Lights	No	Yes	Yes
Runway Centerline Lights	Yes'	Yes	Yes
RVR Transmissometer Requirement; Touchdown Mid Rollout	Required Required Required	Optional Optional Required	Optional Required Required

Runway Visual Range

The basic principle of a transmissometer is illustrated in Figure 11-2. It does away with the need for depending upon someone's estimate of how far he can see at his particular time and place, but probably not when and where the airplane touches down.

The transmissometer transmits a calibrated light beam across a measured distance to a receiver. Older installations used a 500 Ft distance, but newer ones use 250 Ft. The received light is electronically measured, and visibility thus determined.

Two installations are used: one in the touch down zone, and one closer to the other end of the runway. With category III equipped airfields; a third

transmissometer is used at mid-field. This system gives a standardized reliable measure of the lower limits of visibility for safe landings.

No consideration is given to vertical visibility because that is measured by the pilot himself as he makes his approach. The transmissometer gives the control tower operator a reading which is designated "runway visual range" (RVR). The indication illustrated here (control panel in control tower) is more than 6,000 feet, but less than 6,050. The RVR figure is an arbitrary distance through which it is assumed that a human observer could see a light of a given intensity.

Figure 11-2

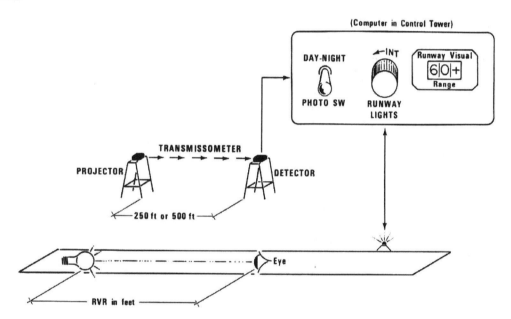

Localizer Geometry

In Figure 11-3, the one dot (75 mV signal) and two dots (150 mV signal) represent the indications appearing on a deviation indicator when the receiver is "one" or "two" dots away from center (0 deviation, reference Figure 6-28).

The "course width" of a localizer beam is measured, either angularly or in feet, from the two-dot point on one side of the beam to the two-dot point on the other side of the beam.

Localizer transmitter antennas are located a nominal 1,000 feet beyond the far end of a runway end used for a Category I, II or III approach.

The course width (beam width) is adjusted at the antenna to be 700 feet wide at the touchdown end of the runway. Most Category II and III runways are at least 10,000 feet long, so this works to a course width of about 4°. One dot is about 1° to one side. This illustration, for explanation purposes, considerably exaggerates the actual angular width of the beam.

Figure 11-3

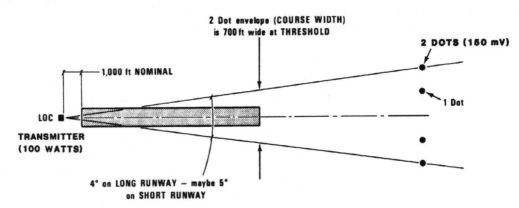

Glideslope Geometry — 2 3/4°

Figure 11-4 shows the geometry of the glideslope transmission. The figures are approximately correct for a 2 3/4° slope on the center of the glidepath. Not all glidepaths use this angle, but it serves as a good nominal figure.

The course width of the glidepath is measured in the same way as the course width of a localizer path, but vertically rather than horizontally. The course width is from the upper second dot to the lower second dot. Each dot represents 75 millivolts change in the signal to the indicator.

At the point where the glidepath is 50 feet above the runway, the course width is 28 feet. This is 14 feet from the center to either "two-dot" deviation. Where the glidepath is 100 feet above the ground, the distance from center to two dots is 28 feet. If the airplane is two dots above the glidepath at the middle marker (MM), it is about 55 feet away from the center.

The angular width of the glidepath course is 1.4° from two dots on one side to two dots on the other. This illustration also exaggerates for illustration purposes.

The narrowing of the glidepath as we approach the runway (called convergence) presents a particular problem in either pilot control or automatic pilot control. The problem is the increased sensitivity of indication as we get closer to the ground. The pilot has to manage that problem mentally. The autopilot design must include an arrangement for taking care of this problem automatically.

The method used in autopilots to overcome this problem is to attenuate the glidepath signal and to increase the attenuation continuously as the airplane approaches the ground. This operation is called "gain programming", or "beam convergence".

Figure 11-4

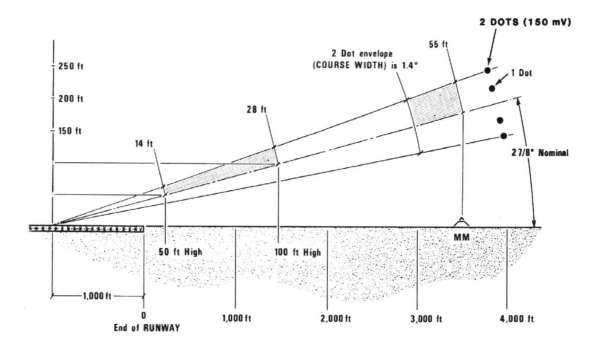

Approach Localizer Indications

Figure 11-5 graphically illustrates, from the pilot's seat point of view, what is happening to his localizer path with respect to convergence as he approaches the runway. In order to simplify the drawing, the actual localizer needle presentation is reversed. For example, if the airplane is to the left of the runway, the needle will actually be to the right in the indicator instead of to the left as represented in the drawing. Calculations are based on 4° localizer with 2 7/8° glideslope.

If the airplane is off course by two dots while on the glidepath 200 feet above the ground, the airplane will be 492 feet off to one side of the center of the runway. If it were off by only 1/8 of a dot, then it would be 82 feet to the center of the runway. If it held the 1/3 dot deviation until it reached a point where the glideslope is 100 feet above the ground, then it would be 70 feet to one side of the center of the runway.

Figure 11-5

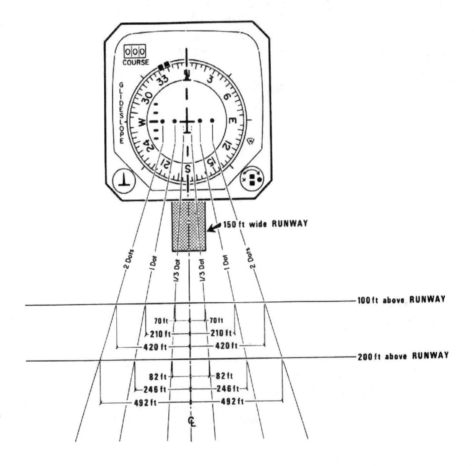

Approach Glidepath Indications

Figure 11-6 is similar to Figure 11-5 except that it concerns the glidepath. Here also, for simplicity of illustration, the needle presentation has been reversed. If, where the glidepath is 200 feet above the ground, an airplane is away from glidepath center by two dots, then it is 56 feet away from glidepath center. Where the glidepath is 100 feet above the ground, two dots deviation represents only 28 feet distance away from center, and one dot deviation at this point represents 14 feet.

Figure 11-6

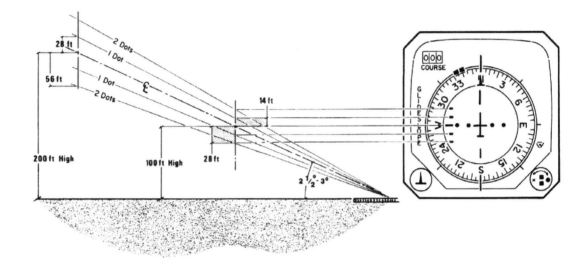

Category II and Category III Performance Criteria

For an autopilot to be certificated by the FAA for use in Category II and Category III approaches, it must pass a rigorous testing program. Among other things involved, at least 300 approaches must be made using several different airports. While making these approaches, the flight crews must monitor and record the autopilot performance. For an approach to be considered satisfactory during this FAA test, the autopilot must keep the airplane within localizer and glideslope limits as indicated in Figure 11-7.

The allowable limit of localizer deviation (top portion of illustration) is a maximum of 1/2 dot where the glideslope is 700 feet high, on down to the point where the glideslope is 300 feet high. From there on down, only 1/3 of a dot is permitted.

In the case of the glideslope (bottom portion of drawing), the airplane must be not farther than one-half dot from glidepath center from the 700 foot point down to the 200 foot point. At 200 feet, the requirement changes to not more than 12 feet from glidepath center. In terms of dots, this means that the limits widen from 1/2 dot to one full dot during the last 100 vertical feet of autopilot control.

In a Category II approach, the height above the runway at which the pilot must be able to see to land is called the decision height. At the decision height, then, the airplane is expected to be not more than 12 feet above or below the glidepath and not more than 70 feet to one side of the localizer center.

A Category III approach (Figure 11-7a) does not use a decision height as in a Category II approach; instead the term alert height is used. Alert height is an advisory in nature and typically is not used to make a go/no go decision as is the case with a category II approach.

In general, during a category II approach the pilot will disconnect the autopilot at decision height and manually land the airplane. In a category III approach and landing, the autopilot will flare, land, the throttles are driven to idle and the rollout function will maintain runway centerline. The autopilot will remain engaged until disconnected by the pilot at some point during the rollout.

Continued on next page ...

Figure 11-7

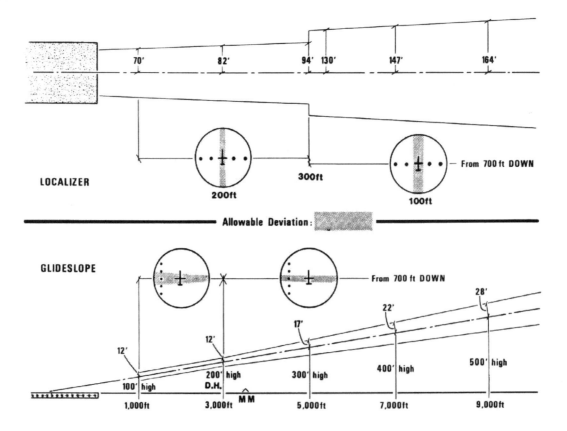

Category II and Category III Performance Criteria (cont'd.)

Figure 11-7a

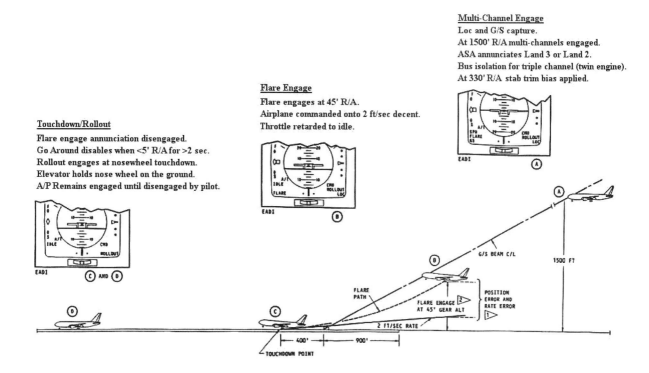

Multi-Channel Engage
Loc and G/S capture.
At 1500' R/A multi-channels engaged.
ASA annunciates Land 3 or Land 2.
Bus isolation for triple channel (twin engine).
At 330' R/A stab trim bias applied.

Flare Engage
Flare engages at 45' R/A.
Airplane commanded onto 2 ft/sec decent.
Throttle retarded to idle.

Touchdown/Rollout
Flare engage annunciation disengaged.
Go Around disables when <5' R/A for >2 sec.
Rollout engages at nosewheel touchdown.
Elevator holds nose wheel on the ground.
A/P Remains engaged until disengaged by pilot.

Localizer Field Pattern

Figure 11-8 illustrates a localizer transmission field pattern. There are, in effect, two transmissions on the same radio frequency in phase with each other. One has a 90 Hz modulation and the other has a 150 Hz modulation. Both are directional transmissions. The 90 Hz modulated transmission is directed a little to one side of beam center, and the 150 Hz modulation transmission is directed to the other side of beam center by the same amount. The strength of the left and right signals are carefully maintained equal.

Consequently, when a localizer signal is received it has two audio signals. If these signals are of equal amplitude, a null signal results which indicates that the receiver antenna is on the center of the localizer beam. The teardrop loops drawn in Figure 11-8 represent the respective field strength patterns of the two transmissions.

You can see that maximum strength for each is off to one side of the center line. Since the signal strengths are equal, only along the runway center is there a line of points where the detected modulations will be equal. A receiver moved to the right of the center line would have a stronger 150 Hz than a 90 Hz signal, and the farther the receiver moves away from the center line, the greater will be the difference. Moving the receiver to the left of the center line would result in a preponderance of 90 Hz signal over 150 Hz signal.

The task of the receiver, consequently, is to separate the two audio signals and to compare their amplitudes. In that way, it determines the angular displacement from the beam center. The receiver measures, not the strength of the radio signal, but the relative strength of the two audio signals.

Figure 11-8

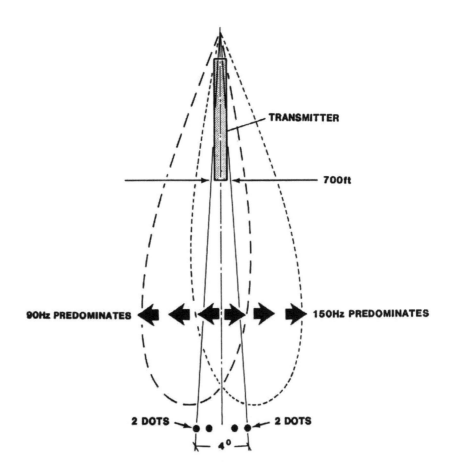

Glideslope Field Pattern

As illustrated in Figure 11-9, the glideslope field pattern is very similar to that of the localizer. The essential difference is that the transmission is rotated through 90°, so that the 90 Hz and 150 Hz audio signals are displaced from each other vertically rather than horizontally. The beam width of 1.4° is considerably narrower than that of the localizer. The principle of the transmission is the same, but a carrier frequency several times that of the localizer frequency is used.

When a particular localizer frequency is selected on the navigation frequency selector panel, the frequency of the associated glideslope transmission is simultaneously selected, although not indicated on the panel.

A typical elevation of the glidepath center is 2 3/4°. This does vary from one installation to another. In the glideslope receiver, as in the localizer receiver, the detected audio signals are compared by having them oppose each other. If those signals are equal in amplitude, a null signal results which indicates that the receiver antenna is on the center of the glidepath. If the 150 Hz signal predominates, the receiver antenna is below the glidepath, and if the 90 Hz signal predominates, the receiver antenna is above the glidepath.

Figure 11-9

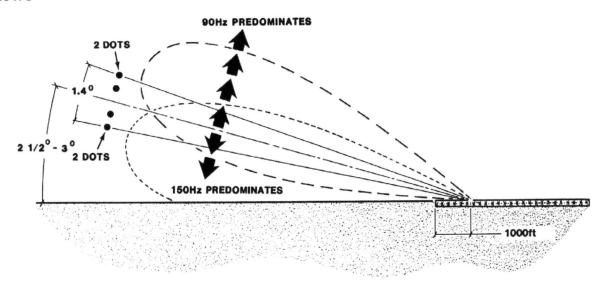

LOC/GS Receiver Principle

Figure 11-10 is a very simplified schematic of the essential elements of either a glideslope or localizer receiver. The receiver, localizer or glideslope, includes conventional RF, IF, and audio stages. The output of the audio detector is the 90 Hz and 150 Hz signals. These are separated in filters.

The two filter outputs are opposed to each other, and schematically represented by the reversed diodes in the "difference" signal. If the 90 Hz and 150 Hz signals have the same amplitude, they cancel each other downstream of the diodes. The deviation meter would then be centered, since it is often a zero-centering DC meter. If a meter of this type is used, it is generally a 1,000 ohm meter.

Sometimes the indicator is driven by a servoed D'Arsonval movement. (The strength of the localizer or glideslope signals, although usually referred to as millivolt (mV) signals, may be referred to as microamp signals; since the numerical equivalent of microamps and millivolts in a 1,000 ohm meter is the same).

When the receiver antenna is located to one side of the center of the beam, either the 90 Hz or the 150 Hz signal is stronger. This results in a DC negative or DC positive half wave signal to the meter. The signal input to the meter is filtered with large condensers to provide a steady DC signal. These condensers also damp the meter movement.

If the difference signal generated has a 75 mV amplitude, the deviation indication is one dot to the left or right on the deviation meter. A 150 mV signal moves the meter movement to the second dot away from center. In some indicators it is more convenient to have a servo motor driven indication, but, in either system, a 75 mV signal causes a one-dot indication and a 150 mV signal causes a two-dot indication.

Since it is very comforting to know whether our radio signals are adequate when we are making a final approach, a warning system is incorporated in these receivers. Earlier warning systems operated just as indicated in this diagram. The output of the two audio filters was combined as indicated in the "sum" circuit. The voltage generated was used to power another meter to which a warning flag was attached.

The warning flag could not be pulled out of view unless the amplitude of the sum of the two filters equaled 250 mV. That same concept is used in present day receivers and the sum of these signals must still equal 250 mV in order to be considered adequate. However, more sophisticated monitoring also includes monitoring of other receiver functions.

Level detectors provide a more positive "valid" output than the old arrangement, which often times resulted in the flag "peeking" in and out of view. The more positive output of the later monitoring circuits is a 28 V DC signal, available to other users of localizer and glideslope receiver information such as autopilots and flight directors, so that they also may know whether the signals they are trying to use are valid.

Older installations may have separate receiver boxes of glideslope and localizer, but the newer ones generally incorporate glideslope, localizer, and very high frequency omni-range receiver all within the same unit.

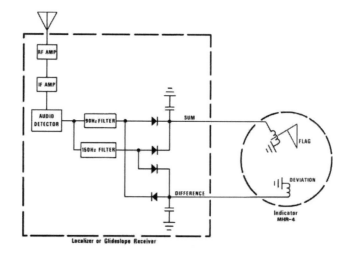

Figure 11-10

Instrument Landing System (ILS)

Figure 11-11 is a pictorial representation of an Instrument Landing System approach.

The combination of limits for the localizer envelope and the glideslope envelope results in a rectangular elongated pyramid (funnel) shaped approach path. The course widths in each case are 150 mV (two dots) deviation on the other side of center.

The course width limits are much greater than would be acceptable for a satisfactory Category II approach (reference Figure 11-7).

Figure 11-11

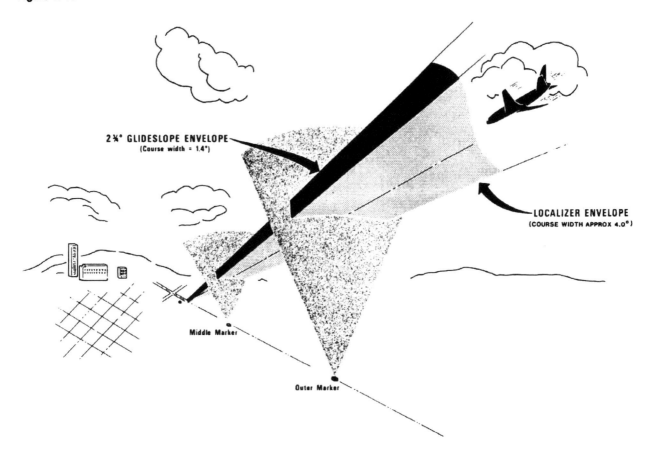

CHAPTER 12 - AIR TRAFFIC CONTROL SYSTEM (ATC)

ATC Transponder Principle

Figure 12-1 illustrates the ATC transponder principle. The ATC is distinguished from the primary surveillance radar. The primary surveillance radar, used by the air traffic control ground station, provides the ground station operator with a symbol on his surveillance radar scope for every aircraft in his area. The primary surveillance radar is a reflection type radar system not requiring any response from the aircraft.

The primary and secondary surveillance radar antennas are mounted on the same rotating mounting, and therefore both always look in the same direction at the same time.

The secondary surveillance radar system uses what is called an "ATC transponder" in the aircraft. The ATC transponder is a transmitter/receiver which transmits in response to an inter-rogation from the ground station secondary surveillance radar system. When no ATC reply is made from the aircraft under surveillance, the indication on the ground radar scope is a single short line, like the one at about 7 o'clock. If an ordinary ATC reply is made, the aircraft indication on the radar scope is like the two close to the sweep line.

The aircraft transponder reply can include a special code which identifies that particular airplane on the scope. If the pilot receives instructions from the ground station to do so, he presses his "ident" button on his control panel. This causes the display on the radar scope to change, so that the ground station operator can be positive of his particular location on the radar scope. The transponder can also transmit aircraft altitude information, which can then be displayed to the ground station operator.

The ATC transponder system is an outgrowth of the war-time identification "friend or foe", and all commercial aircraft presently use it.

The Air Traffic Control (ATC) system allows controllers to track airplane movement on ground radar displays.

The ground station monitors the airplane's altitude and identification, and computes its range, bearing and airspeed.

The altitude and identification of the airplane is transmitted to the ground station by an on-board transponder. This information is sent in response to interrogation signals from the ground station.

Figure 12-1

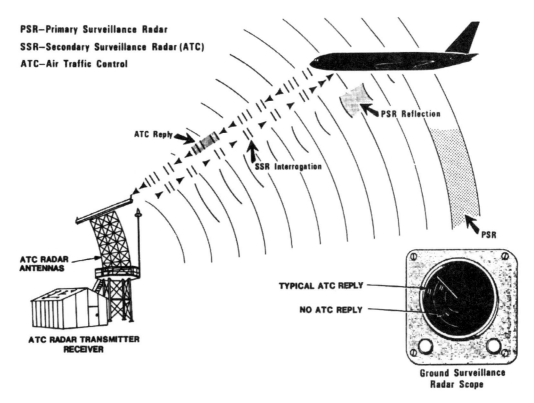

PSR—Primary Surveillance Radar
SSR—Secondary Surveillance Radar (ATC)
ATC—Air Traffic Control

ATC Reply

PSR Reflection

SSR Interrogation

PSR

ATC RADAR
ANTENNAS

TYPICAL ATC REPLY

NO ATC REPLY

ATC RADAR TRANSMITTER
RECEIVER

Ground Surveillance
Radar Scope

ATC Transponder System

Figure 12-5 shows the pilot's control panel for the ATC system. The switch in the lower left corner of the panel marked "altitude source" is a two-position switch by which the pilot can select No. 1 or No. 2 central air data computer to give altitude information to the ATC transponder. Two sets of knobs of two each are below the "code" digital readout. Each knob controls one digit. A total of 4,096 codes are available.

At the request of the ground station, the pilot selects a particular code. The selector knob in the upper left corner of the control panel selects the ATC system desired for use. With the knob in "standby" position, an otherwise long warm-up can be eliminated prior to actual use without putting the ATC system into full-time operation.

The button marked "ident" below the digital readout is the one which the pilot uses at the request of the ground station for identification on the ground radar scope. Each display is unique from other aircraft radar returns.

The upper right switch is a mode select switch. Mode A is ordinarily used for domestic operation and mode B is ordinarily used in Europe. Mode C puts the transponder in condition for transmitting

altitude information alternately with code information. Since mode C is no longer used, the altitude information is transmitted by the aircraft transponder, if the altitude report switch is on and either mode A or mode B is selected.

The ground station transmits on 1,030 MHz. If the pilot has selected mode A (Figure 12-2), his transponder will reply only to an interrogation which consists of two pulses eight microseconds apart. If the pilot has selected mode B, the transponder will reply only if the interrogation consists of two pulses separated by 17 microseconds.

A mode S (TCAS) equipped aircraft control panel will include other functions such as;

- ABV – N – BLW switch selects the upper or lower antenna position. N (normal) uses both upper and lower antennas.
- The ABS/REL switch will indicate an intruder's altitude relative to your aircraft or altitude above sea level.
- The Norm/ALTN will select either the ADIRU or SAARU as the primary altitude source.

Continued on next page ...

Figure 12-2

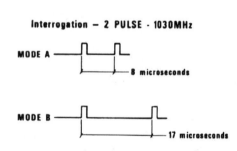

Figure 12-3

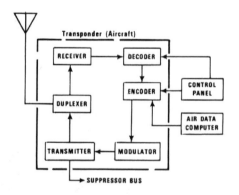

Figure 12-4

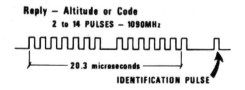

Figure 12-5

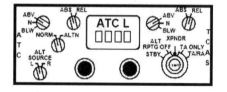

ATC Transponder System (cont'd.)

Figure 12-4 shows a digitized altitude or code reply by the aircraft transponder. This figure shows all 14 possible pulse positions and the identification pulse. The first and the last are framing pulses and will always be present (there is never a pulse in the middle open space). The other pulses will or will not be present in their allotted spaces, depending upon the altitude code (submitted by the central air data computer), or by the aircraft code (submitted by the control panel). By the relative timing of the aircraft transponder transmission pulse groups and identification pulses, the ground station distinguishes between altitude and code pulses.

Figure 12-3 shows a simplified block of the aircraft transponder system. A single antenna is used for both the receiver and the transmitter.

A duplexer arrangement switches the antenna back and forth between the receiver and transmitter as required. The information from the receiver goes to the decoder then to the encoder, which determines whether the transmitter will transmit. The air data computer supplies coded altitude information to the encoder, and the control panel supplies selected code information to the encoder.

The two DME systems and the two transponder systems are interconnected by a suppressor bus which prevents transmission from more than one system at a time.

Figure 12-2

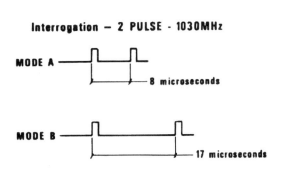

Figure 12-3

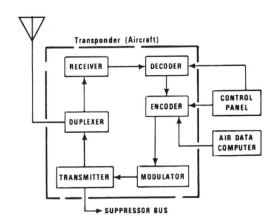

Figure 12-4

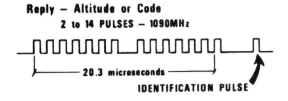

Figure 12-5

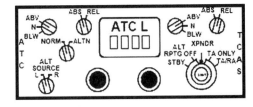

TRAFFIC ALERT AND COLLISION AVOIDANCE SYSTEM (TCAS)

TCAS is an abbreviation for traffic alert and collision avoidance system. TCAS is a system that is designed to alert the pilots to the potential of conflicts with other airplanes equipped with mode C or mode S transponders.

TCAS II provides 2 types of advisories to the pilots. One type of advisory is called a traffic advisory (TA), which informs the pilots that there are other airplanes in the area. The other type of advisory is called a resolution advisory (RA), which advises the pilots that a vertical corrective or preventative action is required to avoid an intruder airplane. TCAS II also provides aural alerts to the pilots.

The dedicated components of a TCAS II system consist of a receiver/transmitter (R/T), a top directional antenna and a bottom antenna which may either be omni-directional or directional.

The TCAS II system interfaces with the ATC system and requires the use of a mode S transponder, a top and bottom ATC antenna, and a mode S/TCAS control panel. Also the TCAS II system interfaces with the EFIS and warning systems to provide visual advisories and aural alerts Figure 12-6).

The TCAS R/T contains the computational processors required to determine if the path of nearby airplanes will intersect the flight path of the TCAS equipped airplanes.

An omni-directional antenna (bottom) and a directional antenna (top), or 2 directional antennas are used by TCAS to determine the bearing and altitude of transponder equipped airplanes responding to the interrogations of the TCAS R/T. Two additional antennas are required for the mode S transponder associated with the TCAS.

Figure 12-6

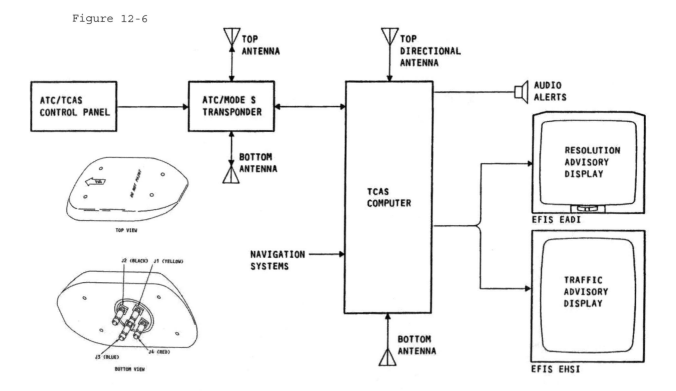

CHAPTER 13 - DISTANCE MEASURING EQUIPMENT (DME)

DME Principles

Figure 13-1 illustrates the principles of a distance measuring equipment system (DME). The distance readout may be in a separate indicator, as drawn, or it could appear in a horizontal situation indicator at the upper left or upper right.

A DME station is typically located at a VOR station or near an instrument landing system (ILS) at an airport. When a VOR or LOC frequency is selected on the VHF navigation control panel, the frequency of the associated DME station (if there is one) is simultaneously selected.

If the DME function knob is in the normal position, distance indications up to 199 miles will be given for VOR stations and up to 50 miles will be given for ILS stations. If the knob is in the override position, distances up to 199 miles will be given for localizer stations. When the function knob is in the test position, flag, blank, and 000.0 indications appear in succession in one type of DME.

In practically all radio equipment using an oscillator, stability is much to be desired and is made as good as possible. The aircraft DME system, however, utilizes one oscillator (to control spacing of pulses) whose stability is deliberately made very low. This is the secret of success for a DME system

since this oscillator provides the spacing of transmitted pulse pairs in a random manner.

Since the random spacing of the pulse pairs is unique to each DME unit, the aircraft is able to recognize its own signal when retransmitted by the ground station, and to distinguish it from other DME transmissions in the same area.

The aircraft DME constantly (although intermittently) transmits its uniquely, randomly spaced pulse pairs to the DME ground station. After a short delay, the DME ground station retransmits these pulse pairs at a frequency either above or below the frequency of the airplane transmitter.

Random spacing of the pulses makes it possible for each DME system to discriminate between its own retransmitted signals and those of other aircraft. The random spacing identification also makes it possible for the aircraft receiver to deter-mine elapsed time between transmission from the aircraft and the receiving of that signal retransmitted from the ground station. Since this time interval is a function of the intervening distance between the airplane and the ground station, the aircraft system can display the distance between the airplane and the ground station.

Figure 13-1

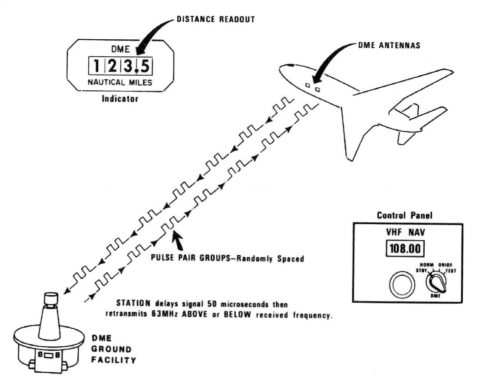

DME System

Figure 13-2 is a block diagram of a DME system. The same antenna is used by the transmitter and the receiver. The antenna is switched back and forth by a duplexer arrangement in the antenna lead-in. The frequency selected at the VHF NAV control panel also controls the receiver and transmitter frequencies selected in the DME interrogator.

Distance and time delay are functionally equivalent, so the "controlled variable delay" block accomplishes two things: It can delay the transmitted signal internally and at the same time give out a signal to the indicator to display the equivalent distance of the delay set up in the controlled delay block.

When the system is first turned on, a controlled variable delay servo motor system begins to run. It runs from zero delay to maximum delay, causing the indicator readout to change from zero to maximum then back, until it matches the delayed transmitted signal to the received signal (usually on the first run).

This running through delay and distance indication is called searching. During this time, the warning flag in the DME indicator is in view, showing that the DME system is not yet operative.

The transmitter is periodically transmitting, and the receiver is periodically receiving, at the same time the delay is changing. The matching circuits furnish the error information controlling the servo motor. When the matching circuits see that the received signal is the same as the delayed transmitted signal, it causes the servo to lock on to that particular delay, which is the same as locking on to a particular distance indication.

"Locking on" means that the servo motor stops its relatively rapid search. It runs slowly after "lock-on" as it sees a signal from the matching circuits calling for a move in one direction or the other to maintain the matched condition.

As long as the delayed transmitter signal matches the signal received, the distance indication is correct. When a difference occurs, the servo motor has to be driven one way or the other to increase or decrease the delay, depending upon whether the airplane is approaching or receding from the ground station.

The transmitter is connected to a suppressor bus connected to other radar type transmitters in the aircraft, such as ATC transponders or another DME transmitter. When one transmitter connected to a suppressor bus is transmitting, other transmitters connected to the bus are suppressed. Continuous transmission is not necessary for any of these systems.

Each DME station on the ground periodically transmits its identification letters in Morse code at an audio frequency of 1350 Hz. An audio detector in the DME receiver sends this information to the flight interphone system in the cockpit so the flight crew can determine whether the DME station which their DME interrogator is using is indeed the correct station for the particular VOR or localizer frequency they have selected.

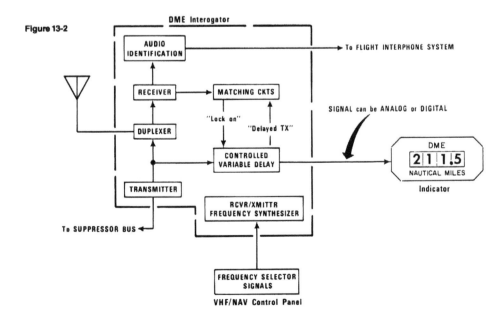

Figure 13-2

Distance Measuring Equipment

The DME (Distance Measuring Equipment) system determines the distance (in nautical miles) between an airplane and a selected VOR ground station. The system measures airplane-to-station slant range distance (line of sight), by timing a signal being sent from the airplane to the ground station, and the ground signal being sent back to the airplane.

The time is converted to a signal and fed into the DME indicator, displaying miles-to-go, in nautical miles.

This distance is measured and determined by the interrogator (receiver/transmitter) unit. When a VOR frequency is selected, the DME frequency is automatically selected. In flight the DME is constantly monitoring itself and computing DME slant range.

The DME unit measures slant range from airplane to ground station. When the plane is near or over the beacon, the DME range is more nearly the altitude of the plane than the true horizontal range from the beacon.

This would normally cause the ground speed shown on an indicator to be slower than actual ground speed when near the beacon.

To overcome this problem, the indicator goes into memory when the airplane is twenty miles from the beacon, and indicates the last calculated ground speed, corrected for the slant range at twenty miles, until the airplane passes twenty miles on the other side of the beacon. If the airplane is making a straight-in approach so that slant measurements will not cause an error in ground speed, the memory feature can be overridden.

Figure 13-3

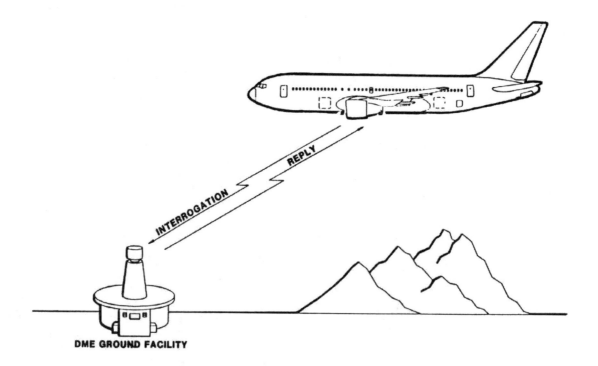

DME GROUND FACILITY

CHAPTER 14 - MARKER BEACON SYSTEM (MB)

Marker Beacon System

Most marker beacons transmit a fan shaped signal. It is a directional signal such that, in three dimensions, it resembles a thick, spread out Japanese hot weather fan. It is oriented so that the intended path of an airplane is through the fan from front to back.

The representation of the outer and middle marker transmissions in Figure 14-2 are shown in a side view. Since inner markers are virtually non-existent as far as Category II runways are concerned, the term "middle marker" seems a bit out of place. The middle marker is placed about 3,500 feet from the near end of the runway, and the outer marker about five miles away.

There is another type of marker transmitter station which is located along airways, and therefore called "airways marker". All marker station transmissions are on a basic radio carrier frequency of 75 megacycles. The thing that distinguishes one from the other is the frequency of the audio tone with which the carrier is modulated. The outer marker transmits a 400 Hertz tone, the middle marker a 1,300 Hertz tone, and the inner and airways markers transmit a 3,000 Hertz tone.

Figure 14-1 shows the receiver with a 75 MHz filter and Rf amplifier and detector. From the detector comes whatever audio tone has modulated the carrier frequency. Three audio filters discriminate audio tones. The audio AC (about 6 volts) is used to light an indicator light. If the audio tone is 400 Hertz, it lights a blue outer marker light, shown on the indicator panel. If it is a 1,300 Hertz tone, it lights an amber middle marker light. And if it is a 3,000 Hertz tone, it lights a white inner/airways marker light.

These tones, along with the Morse code for identifying a station, are made available through an audio amplifier to the cockpit audio interphone system. If equipped, the sensitivity switch is used in the low position if an ILS approach is to be made, and in the high position at cruise altitudes.

The marker beacon system is turned on when the airplane electrical buses are energized, and the applicable MKR BCN circuit breaker is closed. It receives VHF signals consisting of a 75 megahertz carrier; amplitude modulated with 400, 1300, or 300 Hertz, depending on which of the three marker beacon signals is received. These signals are converted to an aural and visual output to indicate passage over a marker beacon transmitter.

Continued on next page ...

Marker Beacon System (cont'd.)

Airway markers are usually associated with specific aids to enroute navigation (or holding points), and provide the operator with an exact position at the time of passing over the associated range station. Airway marker facilities are identified when the white lights come on and a 3,000 Hertz tone is heard.

Outer and middle markers are associated with an instrument landing system. The outer marker is usually located directly below the point where an airplane on a localizer course should intersect the glideslope and start descending. An outer marker is identified when the blue lights come on and a 400 Hertz tone is emitted.

The middle marker is located near the runway, usually under the point on the glidepath where a descent could be discontinued. A middle marker is identified when the amber lights come on and a 1,300 Hertz tone is emitted.

Figure 14-1

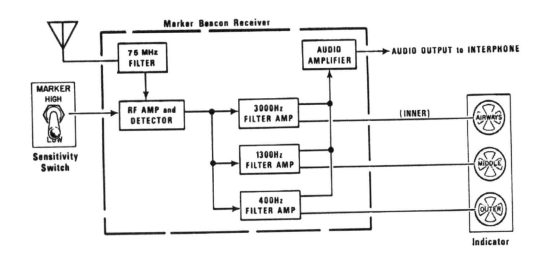

Figure 14-2

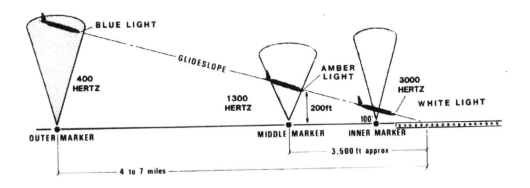

CHAPTER 15 - RADIO ALTIMETER PRINCIPLE (RA)

Radio Altimeter Principle

A radio altimeter is called "radio" rather than "radar" because its transmissions are not pulsed as radar transmissions are. The transmission is a continuous wave, constant amplitude, frequency modulated carrier of 4,300 MHz. The depth of frequency modulation is 50 MHz, so the transmission is continuously varied between 4,250 MHz and 4,350 MHz. The rate of frequency modulation is 100 times per second.

Some systems use a depth of frequency modulation other than 50 MHz. However, even though the number changes, the results will be the same. In an airplane flying above the ground, there is therefore a difference in the frequency of the reflected signals seen by the receiver and the transmitter frequency at the same instant. The difference in frequencies between transmitter and receiver at any given instant is a direct function of the distance that the radio wave had to travel from the transmitter antenna to the ground and back to the receiver antenna.

If you calculate the frequency change per foot of transmitted distance, you will find that for each foot of transmitted distance, there is a frequency change of approximately ten cycles. Since the transmission must travel to the ground and back again to the receiver, the frequency change per foot of airplane altitude is approximately 20 cycles. For example, if the airplane is 1,000 feet above the ground, there will be 20,000 cycles of frequency difference between the transmitter frequency and the frequency received by the receiver.

In the receiver mixer, the transmitted and received frequencies are mixed and the beat frequency (difference) is counted in the counter.

The beat frequency counter converts the frequency difference to an analog DC voltage whose amplitude is a function of aircraft altitude above the ground. A servo system in the indicator drives the indication (here a vertically movable tape) to a position corresponding to the amplitude of the DC analog voltage received from the beat frequency counter.

Continued on next page ...

Radio Altimeter Principle (cont'd.)

The antennas are so designed that as long as the bank angle does not exceed 30°, and the pitch attitude is not more than 20°, the altitude indication remains correct. If the bank angle exceeds 30°, or pitch attitude exceeds 20°, the indicated altitude is excessive. These high figure attitudes would not be maintained very long, so that does not present a problem.

This radio altimeter is called "low range" be-cause it is not intended to operate at airplane altitudes above the ground greater than 2,500 feet. Its principal usage is during the final approach. When making a Category II approach, the crew is notified by the radio altimeter when the airplane is 100 feet above the runway extended. This is the point at which the crew must be able to see to land, called "decision height".

Since the terrain is not necessarily level in front of the runway, each Category II runway chart shows the radio altimeter setting for that particular runway at which the airplane will be 100 feet above the runway itself. It might be more or less than 100 feet. A decision height selection is made with a knob on the front of the indicator which positions an index on the altimeter indicator to the selected point.

In addition, it controls a switch in the indicator so that, when the selected decision height is reached, a "decision height" light is lit by the indicator. A self-test button on the indicator self-tests the transmitter/receiver and indicator.

Figure 15-1

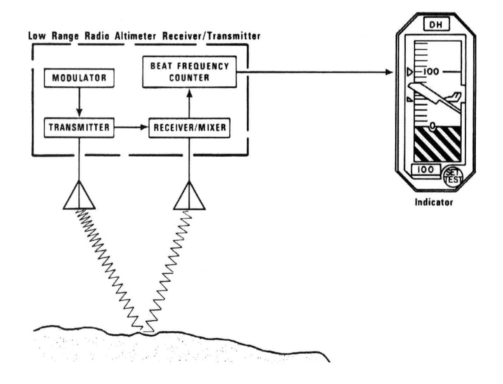

Indicator

Radio Altimeter System

Figure 15-2 calls out altitude trip switches in the receiver transmitter. These are switches driven by the servo motor system which develops the DC analog voltage. Preset to operate at selected altitudes, they are used for such things as initiating gain programming in flight directors and auto-pilots, and for lighting an "alert" light.

The receiver/transmitter is typically a dual unit which develops DC analog altitude voltages in two separate systems, normally the same. A part of the monitoring system in the receiver/transmitter unit is checking on the agreement between these two outputs.

Figure 15-3 illustrates a typical set of radio altimeter antennas. This illustration is for an aircraft with two radio altimeter systems, and therefore, four antennas. The actual antenna locations are usually established experimentally.

Figure 15-4 illustrates radio altitude and decision height indication on EFIS equipped aircraft using and EADI. On aircraft equipped with a Primary Flight Display (PFD), radio altitude will be shown in the center lower portion of the attitude ball and decision height illustrated using a pointer on the left side of the altitude tape. Decision height on both EFIS aircraft types will be set using the EFIS control panel.

Figure 15-2

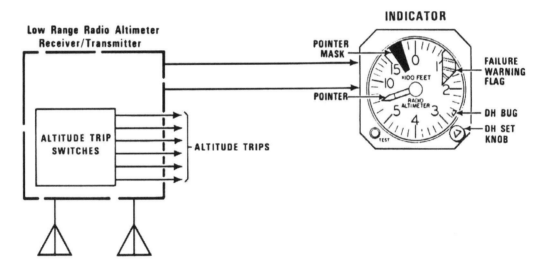

Figure 15-3

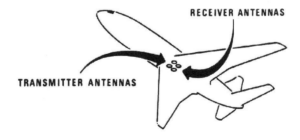

Radio Altimeter System (cont'd.)

Figure 15-4

RST Switch
Resets the DH alert display on the associated ADI. Changes the DH from yellow to green and causes the decision height value to reappear. Also causes the radio altitude to change from yellow to white. Reset occurs automatically on touchdown or climbing through selected decision height plus 75 feet.

Decision Height Knob
Sets decision height on the associated ADI. Setting negative value removes DH display.

DH REF Indicator
Indicates the decision height set with the decision height knob from -20 to 999 feet.

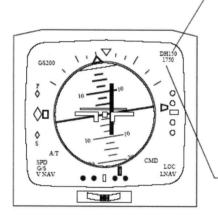

Decision Height Display
Indicates decision height set in the associated DH REF indicator. When descending through selected decision height, the digital indication disappears and the letters DH change from green to yellow, enlarge to the same size as the radio altitude display, flash several times, then remain steady. Replaced by DH flag when input is unreliable.

Radio Altitude Display
indicates radio altitude at 2,500 feet and below. Changes from white to yellow at decision height. Changes from yellow to white when passing selected DH during go-around, after touchdown, or after pushing RST switch. Replaced by RA flag when input is unreliable.

CHAPTER 16 - AIR DATA SYSTEMS

Typical Pitot And Static Sensing

Air data systems and air data instruments depend upon pitot pressure and static pressure sensing, as well as air temperature sensing. Static air pressure is the pressure of the outside air at the location of the airplane. It could be measured exactly in a motionless balloon at that altitude, but is difficult to measure precisely in a moving airplane.

Figure 16-1 on the left side shows a flush skin mounted static port, and in the upper right side illustrates a typical location. The optimum locations for the static ports have to be determined experimentally in the prototype airplanes. The locations chosen are where the errors are least.

Static ports of a different and later kind are located around the outside of the pitot static probe, indicated in the lower part of the illustration. These locations have been found to be as good as any, but flush static ports are still used for alternate or auxiliary sources.

The pitot static probe illustrated has two sets of static ports which have separate static line connections as indicated. The forward end of the probe is an open tube with a knife edge around the circular opening in front. These pitot probes are located on the airplane in a spot where they have the best possible access to undistorted airflow. They are oriented so that in normal flight attitude the probe is pointed directly into the airstream.

In a parked airplane, the pitot and static pressures are equal. In a moving airplane, the pitot pressure is greater because additional pressure is developed at the forward end of the tube by its motion through the air.

Altitude is calculated on the basis of static air pressure, and airspeeds are calculated on the basis of the difference between pitot and static pressures. Since a pitot static probe is, under some conditions, subject to icing, it is necessary to have available a heater to melt the ice which otherwise would block the ports. Flush static ports may also be heated, if necessary.

Figure 16-1

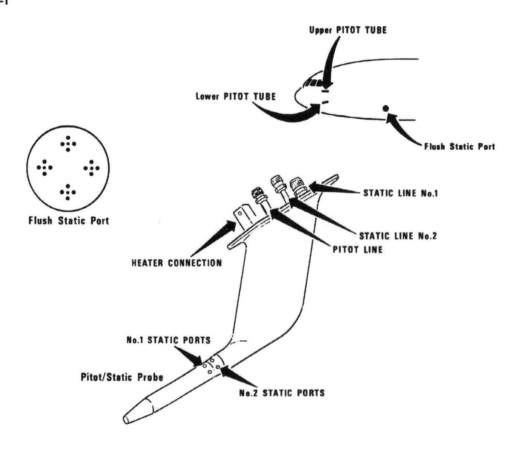

Total Air Temperature Sensing

Figure 16-2 illustrates a total air temperature probe. Total air temperature is the static air temperature plus the rise in temperature due to the pitot effect.

As Boyle's Law pointed out quite a few years ago, increasing the pressure of air also increases its temperature. Pitot pressure is static pressure increased by a pressure factor which results from forward motion of the airplane through the air. Similarly, total air temperature is static air temperature increased by an amount accounted for by the forward motion of the airplane through the air.

Total air temperature is of great importance in setting up the operating conditions of a jet engine since the temperature of the air into the jet engine is static air temperature increased by the pitot factor. It is also possible to derive static air temperature from total air temperature and pitot pressure information.

The total air temperature probe is constructed similar to a pitot pressure probe. Constructing a total air temperature probe is considerably complicated by the need for providing deicing heat. The air whose temperature is being measured must be carefully shielded from deicing heat. That part of the construction has been eliminated from this illustration, but the principle of total air temperature is shown here.

A pitot pressure sensing tube can be dead-ended because there is no need for airflow through the pitot pressure system. A total air temperature probe, however, needs some airflow through it to avoid heating of the measured air by the heating element, and so as always to be measuring ambient, not static air. The amount of airflow through the metered orifices is quite small and, therefore, has only a minor influence on the pitot effect. The total air temperature sensing element is shielded as well as possible from the deicing heat, and therefore senses the temperature of air which has been affected very little by it. Altogether, the results are remarkably satisfactory.

Figure 16-2

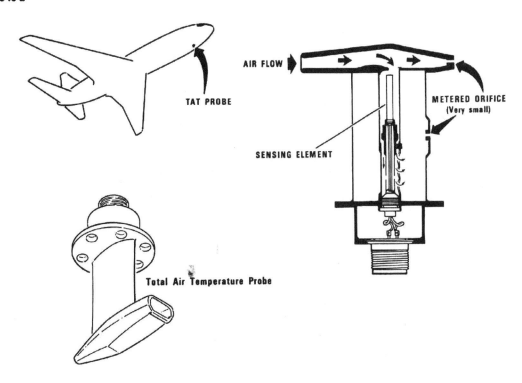

Pitot/Static System No. 1

Figure 16-3 is a chart of a typical pitot static system. Each of the pitot static probes has one pitot pressure line, indicated by a solid line, and two static lines, indicated by dashed lines. For example, the No. 2 alternate pitot in the upper left of the drawing has A, B and C lines. The legend in the center calls out A as pitot, B as first officer static, and C as second auxiliary static.

In order to help compensate for port position errors, as well as to compensate for static errors introduced by airplane yawing, each static system has a port on the left side of the airplane and a port on the right side of the airplane, joined in a common line.

If you follow the B line from the No. 2 alternate pitot, you find it connected to a static source in the captain's pitot on the left side of the airplane. The other three forward static systems are similarly cross-connected.

On the far right side of the drawing are represented two flush static ports at the skin of the airplane. They are connected to each other and are designated "alternate static".

The instruments at the front of the airplane in this drawing are all direct reading mechanical instruments. They all give information derived mechanically at the instrument from static and/or pitot information. Each of the boxes marked "static" are static manifolds, which consist of large tubing from which the various instruments take their static pressure.

The alternate static pressure source is used for comparison checking of the normally connected static pressures. Moving one of the static source selector valves, so that it connects its static manifolds to the alternate static source, should cause a very minimum amount of change in indication of any of the instruments.

By checking between the captain's and first officer's instruments and further checking with the alternate static source, the crew can determine whether a fault exists in a static system, or whether it exists in the instrument itself.

Pitot and static lines must be carefully protected from leaks and from water accumulation. Drain ports are provided where they are periodically drained.

A mach meter/airspeed indicator is a dual indicator which shows both the "indicated" airspeed and the "Mach" speed of the airplane. "Indicated" airspeed is basic information because the flight characteristics of the airplane change as a function of "indicated" airspeeds.

"Indicated airspeed" is differentiated from true airspeed since indicated airspeed is a function of the ratio between pitot and static pressures. At sea level on a "standard" day, indicated airspeed and true airspeed would be the same. If an airplane flies a constant true airspeed while changing its altitude from sea level to 40,000 feet, the indicated airspeed continuously falls off as the altitude increases.

The mach meter gives a reading of the aircraft's airspeed as a fraction of the speed of sound at ambient temperature. The speed of sound at commercial airline altitudes varies only as the temperature of the air, not as an altitude function. The mach meter indication, paradoxically, is derived from indicated airspeed and altitude of the airplane.

The altitude of the airplane is substituted for the temperature function because it is more convenient. However, it is usually not correct at altitudes below about 25,000 feet. The altitude function can be substituted for the temperature function only if the expected temperature is found at a particular altitude. The expected, or standard day, temperatures usually are found at the higher altitudes. Standard day temperatures usually are not found at lower altitudes.

Continued on next page ...

Pitot/Static System No. 1 (cont'd.)

For example, lower altitude temperatures above Miami in the summer are quite different from lower altitude temperatures above Minneapolis in the winter, but above 25,000 feet there is not much difference. Therefore, Mach meter indications below about 25,000 feet are not used.

Only static pressure is needed to find the altitude of the airplane. The static pressure needs to be compensated according to the prevalent barometer reading in the area. Altimeters, therefore, have a barometric correction knob by which corrections are made according to information given from a ground station.

By common agreement, all aircraft set their altimeters to a standard pressure, 29.92" Hg, above 18,000 feet. This makes it easier for cruising aircraft to maintain their assigned altitudes with-out constantly resetting altimeters.

A vertical speed indicator needs only static information to show the rate at which the altitude is changing.

The cabin pressure computer maintains the desired pressurization of the aircraft with respect to outside air pressure. A cabin altitude and differential pressure indicator is used for monitoring aircraft pressurization.

The mach/airspeed warning switch provides a warning if the airplane speed becomes excessive.

Central air data computers are capable of computing any desired air data function.

Figure 16-3

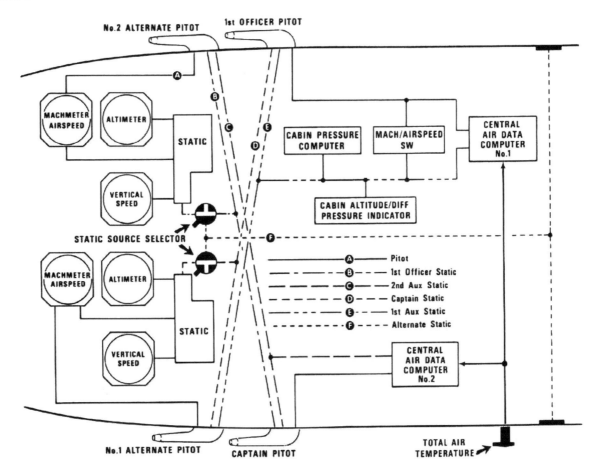

Pitot/Static System No. 2

Figure 16-4 illustrates a centralized computer developing all of the air data information and electrically transmitting it to the flight instruments in the cockpit. This print shows that only the standby altimeter and standby airspeed indicator are mechanical devices using pitot and static information directly.

By more or less common agreement the term "air data sensor" is reserved for an air data computer belonging to and used by only an autopilot.

The term "central air data computer" is used to describe an analog computer, typically giving analog outputs, synchro, or DC voltages.

The term "digital air data computer" refers to one which makes its computations digitally and provides digital output information. In addition, they supply analog information where needed.

Before going into central air data computers, we will show schematically how some of the indications are derived, directly and mechanically, by self contained indicators.

Figure 16-4

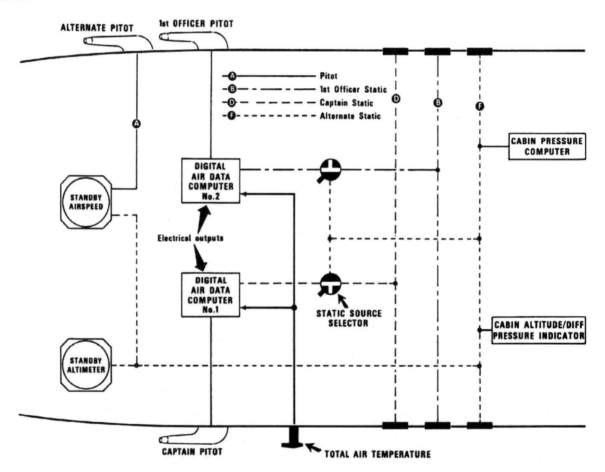

Boeing 777 Air Data System

Boeing 777 air data information is obtained from both the ADIRU (Air Data and Inertial Reference Unit and the SAARU (Secondary Attitude Air Data Reference Unit) illustrated in Figure 16-4a.

The ADIRUs four processors get air data from the air data modules. The ADIRU gives these air data outputs:

- Altitude rate
- Pressure altitude
- Computed airspeed
- Mach number
- True airspeed
- Static air temperature
- Total air temperature
- Impact pressure
- Total pressure
- Static pressure
- Angle-of-attack

The SAARU supplies pitch and roll attitude to the standby attitude indicators. It is also the secondary source of inertial navigation and air data for the Primary Flight Displays (PFDs), primary flight controls system (PFCS), autopilot flight director system (AFDS) and other airplane systems.

The captain or first officer enters altitude barometric correction at the EFIS control panel. If one of the EFIS control panels fail, barometric corrections can come from the CDU.

There are two AIR DATA / ATT source select switches on the flight deck. The switches control the source of display data for the PFD. The primary source of display data is the ADIRU and the secondary source of display data is the SAARU.

The ADIRU sends data on the left and right ARINC 629 flight control buses and gets data from all three ARINC 629 flight control buses The SAARU sends data on the center flight control data bus, and gets data from the left, center and right flight control data buses.

Each pitot probe and static port connects to a 629 air data module (ADM). The ADMs change the air pressure into ARINC 629 digital data. The ADIRU and SAARU get the 629 air data from the ADMs.

The center pitot probe and the standby static ports also have a 429 ADM. The standby altitude and airspeed indicators get the 429 air data from the 429 ADMs.

Three flat panel standby displays show:

- Indicated airspeed
- Altitude data
- Attitude data from the SAARU

The angle of attack (AOA) vanes send analog signals to the AIMS cabinet.

The total air temperature (TAT) probe is a dual element probe with two analog outputs which goes to the AIMS cabinets.

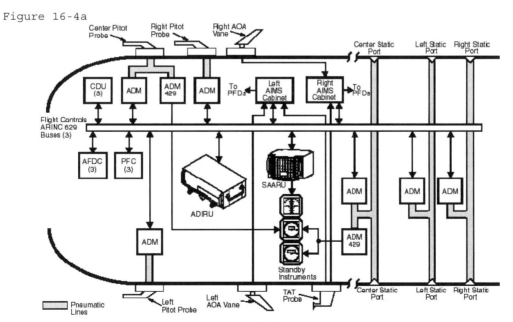

Figure 16-4a

Air Data Inertial Reference System

Air Data Module

The Boeing 777 has 6 air data modules (ADMs) in the air data inertial reference system (ADIRS). Three are for pitot pressure and three are for static pressure (Figure 16-4b) All the ADMs operate in the same manner.

The air data module (ADM) measures absolute pressure. This pressure comes from a pitot probe or static pressure port. The ADM transmits this pressure on a flight control ARINC 629 bus for use by the ADIRU and SAARU. All ARINC 629 ADMs are interchangeable and can be a pitot or a static ADM.

The ADM has one pneumatic connector and two electrical connectors. The pneumatic connector seals itself when removed from the ADM so that you do not need to plug the pneumatic line when you remove the ADM. One electrical connector is for program pins and power. The other electrical connector is for ARINC 629.

The ADM receives pneumatic pressure from a pitot probe or static ports. The pressure sensor changes the pressure value to an analog signal. The analog signal goes through an analog to digital converter. The ADM calibrates the pressure and

compensates for temperature. The ARINC 629 transmitter / receiver sends the data on a flight controls ARINC 629 bus. Program pins define the location of the ADM. The input/output decoder logic sends the program pin ADM location data to the processor. This location information goes to the ADIRU and SAARU on a flight controls ARINC 629 bus.

The pitot ADMs control the heat for the pitot probes and the angle of attack (AOA) sensors. The left pitot ADM controls the heat for the left pitot probe. The center pitot ADM controls the heat for the center pitot probe.

The ADMs supply a ground heat signal to the power management panels when the PFCs report these two conditions are true:

• The airplane is on the ground
• An engine running.

The ADMs supply an air heat signal to the power management panels when any of these conditions is true:

• The airplane is in the air
• Computed airspeed (CAS) is more than 50 kts.

Figure 16-4b

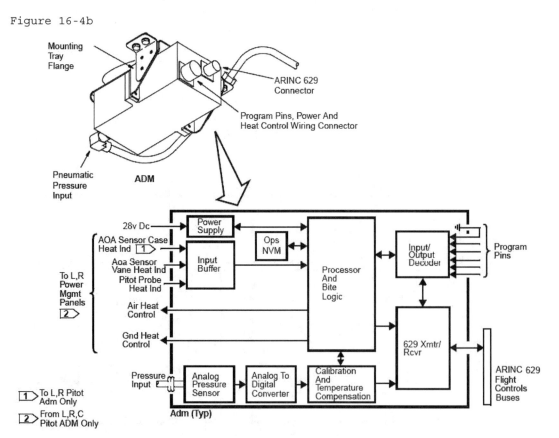

Reduced Vertical Separation Minimum (RVSM)

Reduced Vertical Separation Minimum (Figure 16-4c) is a program designed to increase usable airspace by reducing the vertical separation between aircraft from the present 2,000 foot vertical separation requirements to 1,000-feet at altitudes between 29,000 and 41,000 feet (FL290 and 410). With the implementation of RVSM, airlines experience an increase in operating capacity, improved operating efficiency, and increased controller flexibility. Projected savings, based on FAA calculations, are $176 Million over the next 20 years. All this comes at minimal cost and provides significant operational and economic benefits to operators.

Aircraft equipment required for RVSM is:

- Two independent Air Data Computers and their related Pitot/Static sensors and AOA vane.
- Two Primary Altimeter Systems.
- One Altitude Reporting Transponder.
- An Altitude Alert System.
- An Autopilot System with Altitude Select and Altitude Hold.

Operators who wish to fly RVSM must qualify their fleets of aircraft and maintain them to rigorous standards for the life of the aircraft.

Altitude errors must be minimized, and if certain limits such as skin waviness in the RVSM critical areas or damage to a pitot or static port are exceeded, must be reported to the FAA and the aircraft removed from RVSM qualification until the problem is detected and corrected. Non-RVSM aircraft can be flown through RVSM airspace with prior approval from the FAA and flown with a 2,000 foot separation for the duration of the flight.

Recurrent training is required for maintenance and inspection personnel who maintain RVSM aircraft

Figure 16-4c

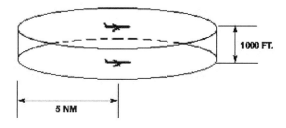

Basic Altimeter

Figure 16-5 illustrates the principle of the mechanically operated, direct reading altimeter. The "31,000" digital readout is not operated by the basic altimeter; it has to be servo motor driven using information from a central air data computer, discussed later.

The instrument actually uses complicated and delicate lever mechanisms to drive indicator needles or discs. Only one needle is shown.

The diaphragm in the lower right is shown in cross section. It is a sealed, partially evacuated, flexible corrugated case whose lower center is attached to structure. It is inside a sealed chamber connected to static pressure.

The rod and gear assembly above the diaphragm does not exist; it merely illustrates a principle. The principle is the lengthening and shortening of the connection between the diaphragm and the indication drive mechanism.

The gears in the rod and gear assembly do not turn unless the "baro" (barometric) set knob is turned; therefore, movement of the diaphragm upper half is directly transmitted to indication as if the rod and gear assembly were replaced by a single solid rod. The wavy dashed line represents freedom of vertical movement between the rod and gear assembly and the "baro" set knob.

Leaving the "baro" knob operations out of it at first, consider the diaphragm connected directly to the indicator needle mechanism. The coiled spring loads the indication toward the low side. The indication is shop adjusted to be zero at sea level.

As the airplane climbs, the static pressure decreases. Therefore the bellows exerts an upward force, changing the indication correspondingly. If the airplane descends, the diaphragm force decreases, and the coil spring changes the indication toward the low side.

Continued on next page ...

Figure 16-5

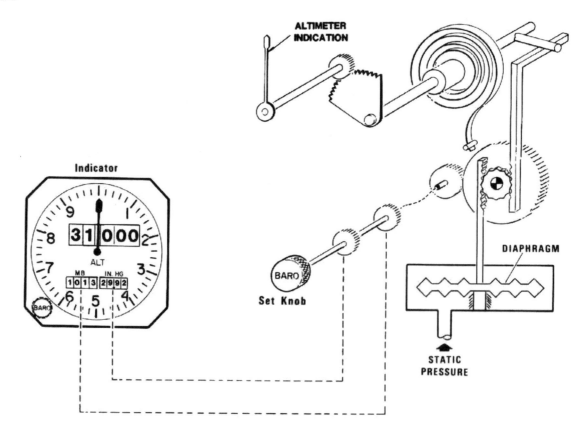

Basic Altimeter (cont'd.)

The baro set knob drives two digital readouts in the indicator. One is marked MB for "millibars" and the other IN. HG for "inches of mercury".

A millibar is l/1000 of a "bar" which is 1,000,000 dynes per square centimeter. It is a unit now commonly in use to measure atmospheric pressure. Another common unit is "inches of mercury", meaning the atmospheric pressure equivalent developed at the bottom of a column of mercury that many inches high. Under conditions of a "standard" day, the atmospheric pressure at sea level is 29.92 inches of mercury, which is the same as 1,013 millibars.

Since the atmospheric pressure changes with the weather, for an altimeter to read correctly, it must be adjusted for the actual atmospheric pressure at the time. This information is given to the flight crew by the tower at the time of departure, and subsequently as may be needed in flight.

The flight crew uses the baro set knob to bring the readings in the "MB" and "IN. HG" to agree with the tower information at departure. In the process, the length of the linkage between the diaphragm and needle mechanism is adjusted by the rod and gear assembly so that the altimeter reads the field elevation.

The information given by the tower is always "corrected to sea level", which means that even if measured at mile high Denver, the figures given are as if measured a mile below Denver.

At flight altitudes of 18,000 feet or above, all aircraft altimeters are set to 29.92 inches of mercury so that constant changes are not necessary enroute, and all aircraft readings are uniformly coordinated.

Figure 16-5

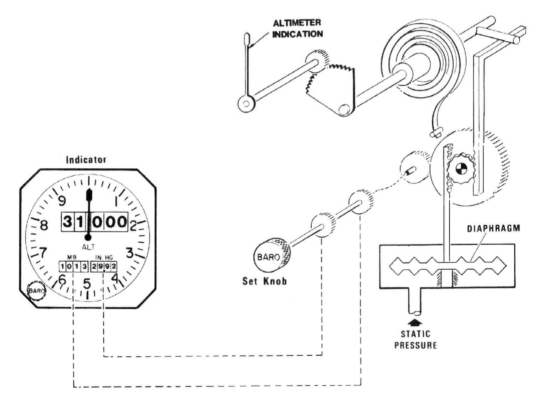

Basic Airspeed Indicator

Figure 16-6 shows the basic principle of the airspeed indicator. The diaphragm has pitot pressure on the inside and static pressure on the outside, so that a difference function governs the indication. The faster the airplane flies, the greater the difference between pitot and static pressure.

Figure 16-6

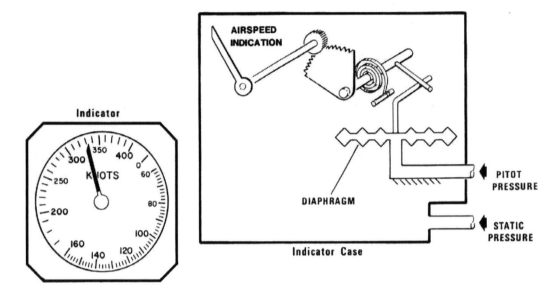

Altitude vs. IAS / Mach

Figure 16-7 illustrates the difference between indicated airspeed, true airspeed, and Mach. True airspeed and indicated airspeed are the same at sea level; however, as altitude increases, holding a constant indicated airspeed results in continually increasing true airspeed. For an example, 400 indicated airspeed at sea level becomes about 450 true at 10,000 feet, and about 550 true at 20,000 feet.

The Mach numbered lines are drawn on the basis of a standard day air temperature chart; .90 times the speed of sound at sea level would be about 600

knots true airspeed, but from about 36,000 feet on up, .90 Mach equals only 525 knots true airspeed.

If a particular airplane is not supposed to fly faster that 390 knots indicated, and not more than .885 Mach, it could fly 390 knots indicated until it got to 21,000 feet. Above that altitude, the indicated airspeed would have to decrease in order not to exceed maximum Mach. Flying an indicated airspeed as high as 390 knots above 21,000 feet would result in transonic or supersonic speeds.

Mach varies as a result of temperature and altitude.

Figure 16-7

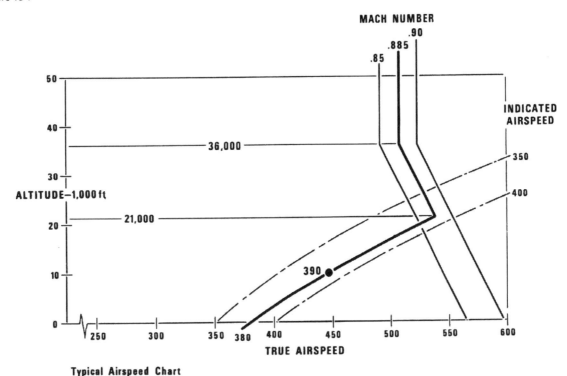

Typical Airspeed Chart

Basic Mach Meter

The speed of sound, contrary to what might seem reasonable, is not at all dependent upon altitude (within the ranges of altitude that commercial aircraft fly). It is dependent only upon the temperature of the air.

Since it would be awkward and difficult to build a mechanical indicator using outside air temperature as one of its inputs, altitude is substituted for temperature in the construction of the instrument. Since above about 25,000 feet the air temperatures encountered are fairly constant, it is possible to build a Mach meter using altitude instead of temperature (and use it only above 25,000 feet).

The speed indicated in a Mach meter is a fraction of the speed of sound. In computing this fraction of the speed of sound, the indicated airspeed of the airplane is first computed, then modified by altitude so that it becomes a fraction of the speed of sound.

In the lower right corner of Figure 16-8 is the airspeed diaphragm with pitot pressure on the inside, static pressure on the outside. If the airplane holds a constant airspeed and changes its altitude, the Mach indication is changed by the altitude diaphragm.

If the airplane holds a constant altitude and changes its airspeed, the Mach indication is changed by the airspeed diaphragm. The two diaphragms work in combination to give the Mach indication for a particular altitude and airspeed.

Airplanes using central air data computers (CADC), provide this information electronically to the pilot and co-pilot Mach/airspeed indicators (reference Figure 16-12).

Figure 16-8

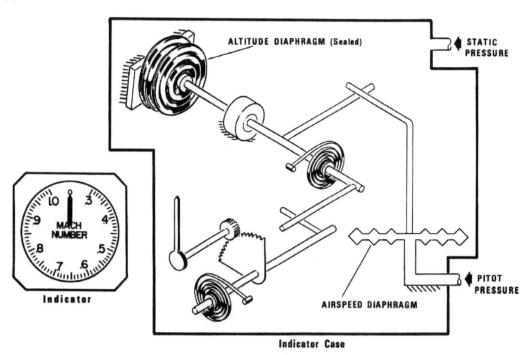

ALTITUDE DIAPHRAGM modifies INDICATED AIRSPEED information to determine MACH NUMBER.

Rate Of Climb Indicator

Older rate of climb indicators do not have the vertical acceleration pump and, therefore, did not give what is sometimes called instantaneous rate of climb, or vertical speed.

In going through this diagram let's ignore the vertical acceleration pump at first. With every-thing at rest and no change in altitude (no change in static pressure), the indicator would show zero rate of climb as drawn. If the airplane is climbing, the static pressure is decreasing and the air trapped in the case is gradually venting through the metering restriction. But the air from inside the diaphragm is leaving at a faster rate through the bypass restriction, and there is much less of it.

So if the airplane is climbing rapidly, there is a large difference in air pressure between inside and outside the diaphragm, and a large indication on the indicator. If the airplane then levels off and holds altitude, the rate of climb indicator continues to show a rate of climb until the air pressures are balanced.

Since it takes a little while to balance the air pressures, the old indicators (without vertical acceleration pumps) give a lagged indication for which the pilot has to mentally compensate.

The acceleration pump makes it possible to give indications which are very nearly instantaneous with vertical speed. First think what would happen if you are holding the indicator in your hand and then move it up. The weighted piston of the acceleration pump descends in its cylinder, causing pressure inside the diaphragm to be lessened, and causing the indicator, therefore, to show a climb. The reverse would happen if you move the indicator down.

If the airplane is climbing at a steady rate, the acceleration pump piston holds steady. Since the static pressure is changing steadily, the rate of climb indication is steady.

If the airplane levels off, the weighted piston of the acceleration pump moves up because of its up-ward momentum. The momentum and inertia of the weighted piston, in combination with the sizes of the restrictions, eliminate the indication lag which formerly existed.

Airplanes, using Central Air Data Computers, provide this information electronically to the pilot and co-pilot indicators (reference Figure 16-12).

Figure 16-9

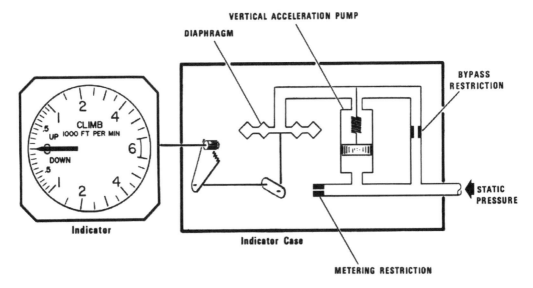

CADC Altitude Module

Figure 16-10 shows an altitude module in a central air data computer using an "E-pickoff" and a servo motor system. A sealed diaphragm in a static pressure chamber is connected to one end of a pivoted armature for the E-pickoff. The other end of the armature is connected through a spring to a servo motor rack and pinion gear.

If the airplane is parked or holding altitude, static pressure is constant. The servo motor will have driven until the force exerted by the spring balances the force exerted by the evacuated bellows, and the E-pickoff armature has been moved to its neutral position. Any position other than neutral causes the servo motor to run in one direction or the other.

If the airplane takes off, static pressure decreases progressively. The armature is no longer at a null. An error signal is developed, causing the servo motor to drive the rack and pinion to counter the changing force of the sealed bellows.

As long as the airplane altitude continues to increase, the servo motor will have to continue to drive, changing the force on the spring. The faster the airplane changes altitude, the faster the servo motor drives to keep up. The faster the servo motor drives, the greater is the altitude rate signal from the tachometer generator.

If the airplane descends, the servo motor drives in the opposite direction, keeping the armature near neutral. The tachometer generator is running in the opposite direction and the phase of its signal is reversed. The phase of the tachometer generator signal tells whether the airplane is climbing or descending. The amplitude of the signal tells how fast (reference Figure 4-7).

The servo motor also drives, through its gear box, a potentiometer wiper arm, and a coarse and a fine synchro pair (reference Figure 4-14).

As the airplane altitude changes, the position of the potentiometer wiper arm changes, and the positions of the coarse and fine synchro rotors change. Airplane altitude can be fed to remote users from any of these outputs. The potentiometer output can be used for other internal functions in the central air data computer itself. For example, it can be one of the inputs in computing Mach.

The altitude error synchro is spring-loaded to a null position so that, when the clutch is disengaged, its signal is a null. If the flight crew wants an autopilot or flight direction in altitude hold mode, they call for that mode on the control panel. This supplies a voltage to the clutch solenoid, engaging the clutch.

From that time on, any change in aircraft altitude develops an error signal. The signal will be one phase for going above the desired altitude, and the opposite phase for going below the desired altitude. The amount of deviation from the desired altitude will determine the amplitude of the error signal.

Figure 16-10

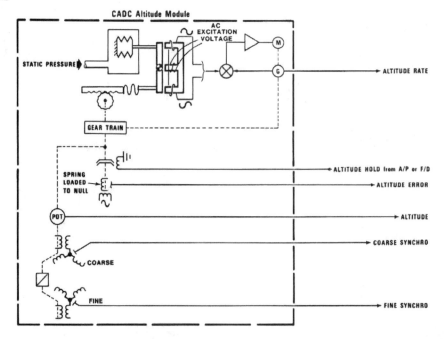

Central Air Data Computer

Figure 16-11 shows four modules of an analog central air data computer. The inputs to the computer are static pressure, pitot pressure, and total air temperature.

The servo motor loop of the altitude module drives: A tachometer generator to produce altitude rate; a potentiometer whose output is to the Mach module; another potentiometer whose output is to the cabin pressure control unit; a pair of coarse and fine synchros; a digital encoder of altitude information used by the ATC transponder; a pair of altitude hold synchros; and an E-pickoff armature.

The dashed lines from the static bellows and the tachometer generator go to a square box with a pair of crossed diagonals. This is a symbol for a differential gear box.

The indicated airspeed module has an indicated airspeed bellows controlling a servo loop which

drives Q-pots for gain control in flight directors and autopilots; a synchro which gives a remote reading of indicated airspeed; and a potentiometer which supplies one of the principal inputs to the Mach module.

The Mach module produces Mach information by combining indicated airspeed and altitude. Its servo motor drives a follow-up potentiometer, a Mach output potentiometer, and a potentiometer supplying Mach information to the true airspeed module.

The true airspeed module servo motor loop drives a transmit synchro for remote indication of true airspeed. It develops true airspeed by combining Mach information and total air temperature information. There are other ways to derive true airspeed information; for example, by combining indicated airspeed and altitude.

Figure 16-11

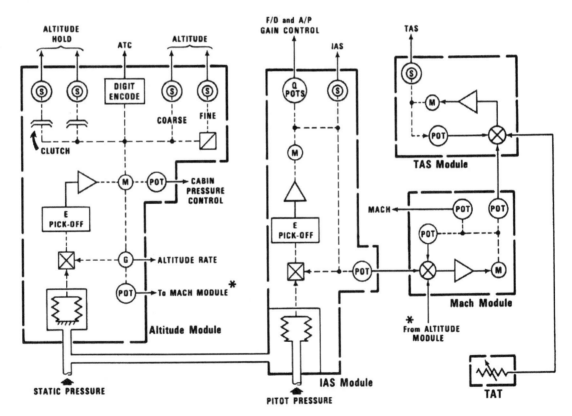

Typical CADC Inputs And Outputs

Figure 16-12 shows a block diagram of typical central air data computer inputs and outputs. The four blocks inside the computer represent separate modules which are capable of supplying the

information and controls indicated on their right. This is essentially the computer of the previous figure.

Figure 16-12

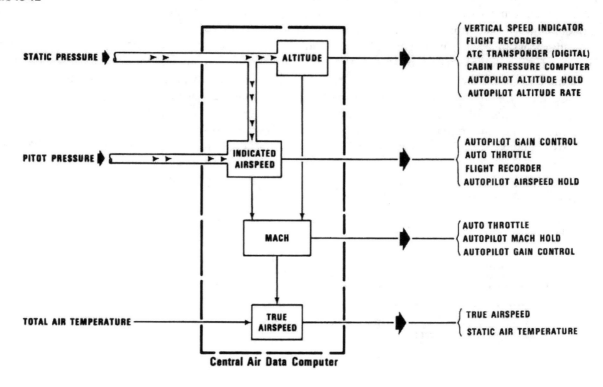

Digital Air Data Computer

Figure 16-13 shows some typical inputs and outputs for a digital air data computer.

A "digital air data computer" uses digital computing and electronic circuits rather than servo motors. Analog inputs are converted to digital for computation. Outputs desired in analog form must be converted from digital.

Many new generation digital air data computers convert pneumatic inputs from pitot and static ports to electrical signals using piezoresistive or capacitive type pressure transducers rather than the mechanical aneroids used with older analog CADC computers.

Because air pressure is analog in nature, the pressure transducer converts the received analog pressure into an electrical signal proportional to the received pressure, then using an analog to digital converter, digitizes the analog signal so that it may processed within the computer.

Figure 16-13

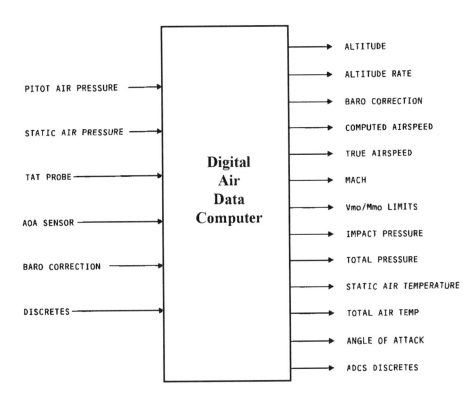

CHAPTER 17 - THEORY OF FLIGHT AND CONTROL SURFACES

Neutral Airfoil

Figure 17-1 shows a neutral airfoil head on into an airstream. It is called a neutral airfoil because the top and bottom contours are symmetrical. The air passing over the top and bottom of the airfoil has to travel farther in the same lateral distance that it would have had to travel if the airfoil were not at that spot.

Mr. Bernoulli (mathematician) figured out quite a while ago that as the speed of a fluid increases, its pressure decreases. Since air is one kind of fluid, the pressures in the air above and below this neutral airfoil are less than they are ahead of and be-hind it. Since the contours of the top and bottom are symmetrical, however, the changes in pressure above and below the airfoil do not lift it up or push it down.

In Figure 17-2 the angle of attack (angle into the wind) of the airfoil has been changed. It is no longer head on into the airstream; the aft end has been dropped quite a bit. Now the airstream divides at a lower point on the leading edge. Air traveling under the airfoil does not go nearly as far as the air traveling over the top. So the air over the top is speeded up quite a bit more than the air under the bottom. And the air over the top has a lower pressure than the air under the bottom. Now that there is a difference, the difference in pressure between the two sides of the airfoil gives a resulting force upward.

Although the force on any separate small area of the airfoil is exerted everywhere perpendicular to the surface at that place, the resultant sum of all the small area forces is very nearly perpendicular to the force and aft line of the airfoil (chord). Total lift is therefore usually shown as a single vector perpendicular to the chord.

The greater the speed of the airfoil through the air, the greater the difference in pressure between the upper and lower surfaces and, therefore, the greater the lift. That is true until the airspeed or angle of attack becomes so great that turbulence results. If the turbulence is great enough, practically all of the lift can be lost.

The functioning of the phenomenon observed and analyzed by Mr. Bernoulli can be seen in many places. It is the operating principle of an automobile carburetor. Another example is that clouds can often be seen hanging near the tops of hills. If the wind is blowing across the top of the hill, the pressure in the air above the hill is less than on either side. A drop in pressure causes a drop in temperature, which in turn causes precipitation of water vapor in the air.

So a cloud can form in that low pressure, low temperature area and seem to just remain there. As it is formed on the up wind side from decrease in pressure and temperature, it is vaporized on the downwind side from the resulting increase in pressure and temperature.

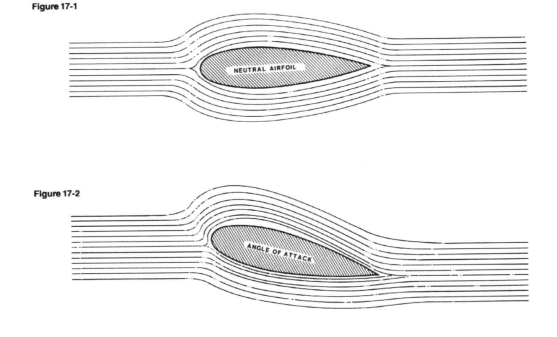

Figure 17-1

NEUTRAL AIRFOIL

Figure 17-2

ANGLE OF ATTACK

Increasing Angle Of Attack

Figures 17-3, 17-4 and 17-5 show cross sections of a typical Davis wing. This type of wing is used on most subsonic airplanes today.

The wing is constructed so that, as the air divides at the leading edge, the portion that travels to the upper side has farther to go than the portion which travels over the lower side. Since it has to travel farther over the upper side, it has to travel faster to get to the trailing edge at the same time as the air passing over the lower side. Because it travels faster, the upper air pressure is less than the lower air pressure. The difference in pressures accounts for the lift of the wing.

Figure 17-3 shows the wing at a low angle of attack. This means that the angle formed by a line drawn from the front of the wing to the back of the wing (chord) forms a small angle with a line representing the direction of airflow approaching the wing.

Assuming the same airspeed in Figure 17-4 as in 17-3, the lift is greater because the angle of attack has been increased. The increase in the angle of attack causes the air to separate at a lower point on the leading edge of the wing, decreasing the distance of travel for the lower air and increasing the distance of travel for the upper air. This increases the pressure difference between the upper and lower surfaces, increasing the amount of lift.

Figure 17-5 illustrates the angle of attack increased to the point where considerable turbulence has developed. This results in loss of lift. And, if there is too much turbulence, the airplane will descend rapidly unless the angle of attack changes.

Figure 17-3

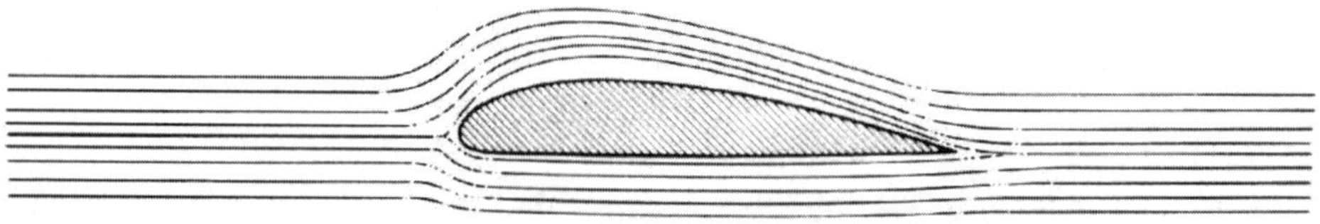

Figure 17-4

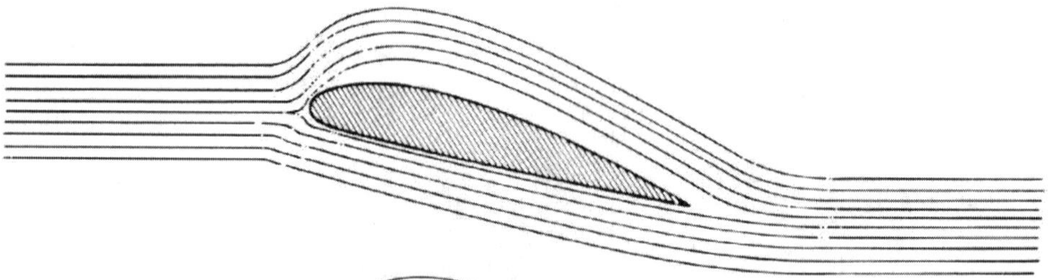

Figure 17-5

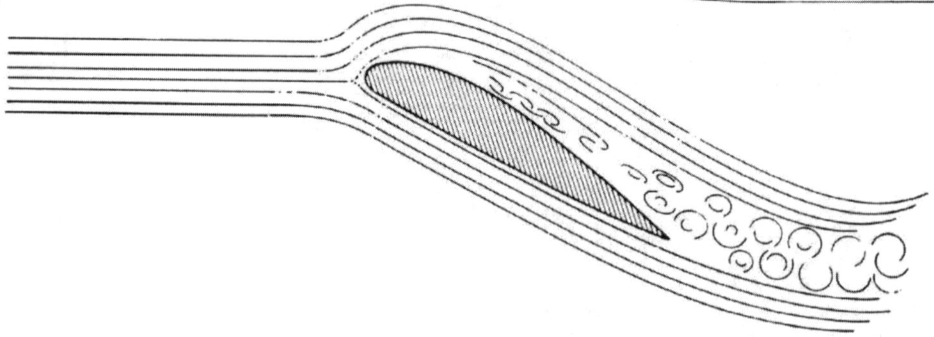

Aileron And Flap Position

Figure 17-6 shows a wing cross section with an aileron. When the aileron is up, the distance of travel over the wing is diminished at the same time that it is increased under the wing. Lift is therefore diminished and the wing tends to drop.

In Figure 17-7, the aileron is shown in a down position, increasing the distance of airflow travel over the wing and decreasing the distance under the wing. The result is an increase in lift and a tendency of the wing to rise.

Aileron movement resulting from control wheel rotation, therefore, causes the airplane to roll in one direction or the other, because when the left ailerons go up, the right ailerons go down, and vice versa.

Figure 17-8 shows a wing with previously recessed and streamlined parts extended. The leading edge device is moved forward and down, and the trailing edge flap is moved aft and down. Trailing edge flaps more often than not come in two or three sections. For simplicity, only one is shown

here. We are not particularly interested in flap design or slot openings. It is the overall effect we are interested in, and it is obvious in this drawing.

Leading edge devices and trailing edge flaps are extended for takeoff and landing. They are extended at takeoff to increase lift so the airplane can leave the ground at a lower airspeed than would otherwise be possible. They are extended during the approach mode to make it possible to fly at a lower speed, and thus avoid high speed landings.

Extending the wing in this manner greatly increases the upper surface airflow distance and greatly decreases the lower surface distance. In addition, direction of the airflow aft of the wing is noticeably changed, being deflected downward.

Extending the flaps greatly increases the lift of the wing, but it also greatly increases the drag, requiring more thrust power than for the same air-speed with the flaps retracted.

Figure 17-6

Figure 17-7

Figure 17-8

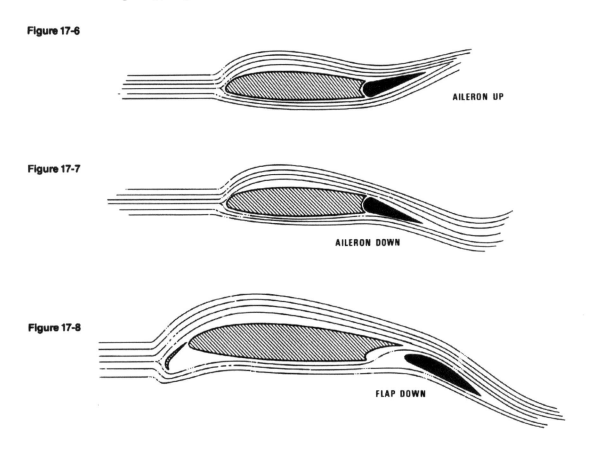

AILERON UP

AILERON DOWN

FLAP DOWN

Angles Of Attack And Incidence

Figure 17-9 illustrates how the angle of incidence is measured. This is the angle that the chord line makes with the fore and aft line of the airplane.

The angle of attack, shown in Figure 17-10, is concerned only with the angle made between the direction of airflow and the chord line. Remember, it is the direction of airflow as compared to the chord line which constitutes angle of attack.

Figure 17-9 shows a rather typical situation with respect to the aspect of the wing on the fuselage. The angle of incidence is such that, at ordinary speeds and altitudes, it is possible to fly the air-plane with the fuselage approximately level and at the same time have approximately the correct angle of attack.

Figure 17-9

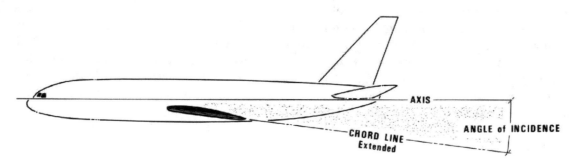

Figure 17-10

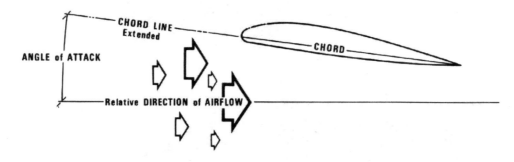

Lift As A Function Of Angle Of Attack

Figure 17-11 illustrates the principle that increasing the angle of attack increases the lift. Assuming the airspeed to be the same for all illustrations, the airplane on the left is climbing rapidly because it has a large lift resulting from a large angle of attack. The middle airplane is climbing less rapidly because the angle of attack is not as great. The airplane on the right has the smallest angle of attack, just enough to hold altitude.

Figure 17-12 illustrates the principle that decreasing the angle of attack decreases the lift. The airplane on the right is holding its altitude. The airplane in the middle is descending because it has diminished its angle of attack. The airplane on the left has the lowest angle of attack, the least lift of all, and is descending most rapidly.

Figure 17-11

Increasing ANGLE of ATTACK Increases LIFT

LINE SHADING INDICATES AIRFLOW RELATIVE TO WING

Figure 17-12

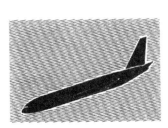

Decreasing ANGLE of ATTACK Decreases LIFT

Effects Of Angle Of Attack

Figure 17-13 illustrates that to avoid climbing; the angle of attack must be decreased as speed increases. Starting with the airplane at the right traveling at only 220 knots, a high angle of attack is required to maintain altitude. The speed increases to 320 knots for the middle airplane; maintaining altitude requires that the angle of attack be diminished. The airplane on the left is travel-ling at 420 knots; the angle of attack has been further diminished in order to hold altitude.

The angles in these examples are exaggerated for better illustration.

Figure 17-14 illustrates that, as the weight decreases, the angle of attack must be diminished to maintain altitude. Starting with the airplane on the right, assume a full load of fuel which requires a high angle of attack at a given airspeed in order to maintain altitude.

A couple of hours later, illustrated by the middle airplane, considerable fuel has been burned out and a much lower angle of attack is required for the same airspeed. Later on (left airplane) the angle of attack is further diminished as the total weight of the airplane is lessened.

Figure 17-13

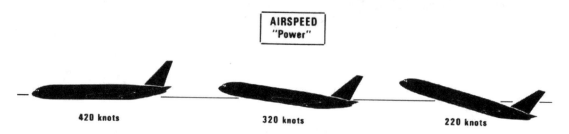

AIRSPEED "Power"

420 knots   320 knots   220 knots

ANGLE of ATTACK must Decrease as speed Increases to avoid climbing

Figure 17-14

LOAD

"LIGHT"   "HEAVY"

ANGLE of ATTACK must Decrease as fuel burns out to avoid climbing

Primary Flight Controls

Airplane flight controls are customarily divided into two groups — primary and secondary.

Figure 17-15 illustrates the primary flight controls. They are considered the primary flight controls because they control the primary functions of yawing (rudder), rolling (ailerons), and pitching (elevators) the airplane.

Figure 17-15

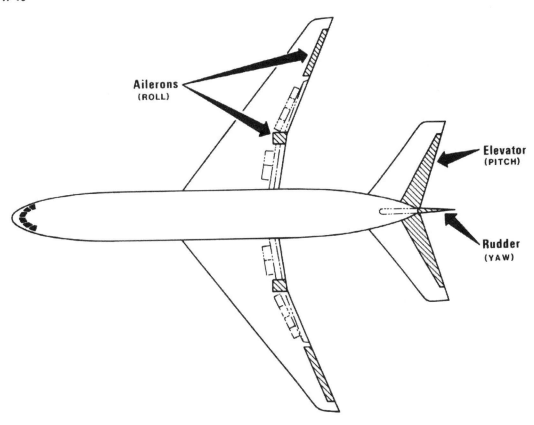

Rudder Action

Figure 17-16 illustrates the rudder causing the airplane to yaw to the right. Yawing, rolling, or pitching of the airplane is always around the center of gravity. So if the airplane is yawing to the right, the nose is moving to the right and the tail is moving to the left (with respect to the center of gravity).

In Figure 17-17 the airplane is yawing to the left. The nose moves to the left and the tail moves to the right with respect to the center of gravity.

The vertical stabilizer, the aft structure on the airplane which supports the rudder, is a neutral airfoil. When the rudder is streamlined, there is no net force to the left or the right. This is the effect illustrated in Figure 17-1.

On the right in Figure 17-16 is a cross section of the vertical stabilizer with the rudder moved to the

right. Air traveling across the left side of the vertical stabilizer must go farther than air traveling across the right side when the rudder is in this position. The result is less pressure on the left side of the stabilizer than on the right, and the resulting force on the vertical stabilizer pushes the tail section to the left, causing the airplane to turn around its center of gravity and yaw to the right.

On the right in Figure 17-17 is a cross section of the vertical stabilizer with the rudder moved to the left. This condition results in a net force exerted to the right on the vertical stabilizer. As the airplane moves around its center of gravity, the nose moves to the left and the airplane yaws to the left.

Figure 17-16

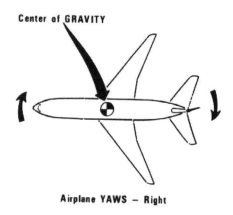

Right RUDDER causes VERTICAL STABILIZER to move LEFT

Airplane YAWS — Right

Figure 17-17

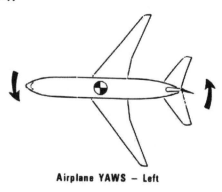

Left RUDDER causes VERTICAL STABILIZER to move RIGHT

Airplane YAWS — Left

Aileron Action

Figure 17-18 shows a rear view of an airplane with the left aileron down and the right aileron up. Increased lift on the left wing and decreased lift on the right wing causes rotation of the airplane around its fore and aft line and around the center of gravity. The airplane rolls to the right.

Figure 17-19 shows the opposite condition, with the left aileron up and the right aileron down. In this case, the airplane rolls to the left.

Figure 17-18

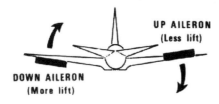

Up AILERON decreases LIFT (Wing DOWN)

Airplane ROLLS — Right

Figure 17-19

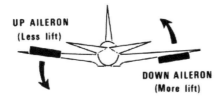

Down AILERON increases LIFT (Wing UP)

Airplane ROLLS — Left

Elevator Action

Figure 17-20 is a side view of the airplane with the elevators up. The horizontal stabilizer, which has the elevators in the aft end, is also a neutral section.

When the elevator is streamlined, there is no net up or down force on the stabilizer, providing that it has a zero angle of attack. Moving the elevators up increases the pressure along the top of the horizontal stabilizer and decreases pressure along the bottom, causing a downward force to be exerted on the horizontal stabilizer.

As the airplane turns around its center of gravity, the nose goes up as the tail goes down and the airplane pitches up.

The opposite condition is shown in Figure 17-21, where the airplane is pitching down.

The horizontal stabilizer is not necessarily a neutral airfoil. It may be designed for some type of lift when the elevator is faired (streamlined). The elevator principle still applies, however, just as in the case of the ailerons.

Figure 17-20

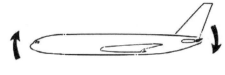

Airplane PITCHES – Up

Up ELEVATOR causes TAIL to move DOWN

Figure 17-21

Airplane PITCHES – Down

Down ELEVATOR causes TAIL to move UP

Secondary Flight Controls

Figure 17-22 shows the usual secondary flight controls.

The leading edge devices are retracted in cruise so that the contour of the wing is smooth and drag is low. The trailing edge flaps are also retracted in cruise for the same reason.

The spoilers are flat panels on the top of the wing which can be hydraulically extended. They are hinged at the front side. The amount of extension of these panels is determined by the amount of movement of a spoiler control handle in the cockpit. As they extend, they disturb the airflow on the wing and diminish the lift. If they are fully extended, they not only diminish the lift on the wing, but they also provide considerable increase in drag so that they can be used as an in flight brake.

They are also used to perform the same roll function as the ailerons. Typically, if the cockpit control wheel is rotated more than 10°, the spoilers begin to operate to assist the aileron action. If the control wheel is turned to the right more than 10°, the spoilers on the right begin to rise.

In cruise, slow rolling motion of the airplane (small control wheel movement) is accomplished by the ailerons. If a more rapid roll maneuver is desired, the spoilers are called upon to assist.

Figure 17-22

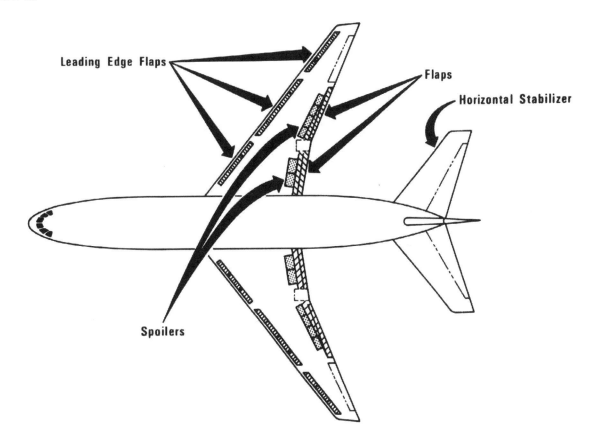

Equilibrium Of Forces In Flight

Figure 17-23 illustrates the four principal forces exerted on an airplane in flight. These are thrust, lift, weight, and drag. They are generally represented vectorially.

In this diagram, thrust equals drag, so speed is constant.

The sum of the weight plus the downward force on the stabilizer equals the vertical lift, so the airplane is holding altitude.

If thrust exceeds drag, airspeed and drag increase until drag equals thrust. If drag exceeds thrust, airspeed and drag decrease until drag equals thrust.

If lift exceeds the weight forces, the airplane rises. If the weight forces exceed the lift, the air-plane sinks.

These four forces are all exerted, in effect, around the center of gravity. The center of gravity is that point in the airplane from which the weight forward equals the weight aft; and the weight to the left equals the weight to the right; and the weight above equals the weight below. The center of gravity moves as the fuel load changes.

The center of lift is that point in the airplane at which all of the lift forces could be exerted (perpendicular to the wing) as a single force without changing the effect. Only the vertical component of the lift force is effective in lifting the airplane. If the lift force is not vertical, a component of lift may be either aiding or opposing thrust.

The center of gravity moves as fuel burns out, and the center of lift position varies as airspeed and angle of attack change. So, the center of lift is seldom at the center of gravity. This results in a torquing force, usually in the direction tending to turn the nose down and the tail up. A downward force on the horizontal stabilizer can balance that torque, resulting in a stable pitch attitude.

If the airplane is maintaining its pitch attitude, any torquing force is counterbalanced by an equivalent but opposite torquing force created by the horizontal stabilizer in combination with the elevators. If the torquing forces are not balanced, the airplane is changing its pitch attitude.

In the illustrated airplane, if we move the elevators up or down, the force on the horizontal stabilizer is changed, causing the airplane nose to go up or down.

In the condition illustrated, it is necessary for the horizontal stabilizer to exert a downward force to overcome the "nose down" torquing caused by the lift center being aft of the center of gravity. This can be accomplished either by moving the elevator up or by rotating the leading edge of the horizontal stabilizer downward. In this case, the stabilizer has been moved down by the amount necessary to balance the torquing forces.

Since moving the elevator out of the streamline position causes increased drag, it is desirable for long term conditions to change the position of the horizontal stabilizer so that the elevator can be streamlined.

Stabilizers are not usually moveable on propeller aircraft because airspeeds are not great enough for elevator drag to be significant. On jet airplanes, elevator drag does have a very noticeable effect on both airspeed and fuel consumption. This changing the position of the horizontal stabilizer to streamline the elevator in the pitch axis is called "trimming".

Figure 17-23

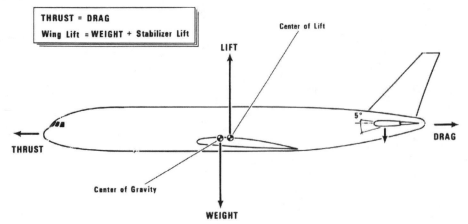

Stabilizer Trimming

Figures 17-24, 17-25, and 17-26 each illustrate two conditions capable of developing the same amount of lift, assuming constant airplane attitude and airspeed. One condition has the elevator not streamlined (faired). The other condition shows the elevator streamlined because the stabilizer is in a different position.

Figure 17-24 shows the stabilizer developing the same amount of upward lift, either at 0° with the elevator down 15°, or at + 6° with the elevator faired.

This will occur only when the weight is aft of the center of lift.

Figure 17-25 shows another pair of equivalent conditions, each developing no lift. In the top case the stabilizer is at 0° and the elevator streamlined. The bottom case shows the stabilizer down 6° and the elevator down 15°.

In Figure 17-26 the same amount of downward lift is generated with the stabilizer at 0° and elevator up 15°, as with the stabilizer at -6° and the elevator faired.

These values and relative positions are only illustrative, not necessarily actual. Actual values and positions depend upon airspeed and airplane attitude.

In each figure, the lift is the same for the two conditions illustrated. However, in each case, the drag is less when the elevator is faired.

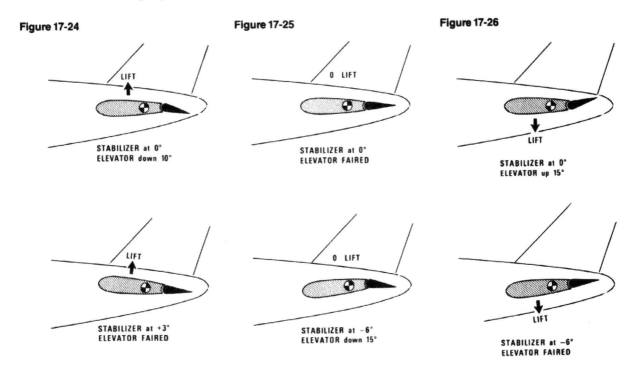

Figure 17-24　　　　**Figure 17-25**　　　　**Figure 17-26**

LIFT

STABILIZER at 0°
ELEVATOR down 10°

0　LIFT

STABILIZER at 0°
ELEVATOR FAIRED

STABILIZER at 0°
ELEVATOR up 15°

LIFT

LIFT

STABILIZER at +3°
ELEVATOR FAIRED

0　LIFT

STABILIZER at -6°
ELEVATOR down 15°

STABILIZER at -6°
ELEVATOR FAIRED

LIFT

Trim Or Control Tab Operation

Figure 17-27 illustrates the function of a trim or control tab. A control tab is operated by pedals, wheel, or control column, while a trim tab is operated by a pedestal trim wheel. This example shows the top view of a control tab on a rudder. We are assuming that the rudder is free to move on its pivots and will take a position dictated by aerodynamic forces.

In the top illustration, the tab has been moved to the left. This causes the rudder to move to the right because pressure is decreased on the right side of the rudder and increased on the left. The net effect on the stabilizer airflow is to increase the distance the air must travel on the left side, and decrease it on the right side. Air pressure difference then pushes the stabilizer toward the left and the air-plane nose toward the right.

In the center drawing, the control tab has been streamlined with the rudder. The rudder assembly then streamlines itself with the vertical stabilizer.

The bottom drawing illustrates a condition similar to the top drawing, but opposite in direction.

Figure 17-27

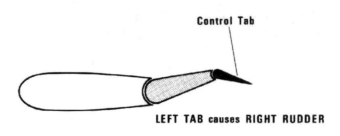

LEFT TAB causes RIGHT RUDDER

STREAMLINED TAB — STREAMLINED RUDDER

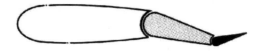

RIGHT TAB causes LEFT RUDDER

Control Surface Actuation

In a small airplane the control column operates the elevator directly because the pilot is capable of exerting sufficient force to push the elevator away from the streamlined position. This direct control operation is illustrated in Figure 17-28.

In larger airplanes the aerodynamic force is too great for the pilot to overcome, so a control tab operation (Figure 17-29) might be used.

In many commercial aircraft the aerodynamic force of operating the elevator is too great for even a control tab operation, so a hydraulic actuator is used.

Figure 17-30 is a typical hydraulic actuator. The hydraulic piston is secured to structure, indicated by the angled lines on the left end of the piston rod. Movement of the actuator body causes movement of the elevator.

When the control column is moved aft, connecting linkage moves the control valve to the left. This opens the left port of the actuating cylinder to pressure and the right port to return. Pressure in the left port moves the actuator body to the left. As it moves to the left, the elevator moves up and the

control valve ports go toward the closed position. The farther the control column is moved aft, the farther the actuator must move toward the left to close off the control valve ports, and the farther the elevator moves up.

The proportioning of elevator movement according to control column movement results from "follow-up" information. "Follow-up" is a term used to describe either the mechanism or the operation of following up on a control valve movement with a corresponding control surface movement.

In this type of actuator, the follow-up action results from actuator body movement and resultant control valve operation. In installations where the body of the actuator is fixed to a structure and the actuator piston and rod cause the movement of the control surface, the follow-up system is not this simple.

When a hydraulic actuator is used, almost no force at all is required to cause the actuator to position the elevator because all that is operated is a control valve.

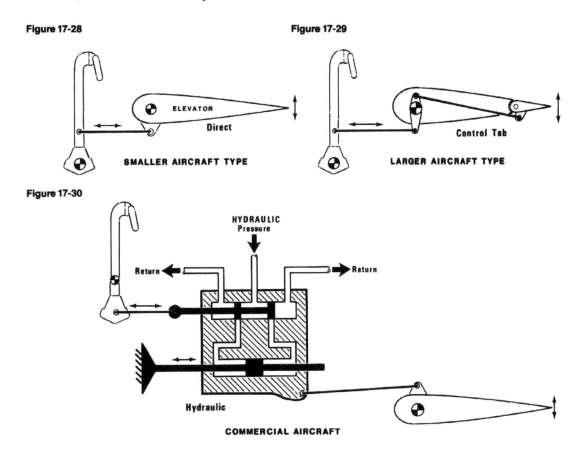

Figure 17-28

Figure 17-29

Figure 17-30

Artificial Feel System

When a hydraulic actuator is used, an artificial feel system must be provided to prevent over controlling by the pilot. In the case of the ailerons, a spring force is usually adequate. However, in dealing with elevators and rudders, it is common to have not only a spring force but also a variable hydraulic force.

Figure 17-31 is a typical artificial feel system using both spring and hydraulic feel. The horizontal stabilizer and elevator system is shown, although this artificial feel could be used in the rudder or aileron systems.

The hydraulic artificial feel is essentially varied as a function of airspeed. Artificial spring feel alone may be adequate at low speeds, but at high speeds greater resistance to cockpit control movement is needed to prevent overstressing the aircraft structure.

Artificial feel systems serve another useful purpose. They position the cockpit controls to a neutral position when the pilot releases the control wheel, column, or rudder pedals. The neutral

position in the case of the elevators is the position where the elevators are faired with the horizontal stabilizer.

The double cam on the aft elevator control quadrant illustrates the tendency of the artificial feel system to put the control column, and naturally the elevators, into the neutral position. If the pilot moves the control column, he must compress the spring and overcome the force exerted on the hydraulic piston.

The schematic of the feel computer shows how the hydraulic pressure on the hydraulic feel piston is varied as a function of airspeed and horizontal stabilizer position. Pitot pressure is delivered to one side of the airspeed bellows and static pressure to the other. As a result, the bellows exerts a force in proportion to aircraft speed. This is exerted against the springs shown — one on top of the stabilizer position cam, the other above the metering valve.

Continued on next page ...

Figure 17-31

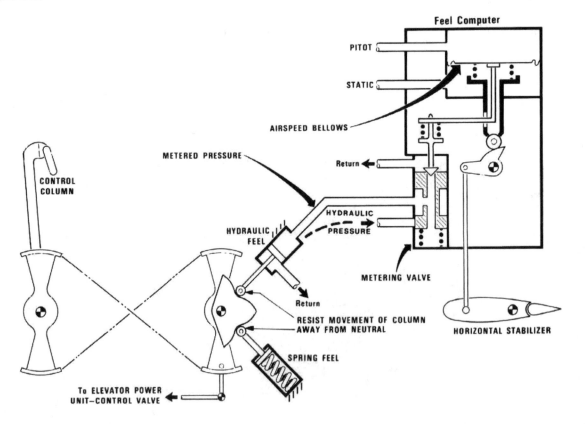

Artificial Feel System (cont'd.)

The metering valve is the shaded portion of the feel computer. Metered pressure forces exerted against the upper and lower interior horizontal surfaces of the metering valve are equal and balance each other. If the metered pressure exerted against the triangular relief is enough to balance the force exerted downward against the metering valve through the spring, then the pressure line is closed off as shown.

If the airspeed increases, the downward force on the metering valve increases and overcomes the metered pressure force. This pushes the metering valve down, opening the interior of the metering valve to the pressure line until the metered pressure balances the downward force on the metering valve. The metering valve is continuously opening slightly to make up for metered pressure leakage.

If the pilot moves the control column, he has to force the hydraulic feel piston up into the cylinder. To do this he must overcome the hydraulic force on the piston and push fluid out through the relief valve (the triangle on top of the metering valve).

The force the pilot has to overcome essentially varies as an airspeed function, since stabilizer position is basically also an airspeed function.

Figure 17-31

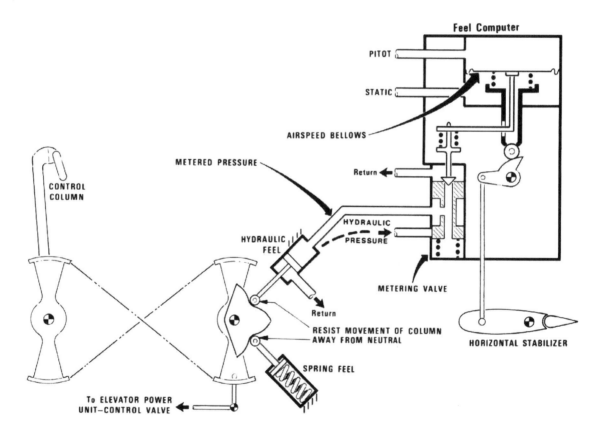

CHAPTER 18 - GROUND PROXIMITY WARNING SYSTEM (GPWS)

Ground Proximity Warning System

The purpose of the Ground Proximity Warning System (GPWS) is to alert the flight crew to the existence of an unsafe condition due to terrain proximity.

The various hazardous conditions that can be encountered in flight are divided into five modes. These are:

Mode 1— Excessive descent rate.

Mode 2 — Excessive closure rate with respect to rising terrain.

Mode 3 — Excessive altitude loss during climb-out (in takeoff or during go-around) when not in landing configuration (landing gear up and/or flaps less than 25°).

Mode 4 — Insufficient terrain clearance when not in landing configuration (landing gear up and/or flaps less than 25°).

Mode 5 — Excessive deviation below glide slope when making a front course approach with the gear down.

The Ground Proximity Warning System modes are annunciated to the flight crew in the flight deck by means of aural messages and visual indications.

The various modes, methods of annunciation and override functions vary on different airplanes.

Figure 18-1

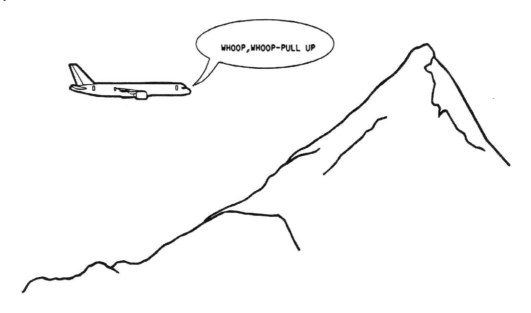

Ground Proximity Warning System Lights and Light Switches

The PULL UP light (red) indicates a mode 1 or mode 2A PULL UP warning condition. It is typically located on the captain's instrument panel.

The GND PROX G/S INHB (Ground Proximity - Glide Slope Inhibit) light switch (amber) is used to indicate advisory modes 1 through 5. When the switch is pressed, the mode 5 (below glide slope) aural and visual indications are inhibited. The switch is a momentary switch. The GND PROX G/S INHB light switch is typically located on the captain's instrument panel.

The GND PROX FLAP OVRD (Ground Proximity Flap Override) light switch and the GND

PROX CONFIG OVRD (Ground Proximity/Configuration Gear Override) light switch serve to simulate flaps down 25° or more, or landing gear down positions, respectively.

They are alternate action push-button switches. The lights (white) illuminate when the override function has been activated. Both override light switches are located on the first officer's instrument panel.

The effect of override switch actuation will be described later.

Figure 18-2

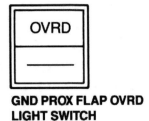

PULL UP LIGHT

**GND PROX FLAP OVRD
LIGHT SWITCH**

**GND PROX-G/2
INHB LIGHT SWITCH**

**GND PROX/CONFIG
GEAR OVRD LIGHT SWITCH**

Ground Proximity Warning System: Mode 1

Mode 1 applies to excessive descent rate with respect to terrain clearance. This mode is independent of landing gear and flap positions.

Mode 1 indications occur below 2,450 feet radio altitude down to 50 feet, when the barometric altitude rate exceeds a threshold value as indicated on the graph.

The Mode 1 envelope is divided into two areas: The initial penetration area ("SINK RATE" area), and the inner warning area ("PULL UP" area). The specific initial penetration area and the inner warning area boundaries are as shown on the graph.

Initial penetration of the Mode 1 envelope is annunciated by the illumination of the GND PROX G/S INHB light switch (amber) and the repeated aural message "SINK RATE".

Penetration of the inner warning area is annunciated by the PULL UP light (red), the Master Warning lights (red), and the repeated aural message "WHOOP! WHOOP! PULL UP!".

Figure 18-3

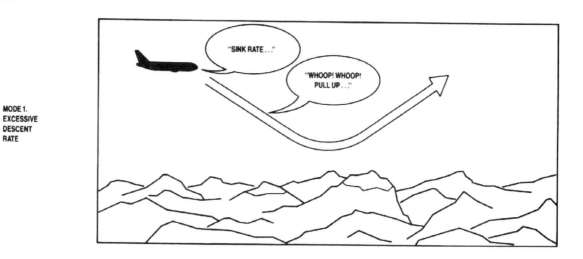

Figure 18-4

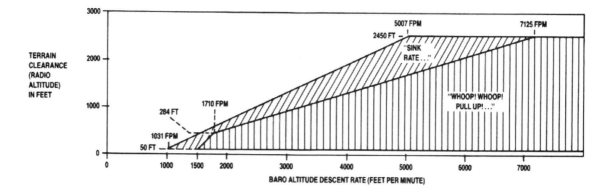

Ground Proximity Warning System: Mode 2A

Mode 2 applies to excessive closure rate with respect to rising terrain. This mode consists of two submodes: If the flaps are not down 25°, mode 2A is annunciated; if the flaps are down 25° or more, mode 2B is annunciated. Mode 2B is described later.

If airspeed is less than 190 knots, the upper boundary is 1,650 feet radio altitude, and the lower boundary is 50 feet. If airspeed exceeds 250 knots, the upper boundary is 2,450 feet radio altitude, and the lower boundary is 50 feet. Between 190 knots and 250 knots, the upper boundary varies according to an airspeed expansion function.

The specific boundary limits for various closure rates is as shown on the graphic (Figures 18-6 and 18-7).

Penetration of the mode 2A envelope can be either on the slope or from the top. The envelope is divided into two areas: the initial penetration area and the inner warning area. The inner warning area is entered after the initial penetration area message "TERRAIN" has been voiced twice upon initial penetration of the mode 2A envelope.

Initial penetration area indications consist of the illumination of the GND PROX G/S INHB light switch (amber), and the aural message "TERRAIN," voiced twice. The indications of the inner warning area are the illumination of the PULL UP light (red), the Master Warning lights (red), and the repeated aural message "WHOOP! WHOOP! PULL UP!"

Upon leaving the inner warning area, due to either terrain drop-off or a pull-up maneuver, the altitude gain function is activated. During this function, the indications change to the GND PROX G/S INHB light switch (amber) and the repeated aural message "TERRAIN". The indications continue until the aircraft has gained 300 feet of barometric altitude, or when the landing gear is lowered.

Figure 18-5

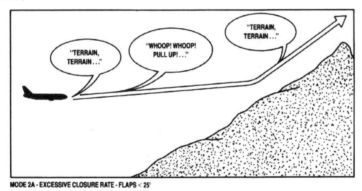

MODE 2A - EXCESSIVE CLOSURE RATE - FLAPS < 25°

ALTITUDE GAIN FUNCTION:
THE AURAL MESSAGE "WHOOP WHOOP PULL UP" CHANGES TO "TERRAIN, TERRAIN, TERRAIN . . ." WHEN AIRPLANE NO LONGER APPROACHES THE TERRAIN.
"TERRAIN, TERRAIN . . ." ANNUNCIATION CEASES AFTER THE AIRPLANE HAS GAINED 300 FT OF BAROMETRIC ALTITUDE OR THE LANDING GEAR HAS BEEN LOWERED.

Figure 18-6   **Figure 18-7**

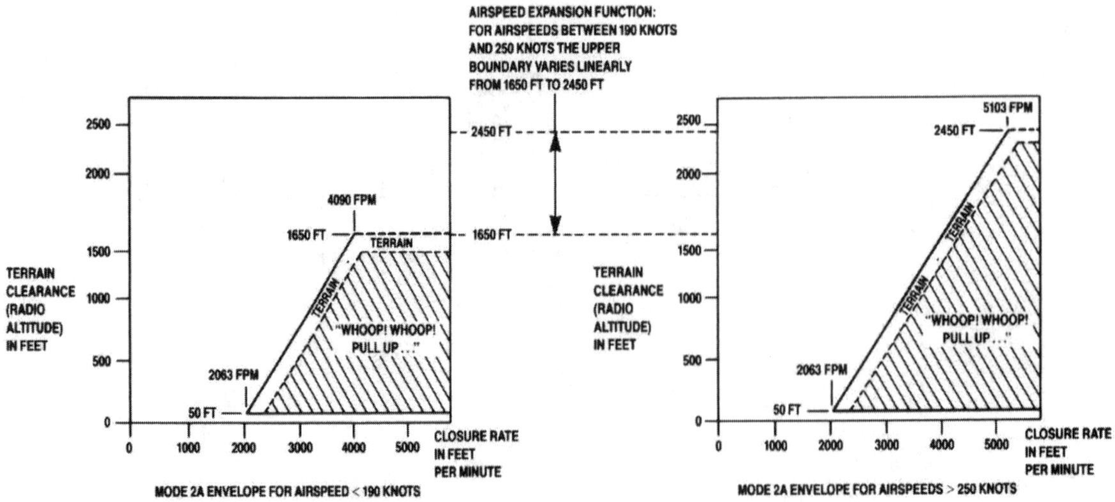

AIRSPEED EXPANSION FUNCTION:
FOR AIRSPEEDS BETWEEN 190 KNOTS AND 250 KNOTS THE UPPER BOUNDARY VARIES LINEARLY FROM 1650 FT TO 2450 FT

MODE 2A ENVELOPE FOR AIRSPEED < 190 KNOTS

MODE 2A ENVELOPE FOR AIRSPEEDS > 250 KNOTS

Ground Proximity Warning System: Mode 2B

Mode 2B applies to excessive closure rate with respect to rising terrain with the flaps down 25° or more. It is an advisory-only mode.

The Mode 2B indications occur below 789 feet radio altitude and down to 200-600 feet, depending upon the barometric rate of descent, when the closure rate exceeds threshold values as shown on the graph (Figures 18-9 and 18-10).

Mode 2B indications consist of the illumination of the GND PROX G/S INHB light switch (amber), and the repeated aural message "TERRAIN".

Figure 18-8

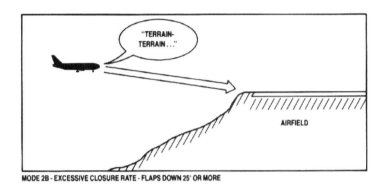

MODE 2B - EXCESSIVE CLOSURE RATE - FLAPS DOWN 25° OR MORE

Figure 18-9

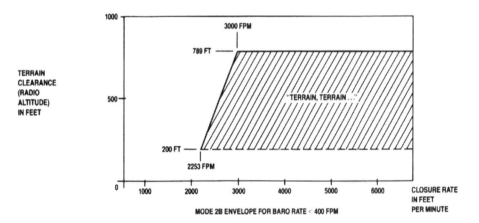

MODE 2B ENVELOPE FOR BARO RATE < 400 FPM

NOTE: FOR BAROMETRIC ALTITUDE RATES BETWEEN 400 FPM AND 1000 FPM,
THE LOWER BOUNDARY VARIES LINEARLY FROM 200 FT TO 600 FT.

Figure 18-10

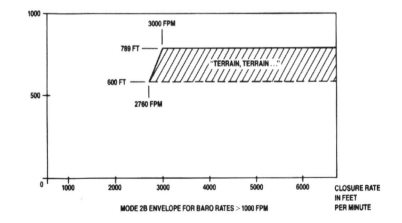

MODE 2B ENVELOPE FOR BARO RATES > 1000 FPM

Ground Proximity Warning System: Mode 3

Mode 3 applies to excessive barometric altitude loss after takeoff, if the flaps are less than 25° or the landing gear is raised; or below 200 feet during a missed approach. It is an advisory-only mode. In general, the altitude loss threshold value is about 10% of the current altitude.

During landing approach, mode 3 is armed after the aircraft has descended below 200 feet in landing configuration (flaps down more than 25° and landing gear down).

Modes 3 and 4 are mutually exclusive: In climb-out, Mode 3 is disabled and Mode 4 is armed after the aircraft has climbed above 700 feet.

Mode 3 indications occur below 700 feet radio altitude and down to 50 feet, when the barometric altitude loss exceeds the threshold values as shown on the graphic (Figure 18-12).

Mode 3 indications include illumination of the GND PROX GIS INHB light switch and the repeated aural message "DON'T SINK"

Figure 18-11

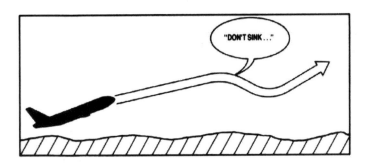

MODE 3 -
ALTITUDE LOSS
DURING CLIMB-OUT
(AFTER TAKE-OFF
OR DURING
MISSED APPROACH)

Figure 18-12

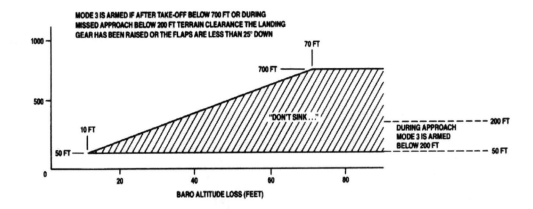

Ground Proximity Warning System: Mode 4

Mode 4 usually applies during the landing phase of flight. It is armed above 700 feet after takeoff. It is annunciated in the event of insufficient terrain clearance when the aircraft is not in the proper landing configuration.

Mode 4 consists of two submodes. When the landing gear is up, mode 4A is annunciated. When the landing gear is down, but the flaps are less than 25°, mode 4B is annunciated.

Each mode envelope, 4A and 4B, is divided into two advisory areas: one for low airspeeds, and another for high airspeeds. In the event of high airspeeds, the boundaries are the same as the ones for low airspeeds, except that an airspeed expansion function has been added, as shown in Figure 18-15.

If airspeed exceeds 190 knots, the mode 4A threshold radio altitude increases from 190 knots to 250 knots. Above 250 knots, the threshold radio altitude is 1,000 feet. If the airspeed is less than 190 knots, the threshold radio altitude is 500 feet. The lower boundary of the mode 4A envelope is 50 feet radio altitude at all airspeeds.

Continued on next page ...

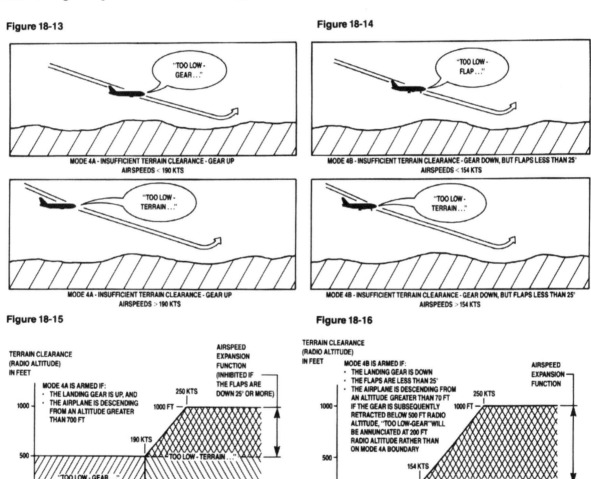

Figure 18-13

MODE 4A - INSUFFICIENT TERRAIN CLEARANCE - GEAR UP
AIRSPEEDS < 190 KTS

MODE 4A - INSUFFICIENT TERRAIN CLEARANCE - GEAR UP
AIRSPEEDS > 190 KTS

Figure 18-14

MODE 4B - INSUFFICIENT TERRAIN CLEARANCE - GEAR DOWN, BUT FLAPS LESS THAN 25°
AIRSPEEDS < 154 KTS

MODE 4B - INSUFFICIENT TERRAIN CLEARANCE - GEAR DOWN, BUT FLAPS LESS THAN 25°
AIRSPEEDS > 154 KTS

Figure 18-15

MODE 4A ENVELOPE

Figure 18-16

MODE 4B ENVELOPE

Ground Proximity Warning System: Mode 4 (cont'd.)

In the high-airspeed advisory area, the aural message is the repeated "TOO LOW — TERRAIN." In the low-airspeed advisory area it, is the repeated "TOO LOW — GEAR." The visual indication in both advisory areas is the illumination of the GND PROX G/S INHB light switch (amber).

In mode 4B, if airspeed exceeds 154 knots, the threshold radio altitude increases linearly as the airspeed increases from 154 knots to 250 knots. Above 250 knots, the threshold radio altitude is 1,000 feet, as in the case of mode 4A.

If airspeed is less than 154 knots, the threshold altitude is 200 feet radio altitude. The lower boundary of the mode 4B envelope is 50 feet radio altitude at all airspeeds.

In the high-airspeed advisory area, the aural message is the repeated "TOO LOW — TERRAIN." In the low-airspeed advisory area, it is the repeated "TOO LOW — FLAP." The visual indication in both advisory areas is the illumination of the GND PROX G/S INHB light switch (amber).

Figure 18-13

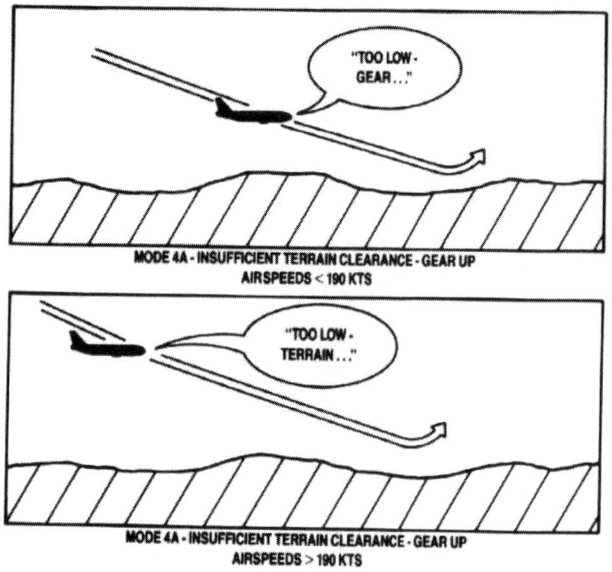

Figure 18-14

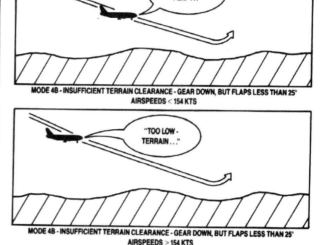

Figure 18-15

Figure 18-16

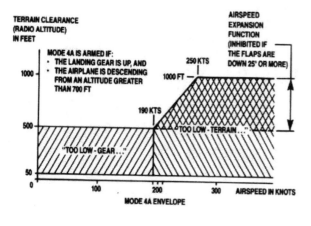

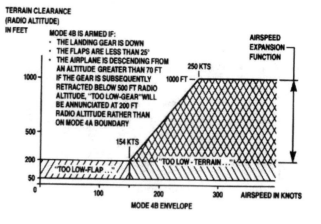

Ground Proximity Warning System: Mode 5

Mode 5 applies to excessive deviation below the glide path when making a front-course approach with the gear down. In a back-course landing condition, mode 5 is inhibited. Initial penetration of mode 5 envelope results in a low-level aural annunciation. A penetration of the inner advisory area is indicated by a normal-level aural annunciation.

The mode 5 envelope is divided into two advisory areas: the low-level (initial penetration) advisory area, and the normal-level (inner) advisory area.

The low-level advisory area indications occur below 1,000 feet of radio altitude and down to 50 feet, when the glide slope deviation exceeds 1.3 dots (0.46 degrees). The specific area boundaries are as shown in Figure 18-18.

Normal-level advisory area indications occur only below 300 feet radio altitude and down to 50 feet, and when the glide slope deviation exceeds 2.0 dots (0.7 degrees). The specific area boundaries are as shown in Figure 18-18.

Mode 5 is annunciated by the illumination of the GND PROX G/S INHB light switch (amber) and the repeated aural message "GLIDE SLOPE." The sound level in the normal-level advisory area is the same as in modes 1 through 4, and it is six decibels lower in the low-level advisory area. The "GLIDE SLOPE" message is repeated more rapidly as the terrain clearance decreases and/or the glide slope deviation increases.

Mode 5 indications may be cancelled by pressing the GND PROX GIS INHB light switch when the radio altitude is less than 1,000 feet. Both the aural message and the light indication are cancelled. The mode is automatically rearmed when climbing above 1,000 feet radio altitude, or descending below 50 feet radio altitude.

Additional boundary variations below 150 feet are provided to allow for normal beam variations near the threshold.

Figure 18-17

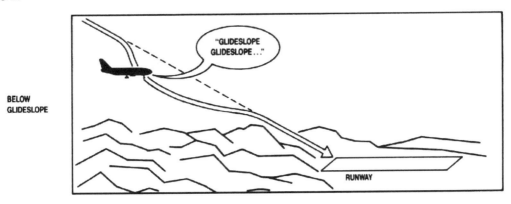

Figure 18-18

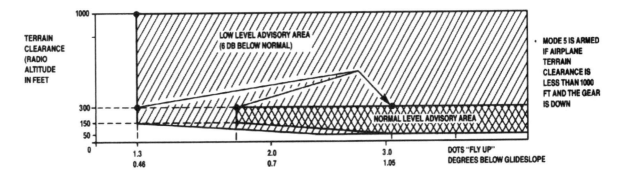

Ground Proximity Warning System Inputs/Outputs

For proper operation of the Ground Proximity Warning System, several different inputs are required in order for the GPWS computer to produce the required output signals.

The various inputs are shown in Figure 18-19. Also shown are the various outputs.

Figure 18-19

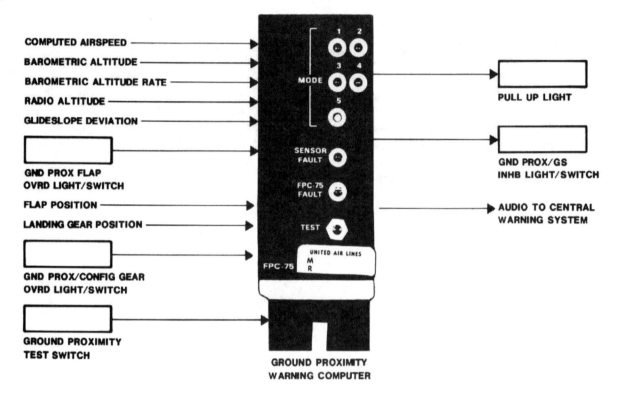

COMPUTED AIRSPEED

BAROMETRIC ALTITUDE

BAROMETRIC ALTITUDE RATE

RADIO ALTITUDE

GLIDESLOPE DEVIATION

GND PROX FLAP
OVRD LIGHT/SWITCH

FLAP POSITION

LANDING GEAR POSITION

GND PROX/CONFIG GEAR
OVRD LIGHT/SWITCH

GROUND PROXIMITY
TEST SWITCH

MODE

1 2
3 4
5

SENSOR
FAULT

FPC-75
FAULT

TEST

UNITED AIR LINES
M
R

FPC-75

PULL UP LIGHT

GND PROX/GS
INHB LIGHT/SWITCH

AUDIO TO CENTRAL
WARNING SYSTEM

GROUND PROXIMITY
WARNING COMPUTER

Ground Proximity Modes And Annunciators

Figure 18-20

MODES	CONDITION			AURAL MESSAGE	PULL UP LIGHT	GND PROX G/S INHB LIGHT SWITCH	CAPT & F/O MASTER WARNING LIGHTS
1	INITIAL PENETRATION AREA			"SINK RATE..."		ON	
	INNER WARNING AREA			"WHOOP, WHOOP, PULL UP..."	ON		ON
2A	FLAPS UP	INITIAL PENETRATION AREA		"TERRAIN" (VOICED TWICE)		ON	
		INNER WARNING AREA		"WHOOP, WHOOP, PULL UP..."	ON		ON
		ALTITUDE GAIN FUNCTION: • STARTS WHEN THE AIRPLANE EXITS THE WARNING AREA • ENDS WHEN 300 FT IN BAROMETRIC ALTITUDE HAVE BEEN GAINED • IS INHIBITED WHEN THE GEAR IS EXTENDED		"TERRAIN..."		ON	
2B	FLAPS 25° OR MORE			"TERRAIN..."		ON	
3	FLAPS LESS THAN 25° OR GEAR UP			"DON'T SINK..."		ON	
4A	GEAR UP	AIRSPEED < 190 KNOTS		"TOO LOW—GEAR..."		ON	
		AIRSPEED > 190 KNOTS		"TOO LOW—TERRAIN..."		ON	
4B	GEAR DOWN BUT FLAPS LESS THAN 25°	AIRSPEED < 154 KNOTS		"TOO LOW—FLAP..."		ON	
		AIRSPEED > 154 KNOTS		"TOO LOW—TERRAIN..."		ON	
	NOTE: WHEN GEAR IS RETRACTED AFTER BEING EXTENDED, "TOO LOW-GEAR" WILL BE ANNUNCIATED ON THE MODE 4B BOUNDARY					ON	
5	GEAR DOWN			"GLIDE SLOPE..."		ON	

Enhanced Ground Proximity Warning System

The Enhanced GPWS incorporates the functions of the basic Ground Proximity Warning System (GPWS) modes 1 through.5 and adds addition modes, callouts as well as display of terrain.

The EGPWS uses aircraft inputs including geographic position, attitude, altitude, airspeed, and glideslope. These are used with internal terrain, obstacles, and airport databases to predict a potential conflict between the aircraft flight path and terrain or an obstacle. A terrain or obstacle conflict will result in the EGPWS providing a visual and audio caution or warning alert.

Additionally, the EGPWS provides (Mode 6) alerts for bank angle and altitude callouts based on system program pin selection.

Detection of severe windshear (Mode 7) conditions is also provided for selected aircraft types when enabled. This feature can use Doppler enabled "X" band weather radar to predict windshear (Predictive Windshear) in addition to the reactive windshear function in which the windshear is detected as the aircraft flies through the condition.

The EGPWS incorporates several "enhanced" features:

- Terrain Alerting and Display (TAD) provides a graphic display of the surrounding terrain on the Weather Radar Indicator, EFIS, or a dedicated display.
- "Peaks" is a TAD supplemental feature providing additional terrain display features for enhanced situational awareness, independent of the aircraft's altitude. This includes digital elevations for the highest and lowest displayed terrain, additional elevation (color) bands, and a unique representation of 0 MSL elevation (sea level and its corresponding shoreline).
- "Obstacles" is a feature utilizing an obstacle database for obstacle conflict alerting and display. EGPWS caution and warning visual and audio alerts are provided when a conflict is detected. Additionally, when TAD is enabled, Obstacles are graphical displayed similar to terrain.
- "Terrain Clearance Floor" feature adds an additional element of protection by alerting the pilot of possible premature descent. This is intended for non-precision approaches and is based on the current aircraft position relative to the nearest runway.

- Windshear alerting (Mode 7) is provided for specific aircraft types. Mode 7 provides windshear caution and/or warning alerts when an EGPWS windshear threshold is exceeded.

The EGPWS adds to these 7 basic functions the ability to compare the aircraft position to an internal database and provide additional alerting and display capabilities for enhanced situational awareness and safety (hence the term "Enhanced" GPWS).

The EGPWS internal database consists of four sub-sets:

- A worldwide terrain database of varying degrees of resolution.
- An obstacles database containing cataloged obstacles 100 feet or greater in height located within North America and portions of the Caribbean
- A worldwide airport database containing information on hard-surface runways 3500 feet or longer in length.
- An Envelope Modulation database to support the Envelope Modulation feature discussed later.

Notification of a Database update is accomplished by Service Bulletin. Database updates are distributed on PCMCIA data cards and downloaded via a card slot in the front panel of each EGPWS computer.

With the use of accurate GPS or FMS information, the EGPWS is provided present position, track, and ground speed. With this information the EGPWS is able to present a graphical plan view of the aircraft relative to the terrain and advise the flight crew of a potential conflict with the terrain or obstacle. Conflicts are recognized and alerts provided when terrain violates specific computed envelope boundaries on the projected flight path of the aircraft.

Alerts are provided in the form of visual light annunciation of a caution or warning, audio enunciation based on the type of conflict, and color enhanced visual display of the terrain or obstacle relative to the forward look of the aircraft. The terrain display is provided on the Weather Radar Indicator, EFIS display, or a dedicated EGPWS display and may or may not be displayed automatically.

Continued on next page ...

Enhanced Ground Proximity Warning System (cont'd.)

Enhanced Functions Include:

Envelope Modulation

Due to terrain features at or near certain specific airports around the world, normal operations have resulted in nuisance or missed alerts at these locations in the past. With the introduction of accurate position information and a terrain and airport database, it is possible to identify these areas and adjust the normal alerting process to compensate for the condition.

Modes 4, 5, and 6 are expanded at certain locations to provide alerting protection consistent with normal approaches. Modes 1, 2, and 4 are desensitized at other locations to prevent nuisance warnings that result from unusual terrain or approach procedures. In all cases, very specific information is used to correlate the aircraft position and phase of flight prior to modulating the envelopes.

Terrain Clearance Floor (Figure 18-21)

The Terrain Clearance Floor (TCF) function (when enabled) enhances the basic GPWS Modes by alerting the pilot of descent below a defined "Terrain Clearance Floor" regardless of the aircraft configuration. The TCF alert is a function of the aircraft's Radio Altitude and distance (calculated from latitude/longitude position) relative to the center of the nearest runway in the database.

TCF alerts result in illumination of the EGPWS caution lights and the aural message **"TOO LOW TERRAIN"**. The audio message is provided once when initial envelope penetration occurs and again only for additional 20% decreases in Radio Altitude. The EGPWS caution lights will remain on until the TCF envelope is exited.

Continued on next page ...

Figure 18-21

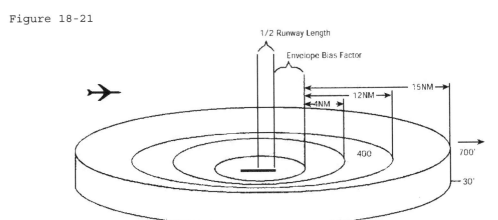

Terrain Clearance Floor Alert Envelope

Enhanced Ground Proximity Warning System (cont'd.)

Terrain Look Ahead Alerting

Another enhancement provided by the internal terrain database, is the ability to look ahead of the aircraft and detect terrain or obstacle conflicts with greater alerting time. This is accomplished based on aircraft position, flight path angle, track, and speed relative to the terrain database image forward the aircraft.

Through sophisticated look ahead algorithms, both caution and warning alerts are generated if terrain or an obstacle conflict with "ribbons" projected forward of the aircraft (see illustration).

A terrain conflict intruding into the caution ribbon (Figure 18-22) activates EGPWS caution lights and the aural message **"CAUTION TERRAIN, CAUTION TERRAIN"** or **"TERRAIN AHEAD, TERRAIN AHEAD"**. An obstacle conflict provides a **"CAUTION OBSTACLE, CAUTION OBSTACLE"** or **"OBSTACLE AHEAD, OBSTACLE AHEAD"** message. The caution alert is given typically 60 seconds ahead of the terrain/obstacle conflict and is repeated every seven seconds as long as the conflict remains within the caution area.

When the warning ribbon is intruded (typically 30 seconds prior to the terrain/obstacle conflict), EGPWS warning lights activate and the aural message **"TERRAIN, TERRAIN, PULL UP"** or **"OBSTACLE, OBSTACLE, PULL UP"** is enunciated with **"PULL UP"** repeating continuously while the conflict is within the warning area.

When a compatible Weather Radar, EFIS, or other display is available and enabled, the EGPWS Terrain AND Alerting and Display (TAD) feature provides an image of the surrounding terrain represented in various colors and intensities. There are actually two types of TAD displays depending on the options selected. The first provides a terrain image only when the aircraft is 2000 feet or less above the terrain ("standard"). A second feature called "Peaks" enhances the standard display characteristics to provide a higher degree of terrain awareness independent of the aircraft's altitude.

Continued on next page ...

Figure 18-22

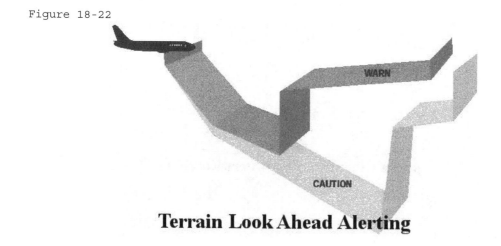

Terrain Look Ahead Alerting

Enhanced Ground Proximity Warning System (cont'd.)

Standard Terrain Alerting and Display

The Standard TAD provides a graphical plan-view image of the surrounding terrain as varying density patterns of green, yellow, and red as illustrated in Figure 18-23. The selected display range is also indicated on the display, and an indication that TAD is active is either indicated on the display (e.g., "TERR") or by an adjacent indicator.

Each specific color and intensity represents terrain (and obstacles) below, at, or above the aircraft's altitude based on the aircraft's position with respect to the terrain in the database. If no terrain data is available in the terrain database, then this area is displayed in a low-density magenta color. Terrain more than 2000 feet below the aircraft, or within 400 (vertical) feet of the nearest runway elevation, is not displayed (black).

When a caution alert is triggered, the terrain (or obstacle) that created the alert is changed to solid yellow (100% density).

When a warning alert is triggered, the terrain (or obstacle) that created the alert is changed to solid red (100% density).

TAD "Pop-Up" and "Auto-Range"

Based on the display system used, there may be additional terrain display features. These are defined as installation options and allow for:

Automatic display of terrain on the cockpit display ("TAD pop-up") in the event that a caution or warning alert is triggered as described in Terrain Look Ahead Alerting. In some cases, an active display mode must be selected first.

"Auto-range" when Pop-up occurs. This provides for the automatic range presentation for terrain as defined for the display system configuration (typically 10 nm). In this case, if the terrain auto-range is different than the display system selected range, the displayed range value on the cockpit display is flashed or changed color until the range is manually reselected or terrain display is deselected.

Continued on next page ...

Figure 18-23

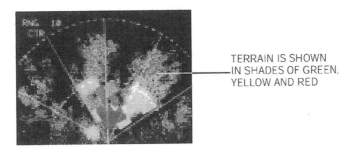

TERRAIN IS SHOWN IN SHADES OF GREEN, YELLOW AND RED

Enhanced Ground Proximity Warning System (cont'd.)

The following table indicates the TAD colors and
elevations (standard and Peaks).

Figure 18-24

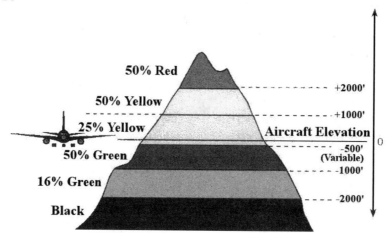

Color	Indication
Solid Red	Terrain/Obstacle Threat Area — Warning.
Solid Yellow	Terrain/Obstacle Threat Area — Caution.
50% Red Dots	Terrain/Obstacle that is more than 2000 feet above aircraft altitude.
50% Yellow Dots	Terrain/Obstacle that is between 1000 and 2000 feet above aircraft altitude.
25% Yellow Dots	Terrain/Obstacle that is 500 (250 with gear down) feet below to 1000 feet above aircraft altitude.
Solid Green (Peaks only)	Shown only when no Red or Yellow terrain /Obstacle areas are within range on the display. Highest terrain/Obstacle not within 500 (250 with gear down) feet of aircraft altitude.
50% Green Dots (Peaks only)	Terrain/Obstacle that is 500 (250 with gear down) feet below to 1000 below aircraft altitude.
	Terrain/Obstacle that is the middle elevation band when there is no Red or Yellow terrain areas within range on the display.
16% Green Dots (Peaks only)	Terrain/Obstacle that is 1000 to 2000 feet below aircraft altitude.
	Terrain/Obstacle that is the lower elevation band when there is no Red or Yellow terrain areas within range on the display.
Black	No significant terrain/Obstacle.
16% Cyan (Peaks only)	Water at sea level elevation (0 feet MSL).
Magenta Dots	Unknown terrain. No terrain data in the database for the magenta area shown.

Enhanced Ground Proximity Warning System – Mode 6

ALTITUDE CALLOUTS

EGPWS Mode 6 includes both altitude callouts and excessive bank angle callouts. Altitude callouts (Figure 18-25), which have become a popular feature with many airlines on new aircraft, are available in the system. They are pin-selectable at the rear connector and there are 32 menus available for the purpose of increasing altitude awareness on final approach.

Tones are also available. Automatic audio level increase is available when windshield rain removal is in use. AlliedSignal recommends the use of a few automatic callouts near the runway and a "smart" callous, which would rarely be heard, for most ILS landings. AlliedSignal recommends that the minimums callout be utilized.

EXCESSIVE BANK ANGLE ALERT

Bank Angle can be used to alert crews of excessive roll angles (Figure 18-26). The bank angle limit tightens from 40 degrees at 150 feet AGL to 10 degrees at 30 feet AGL to help alert the crew on landing of excessive roll corrections which might result in wing tip or engine damage. Bank Angle is also useful to help alert the pilot of severe over banking which might occur from momentary disorientation during initial climb out.

Figure 18-25

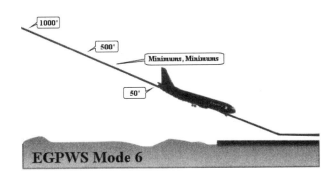

Figure 18-26
EGPWS Mode 6 Bank Angle

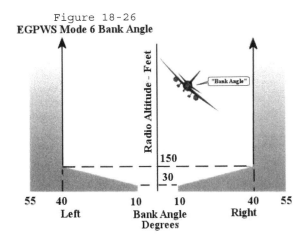

Enhanced Ground Proximity Warning System – Mode 7

Visual and aural windshear warnings are given for windshear that significantly degrade the performance of the aircraft. Optional visual and aural caution alerts can be given for windshear that may be a signature of micro-bursts (Figure 18-27, 18-28).

The detection level is automatically varied by outside air temperature and lapse rate, change in flight path, relationship to glideslope, height above ground, bank angle and approach stall margin. This advances the alert/warning time and improves the margin against unwanted alerts or warnings.

WINDSHEAR PERFORMANCE FEATURES

- Detection and Alerting.
- Senses aircraft performance and atmospheric conditions for advanced alerting.
- Provides caution alerting for excessive "Increasing Energy" Shears.
- Provides aural and visual warning for "Decreasing Energy" Shears.
- Coordination with GPWS allows intelligent alerting.

Figure 18-27

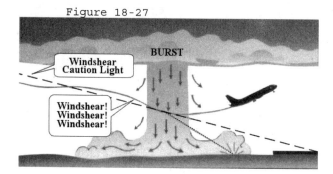

Figure 18-28

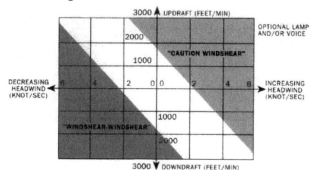

Weather Radar and Predictive Windshear

Description

As discussed in a previous chapters, ground-based ATC primary surveillance radar systems provide information concerning the location of aircraft within the control area covered by the radar. Similar systems are carried onboard aircraft and are used for locating weather systems ahead of the aircraft. If impending weather is identified and located, the flight plan can be altered in order to avoid the areas of turbulence associated with these storms.

A secondary function of the airborne radar is to provide ground mapping of the terrain below the aircraft. Large landmarks, such as mountains and the edges of lakes and rivers, can be seen on the radar return and then used for navigation.

A third usage has been developed in the last few years whereby the return signal reflected back from the precipitation within a storm is analyzed and the lateral wind velocity and direction can be estimated. This information about the lateral winds surrounding a storm can be used to predict turbulence.

While the radar system will show a reflection from other aircraft, it is far too slow to be used for traffic avoidance- modern transport aircraft simply move to fast. As such, airborne radar systems are termed Weather Radars (or simply WXR) since their primary function is weather detection.

Both ATC radar and airborne weather radar operate on the echo principle. A high-energy beam made up of radio frequency pulses is emitted by a directional antenna. The pulses travel at the speed of light outward from the antenna, and when they hit anything of substance, a percentage of the beam reflects back towards the antenna. In the case of the weather radar, the antenna sweeps back and forth along the horizon, ahead of the aircraft.

The raindrops within a storm are substantial enough to reflect the signal back to the antenna. The strength of the return signal, the time it took to go out and back, along with the position of the antenna when the reflection occurred, is displayed on a screen for the flight crew to view (Figure 18-29).

The strength of the return shows the amount of precipitation, the time out and back shows the distance, and the antenna position shows the bearing to the storm. The flight crew has the option to tilt the radar scan up and down in order to get a sense of the height (vertical size) of the storm ahead.

Continued on next page ...

Figure 18-29

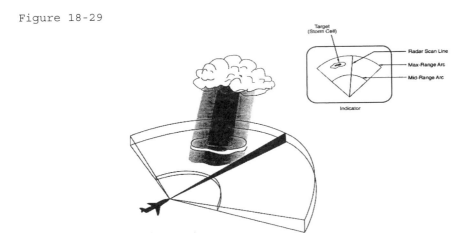

Weather Radar and Predictive Windshear (cont'd.)

Weather Radar System

A typical weather radar system consists of three sub-systems. The three sub-systems are: the antenna, the receiver/transmitter unit (RT unit), and the Control/Display system (figure 18-30). Most modern weather radar systems are designed to operate in the C-Band (5.2 to 5.9 GHz), or the X-Band (8.5 to 10 GHz). The operating range of C-Band radars extends about 200 nautical miles while the X-Band radars have a range extending about 320 nautical miles.

Most aircraft will have two systems installed for redundancy, a Left system and a Right system,. The two systems will share one antenna assembly. The RF signal from the R/T units will both be fed to a relay, which chooses one signal and feeds it to the antenna and feeds the return back to the same R/T unit.

The RF signals usually travel to and from the antenna through a hollow tube called a waveguide. Some systems (usually C-Band systems) use coax cable in place of the waveguide.

The receiver/transmitter unit, or R/T unit, provides the high-energy RF signal that is sent out the antenna. The R/T unit also takes in the reflected return and processes it for display on the flight deck.

Continued on next page ...

Figure 18-30

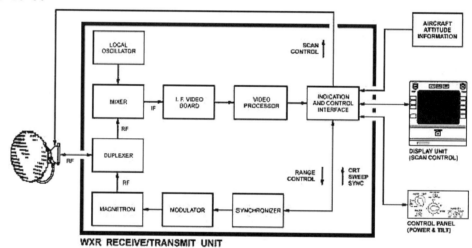

WXR RECEIVE/TRANSMIT UNIT

Weather Radar and Predictive Windshear (cont'd.)

The antenna is mounted on the forward pressure bulkhead and is covered by a special fiberglass fairing called a radome. Most modern systems use antennas consisting of a flat plate array, but parabolic reflectors are also used. The antenna is provided pitch and roll information from the aircraft's attitude source. This information allows the scan of the antenna to be constantly adjusted so it scans the same horizontal plane regardless of changes in the aircraft attitude.

The Control/Display subsystem sets the radar range and signal strength, as well as the antenna elevation. It also displays the information gained from the reflected return. In some installations this subsystem is all one unit, and in others it is several individual components (Figure 18-31).

On "Glass Cockpit" aircraft (aircraft with CRT's or LCD's to display flight information) the weather radar display is usually made in conjunction with the other navigational information. The WXR range controls, and the on/off controls, are usually with the other controls relating to the flight instrumentation displays. Typically, a separate panel controls antenna elevation and left/right system selection.

Continued on next page ...

Figure 18-31

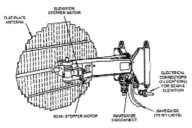

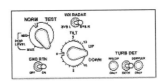

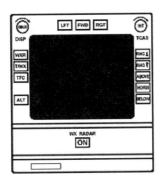

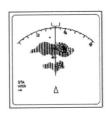

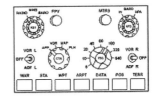

Weather Radar and Predictive Windshear (cont'd.)

PREDICTIVE WINDSHEAR

When the R/T unit generates the RF signal for transmission, that signal has a particular frequency. If the transmitter and the substance the signal reflects off of are both stationary, the reflected signal will have the same frequency as the transmitted signal. Since the aircraft is moving forward during flight, the reflected signal bouncing back from a stationary object will show a frequency shift equal to the speed of the aircraft. If, at the same time, the material causing the reflection is moving toward or away from the aircraft then the frequency shift will include both the speed of the aircraft and the speed of the material. Since the WXR system is made aware of the speed of the aircraft, it can calculate the relative speed of the material causing the reflection. It is this principle, known as Doppler Shift, which allows the WXR system to predict areas of turbulence around storms (Figure18-32).

Since it is the raindrops that reflect the radar signal, the radar system cannot "see" clear air turbulence. There is nothing in the air to reflect the RF energy. When precipitation is present, it will reflect the RF energy, and it will shift its' frequency due to any lateral speed the precipitation has. Once the shift exceeds a preset limit (around 12 mph) the radar system indicates turbulence in that area. Since these frequency shifts are quite small, the WXR system can only detect turbulence for a range of about 40 nautical miles.

Windshear is a condition with very unstable, rapidly changing winds. It may be caused by low-level temperature inversions, or microburst activity on the frontal zone of a thunderstorm. It generally consists of a column of air moving vertically downward. When this happens at a low altitude it becomes especially hazardous to aircraft because as the wind hits the ground it "mushrooms" outward. This causes a headwind (and a corresponding increase in lift) as the aircraft approaches the downdraft, and a tailwind (along with a corresponding loss of lift) as it exits the downdraft. If the flight crew is not prepared for this the results can be tragic.

When the aircraft is below 1200 feet above ground level (AGL) and the WXR system sees turbulence within 3 nautical miles ahead of the aircraft in a pattern where there is a positive shift on the near side of a storm and a negative shift on the far side, (the "mushroom" effect), it will issue a "Windshear Caution". When that distance closes to 1.5 nautical miles a "Windshear Warning" is issued. These warnings give the flight crew time to prepare for what their aircraft is about to experience as it flies through the downdraft ahead. Just how much time does it give the flight crew? Well, at 150 knots, 3 miles goes by in only 72 seconds!

Figure 18-32

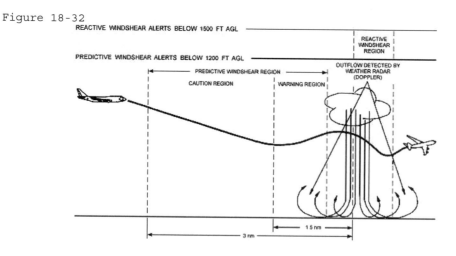

Enhanced Ground Proximity Warning System Aural Priority

AURAL MESSAGE PRIORITY
Two or more messages may be activated simultaneously, so a message priority has been established. The following table reflects the priority for these message callouts. Messages at the top of the list will start before or immediately override a lower priority message even if it is already in progress.

Table 18-28

Message	Notes	Mode
"Windshear, Windshear, Windshear"	d, j	7
"Pull Up"	h, l, k	1, 2, TA
"Terrain, Terrain"	--	2, TA
"Obstacle, Obstacle"	c	TA
"Terrain"	--	2
"Minimums"	a, c	6
"Caution Terrain, Caution Terrain"	c	TA
"Caution Obstacle, Caution Obstacle"	g	TA
"Too Low Terrain"	--	4, TCF
Altitude Callouts	--	6
"Speed Brake, Speed Brake"	c	6
"Too Low Gear"		4A
"Too Low Flaps"		4B
"Sink Rate, Sink Rate"		1
"Don't Sink, Don't Sink"		3
"Glideslope"		5
"Approaching Minimums"	b, c	6
"Bank Angle, Bank Angle"		6
"Caution Windshear"	c, d, e	7
"Autopilot"		6
"Flaps, Flaps"	c	6

Notes:
a) May also be "Minimums, Minimums" or "Decide".
b) May also be "Approaching Decision Height", "Fifty Above", "Plus Hundred".
c) Message is dependent on aircraft type or option selected.
d) Windshear detection alerts provided for some aircraft types.
e) Caution alert if not disabled.
f) May also be "Terrain Ahead, Terrain Ahead".
g) May also be "Obstacle Ahead, Obstacle Ahead"
h) May also be "Terrain Ahead Pull Up"
i) May also be "Obstacle Ahead Pull Up"
j) May be preceded by siren.
k) May be proceeded by "Whoop, Whoop"
TA= Terrain Look-Ahead Alert
TCF= Terrain Clearance Floor

Enhanced Ground Proximity Warning System Inputs and Outputs

SYSTEMS INPUTS

The EGPWS uses various input signals from other on-board systems (Figure 18-29). The full compliment of these other systems is dependent on the EGPWS configuration and options selected. Systems providing Altitude, Airspeed, Attitude, Glideslope, and position are required for basic and enhanced functions. Accelerations, Angle-of-Attack (AOA), and Flap position is required for Windshear. Inputs are also required for discrete signal and control input.

The EGPWS utilizes signals from the following systems:

- AIR DATA
- RADIO ALTITUDE
- FMS, IRS, AHRS, ACCELEROMETER
- GLOBAL POSITIONING SYSTEM
- VHF NAV RECEIVER
- TERRAIN DISPLAY SYSTEM
- AOA VANE
- DISCREETS

SYSTEM OUTPUTS

The EGPWS provides both audio and visual outputs.

Audio outputs are provided as specific alert phrases, and altitude callouts or tones provided by an EGPWS speaker and via the cockpit Interphone system for headset usage. Several audio output levels are available, established during the installation of the EGPWS. These EGPWS audio outputs can be inhibited by other systems having higher priority (i.e., windshear) or cockpit switches in some cases. The EGPWS also has the ability to inhibit other system audio such as TCAS. Visual outputs provide discrete alert and status annunciations, and display terrain video when a compatible display system is available and enabled.

Status annunciations provide information to the flight crew about the status of the EGPWS (e.g., GPWS INOP) or activation of selected functions. Terrain video is generated by the EGPWS computer based on the aircraft's current position relative to the surrounding terrain. This video is presented to a Weather Radar indicator, EFIS display, or a dedicated display unit.

Figure 18-29

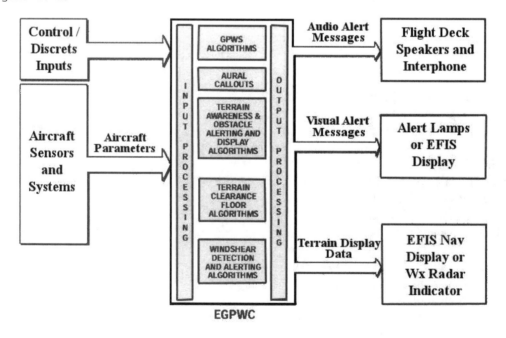

CHAPTER 19 - AUTOPILOTS AND FLIGHT DIRECTORS

Autopilot/Flight Director Comparison

Figures 19-1 and 19-2 compare an autopilot system with a flight director system. In either system, the computers have many different inputs available, some of which are indicated. The gyro inputs are always in use, since they are basic to the operation.

The inputs in use by an autopilot or flight director depend upon its mode of operation. Modes of operation in the roll channel could be following a selected heading, or a selected radio path, or merely maintaining wings level. A pitch channel mode might maintain a certain pitch attitude, certain vertical speed, or follow a glide path.

Whatever the mode of autopilot operation, there is a set of servos of some kind operating the control surfaces as required to achieve the needed aircraft attitude. The computers compute the necessary servo commands which operate the control surfaces. Airplane attitude changes are reflected in the gyro output signals. This is called aerodynamic response. Gyro signals to the computers are follow-ups to commands. Control surface position information is also follow-up (feedback).

The same set of computers could be used as a flight director. The difference is that the flight director does not have a set of servos to operate the aircraft control surfaces. All the flight director can do is command the pilot to operate the control surfaces.

If the flight director wants the elevators raised, it moves the horizontal command bar, called pitch command bar, up. If it wants the elevators lowered, it moves the pitch command bar down. The pilot is then supposed to operate the control column accordingly.

If the flight director wants the right wing moved down, it moves the vertical command bar, called roll command bar, to the right. If it wants the left wing down, it moves the command bar to the left. The pilot then operates his control wheel in accordance with that command.

The job of the pilot is to keep the two command bars centered. Whenever one moves away from the center position, he operates the control column or the control wheel as directed by the command bars.

In earlier airplanes, the autopilot system and the flight director system have separate computers and the flight director computers are generally less sophisticated than the auto-pilot computers.

In airplanes built today, the autopilot computers are also used as flight director computers. The pilot can, at his discretion, allow the autopilot to operate the controls, or he can follow the commands of the flight director. Typically, he uses the autopilot to control the airplane, and the flight director becomes a monitor of autopilot operation.

Standard Body Auto Flight Guidance

Figure 19-1

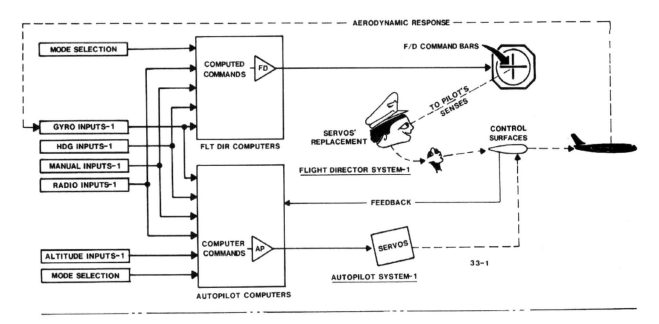

Figure 19-2

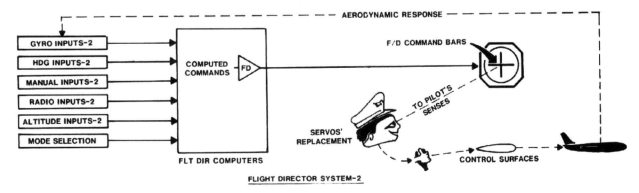

B-767 Auto Flight Guidance

Figure 19-3

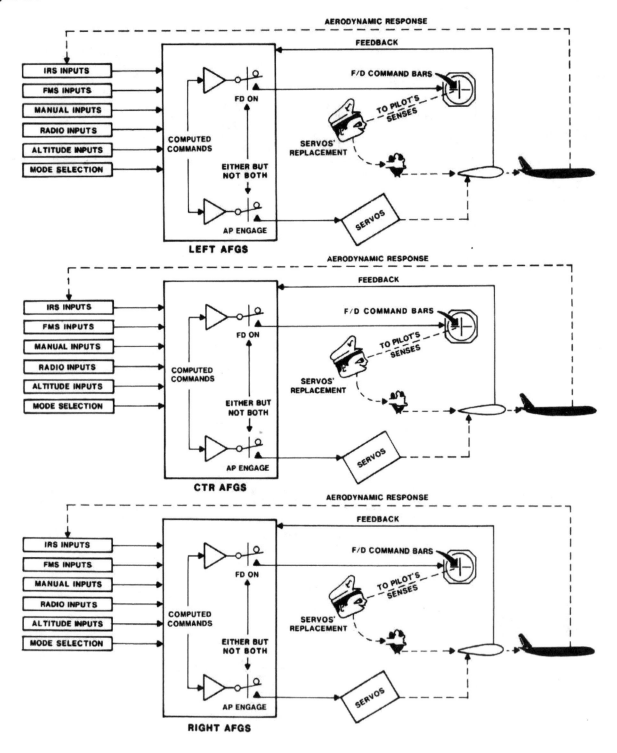

B-747 Auto Flight Guidance

Figure 19-4

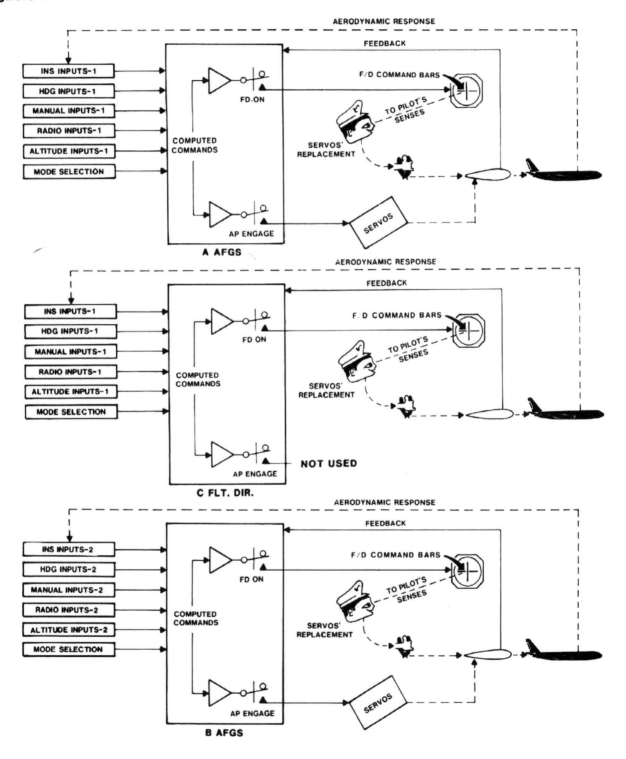

DC-10 Auto Flight Guidance

DC-10 Auto Flight Guidance

Figure 19-5

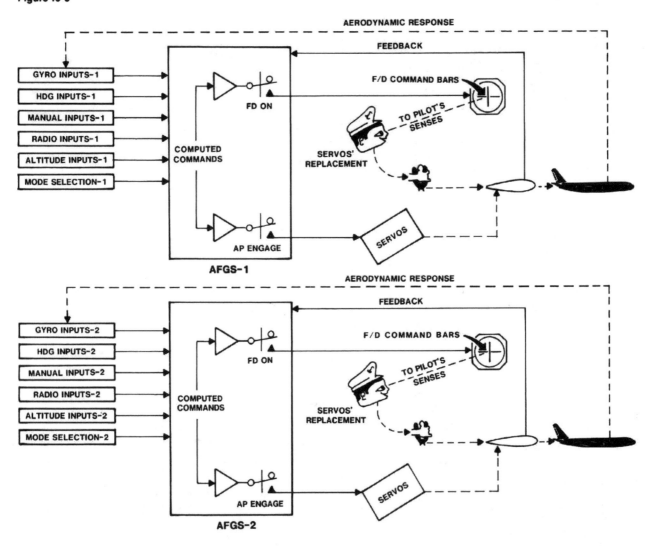

Autopilot Flight Director System – Fly By Wire (Typical Boeing 777)

The autopilot flight director system (AFDS) typically has three channels that supply automatic control of the airplane and flight director guidance. When selected, the system controls the airplane on the selected flight path and at the selected speed. The 777 autopilot flight director system is similar to the auto flight system on Boeing 757/767 and 747-400 airplanes. However, there are differences in the way the AFDS interfaces with the flight control system. The autopilot flight director system (AFDS) has two purposes. These purposes are:

To automatically control the airplane attitude

To supply indications so the flight crew can manually control the airplane attitude.

The autopilot controls the airplane attitude through:

- Takeoff (flight director only)
- Climb
- Cruise
- Descent
- Approach
- Go-around
- Autoland

The flight crew uses the mode control panel (MCP) to select a mode of operation. The autopilot commands are generated in the AFDS computers. They are then sent to the primary flight computers (PFCs) to operate the flight control surfaces.

The autopilot automatically controls airplane heading, track, speed, altitude, navigation paths, and go-around. The flight director provides guidance commands for these functions plus guidance commands for takeoff (Figure 19-5a).

The AFDS has these components:

- One mode control panel (MCP)
- Three autopilot flight director computers (AFDCs)
- Six backdrive actuators
- Two control wheel disengage switches
- Two takeoff/go-around switches (TO/GA)

The AFDS does not have servos to move the primary flight control surfaces. The primary flight computers (PFCs), the actuator control electronics (ACEs) and power control units (PCUs) control movement of the surfaces.

There are two autopilot engage switches on the MCP. All available autopilot channels engage when either switch is pushed.

When the flight director switches are on, the flight director command bars show on the primary flight displays (PFDs). The flight crew uses the flight director bars as guides to control the attitude of the airplane. Autopilot Flight Director Computer

The autopilot flight director computer (AFDC) calculates:

- Flight director (F/D) commands
- Autopilot commands
- Backdrive actuator commands.

The AFDCs must have operational software loaded in them. Shop personnel can load the software, or you can use the Maintenance Access Terminal (MAT) to do a software load.

Continued on next page…

Autopilot Flight Director System – Fly By Wire (Cont'd)

Figure 19-5a

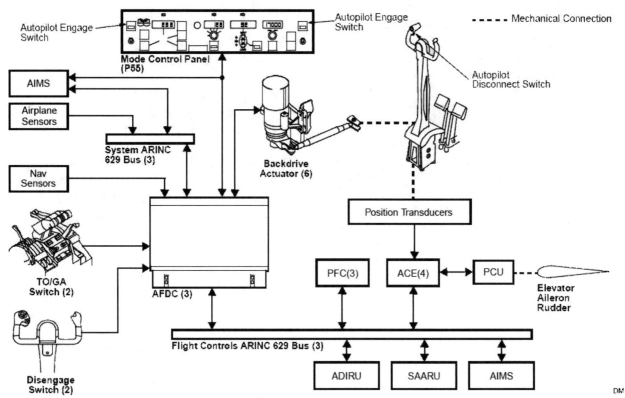

Auto Flight Director Control System

Autopilot Servo Types

Figure 19-6 illustrates the function of a DC servo motor when used to operate an airplane control surface. If the autopilot is engaged, the servo motor gear train is mechanically clutched to the control cables which connect the cockpit control to the control surface actuator. The clutch is operated by an electric solenoid. The servo motor drives a tachometer generator to provide inverse feedback to the amplifier for speed limiting and smoothing.

A sine winding follow-up synchro is driven by the servo motor to a null prior to engaging the autopilot. The null results because there is no command signal to the amplifier while the autopilot is disengaged. When the autopilot is engaged, the null signal from the follow-up synchro must represent the neutral position of the control surface, so the pilot must move his control surfaces to the neutral position prior to engaging the autopilot. That is normally accomplished by the artificial feel system.

The torque-limiting resistor is adjusted so that the maximum possible output of the amplifier does not exert more than the maximum desirable force on the control surface.

Providing that the autopilot was engaged with the control surface in a neutral position, the follow-up synchro signal is in proportion to control surface displacement and of a phase indicating the direction of displacement.

If the autopilot computer determines that the control surface should be moved up, the computed command calls for control surface movement up. The signal from the follow-up synchro is of a phase opposite to that of the computed command. Therefore, control surface movement will cease when the follow-up signal equals the computed command.

Figure 19-7 illustrates a hydraulic power unit which can be operated directly by the autopilot (detailed explanation in following illustrations). The servo amplifier has a low power output on the order of 100 millivolts.

The transfer valve is an electrically controlled hydraulic valve which operates a piston assembly called the autopilot actuator, which in turn operates the main control valve for the actuating cylinder.

The amount of movement of the autopilot actuator is indicated by the output of the linear voltage differential transducer, abbreviated LVDT. An LVDT is a cylindrical E-pickoff. This becomes the follow-up signal to the computed command.

The control surface position may also be used as a follow-up signal. Direct operation of the hydraulic power unit by the autopilot has two main advantages. One is the very low power computer output. The other is that control is effected directly at the hydraulic power unit, bypassing cable slack, stretch and drag. The control is therefore more sensitive and more accurate.

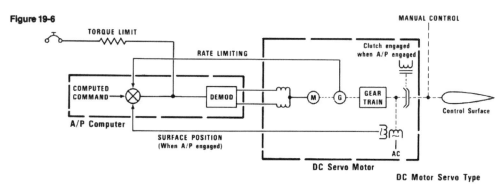

Figure 19-6

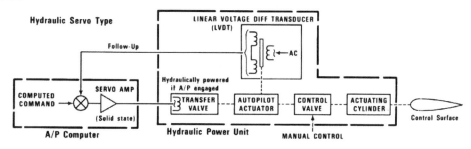

Figure 19-7

Transfer Valve

Figure 19-8 details schematically the construction of the transfer valve used in Figure 19-7. In the upper left is a cross section of coil windings around the "C" shaped core. If the "C" core is magnetized by a signal to the electric coil, it can move the permanent magnet armature up or down about its pivot point. Hydraulic fluid comes in at the lower right, passes through the flex tube and splits across the pointed divider just under the flex tube. This fluid flows all the time that the autopilot is engaged.

If there is no electric signal to the coil, the flex tube is spring-loaded to the neutral position as shown. In this position, the combination spool valve, against whose ends the hydraulic flow is directed, sees equal pressures at both ends. It, therefore, takes up the neutral position. In this position, both of the control ports are closed off by the spool valve.

If the autopilot develops a command signal calling for control surface movement, the first action is the electrical signal in the coil windings polarizing the core, causing the permanent magnet to rotate slightly in one direction or the other.

If the signal develops a north pole at the top and a south pole at the bottom of the core ends, the permanent magnet will rotate slightly counter-clockwise, moving the bottom of the flex tube to the right. That causes greater pressure at the right end of the spool valve than at the left end. The spool will therefore move toward the left. It moves to the left until the force from the feedback springs is sufficient to bring the flex tube back almost to the neutral position.

A stronger signal from the autopilot would cause a greater movement of the spool to the left in order to develop a greater force from the feedback spring to overcome the greater magnetic force on the permanent magnet. With the spool moved to the left, the right control port is connected to hydraulic pressure, and the left control port is connected to return, moving the autopilot actuator shown in the next drawing (Figure 19-9).

If the electrical signal is of the opposite polarity, the spool will move right instead of left, reversing the hydraulic connections to the control ports.

Figure 19-8

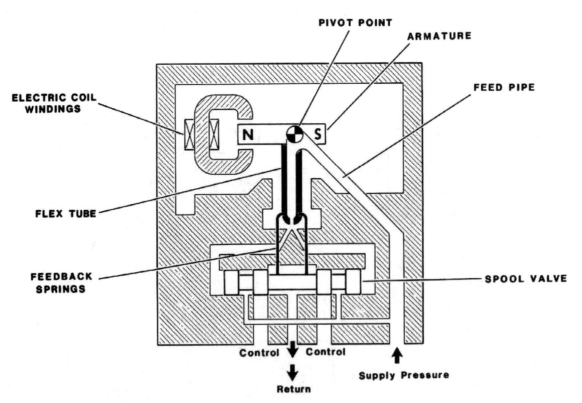

Control Surface Actuator: Manual/Autopilot

Figure 19-9 is a schematic drawing of a hydraulically operated control surface actuator, operable either by the pilot directly or by the autopilot electrically.

The lower portion of the transfer valve, shown in the upper right, is connected to the autopilot actuator. Just below it is an on/off solenoid which, when the autopilot is engaged, opens its valve, supplying pressure to the transfer valve.

Two LVDTs are shown, one at the top middle, one at the bottom left. The signal from the autopilot LVDT, at the top, will indicate how far the autopilot actuator has been moved by the operation of the transfer valve. The LVDT at the lower left is attached to the piston rod of the main actuator, which actually moves the control surface. The signal from that LVDT, therefore, is an indication of the position of the control surface itself.

The body of this actuator is secured to structure, and it is the piston and piston shaft which move the control surface.

Let's operate it first with a manual input from the cockpit. The cockpit control cables move the quadrant at the top of the drawing. Assume that it rotates clockwise. As it turns, it moves the arm attached to it to the left. That moves the autopilot

actuator to the left, compressing the left springs. This is of no consequence since the autopilot is not engaged. The long lower arm then pivots around the ball at its lower end.

As the top end of the long lower arm moves to the left, it moves the control valve to the left. This connects the left side of the main actuator piston to pressure. The right side of the main actuator piston is, at the same time, connected to return.

The main actuator piston then moves to the right, carrying the bottom end of the long arm to the right, and moving the control valve back toward its neutral, or shutoff position. The main actuator piston will move the control surface until the control valve has closed off its ports.

Further movement of the cockpit control in the same direction would again move the control valve to the left, causing the main actuator piston to move to the right until the control valve is once again shut off. The amount of control surface movement is a function of the amount of movement of the cockpit control. The direction of control surface movement is a function of the direction of cockpit control movement.

Continued on next page ...

Figure 19-9

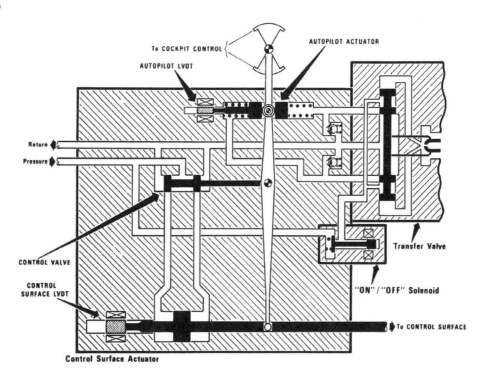

Control Surface Actuator: Manual/Autopilot (cont'd.)

The autopilot is capable of performing the same operation. If the transfer valve has a signal, the black vertical spool on the right moves in one direction or the other. Let's say that it moves down. Hydraulic system pressure is then ported to the right side of the autopilot actuator, and return is ported to the left.

Positioning springs are shown on each side of the autopilot actuator. As long as the transfer valve ports remain open, the autopilot actuator continues to move to the left.

As it moves to the left, however, it develops a follow-up signal in the autopilot LVDT. This will cause the transfer valve signal to become null when the computed command has been equaled by the follow-up signal (Figure 19-7).

As the autopilot actuator moves to the left, it carries the arm above it to the left, rotating the quadrant to which the cables are attached, and moving the associated cockpit control. This could be either the control wheel rotating for aileron movement, or the control column moving for elevator movement (the rudder actuator operates differently).

The amount of movement developed in the autopilot actuator determines how far the control surface moves. When the control surface moves, an aerodynamic follow-up signal shows up in the autopilot as a changed gyro attitude signal.

The autopilot can be overpowered at any time by the pilot if he moves his cockpit control with enough force. A typical overpowering force would be 25 to35 pounds of turning force on the control wheel, or 40 to 50 pounds of force on the control column.

If the pilot overpowers the autopilot, in the preceding example, he would be pushing the autopilot actuator back toward the right. In doing so he would increase the hydraulic pressure in the pressurized side of the actuator enough to open the top relief valve (just below and to the right of the auto-pilot actuator). The relief valve would then dump the excess pressure into the return line.

The operating pressure of the relief valve determines the amount of force required to overpower the autopilot. The actual overpowering operation is considerably more complicated than indicated here, but this is the basic principle involved. It is actually arranged so that if the autopilot is over-powered, a portion of the autopilot actuator moves the LVDT center slug hard over, developing a high LVDT signal. The autopilot LVDT and the control surface LVDT are so arranged that, in normal operation of the actuator by the autopilot, these two signals are equal.

But, if the autopilot has been overpowered, the autopilot LVDT signal becomes very high, and it is then quite easy to detect electrically that the auto-pilot has been overpowered. This fact is of use in aircraft where two or more autopilots are used simultaneously for automatic landing. The over-powering of one autopilot by another would be electrically signaled by the difference in signal levels of these two LVDTs.

Figure 19-9

Autopilot Control Servos

Each A/P servo contains two solenoid valves, an electrohydraulic servo valve (EHSV), two linear variable differential transducers (LVDTs), and a pressure regulator/relief valve. Hydraulic power is applied through pressure and return ports at one end of the servo (Figure 19-9a).

Solenoid Valves

Each solenoid valve (SV1 and SV2) is an electrically operated open-close valve which completes hydraulic pressure through the servo when autopilot arm and engage logic circuits are completed.

Solenoid valve number 1 (SV1) opens when the autopilot is armed. It ports hydraulic pressure to solenoid valve number 2 (SV2) and the electrohydraulic servo valve (EHSV). Solenoid valve number 2 opens when the autopilot is engaged, and ports hydraulic pressure to the detent pistons. The detent pistons clamp the output linkage crank and transmit actuator piston position to the output linkage.

Electrohydraulic Servovalve (EHSV)

The EHSV is controlled by the output command signal from the FCC. The EHSV contains a sealed torque motor, a feedback spring, a projector jet and a piston. Hydraulic pressure through the valve can be applied to either of two output ports. When no command signal is applied to the torque motor, a small amount of hydraulic fluid flows through a flexpipe attached to the torque motor armature and out of the projector jet. From the jet, equal pressure is applied to opposite ends of the piston holding it at center.

When a command signal is applied, the motor armature rotates in proportion to the magnitude and direction of the input signal and moves the projector jet accordingly. The jet directs more hydraulic pressure to one end of the piston than the other causing it to move and open the corresponding output port to complete hydraulic pressure through the servo. When the command signal is nulled, the motor armature and jet return to center. This equalizes the pressure on the piston and, with the aid of the feedback spring, causes the piston to re-center and close both output ports.

In response to the FCC output command, the EHSV ports hydraulic pressure to the right or left side of the actuator piston. Feedback from the actuator piston LVDT (linear variable differential transducer) nulls the command signal at the FCC and piston movement stops.

The EHSV is installed with four bolts and sealed with a gasket plate. Electrical pins are mated when the EHSV is bolted in position.

Continued on next page ...

Figure 19-9a

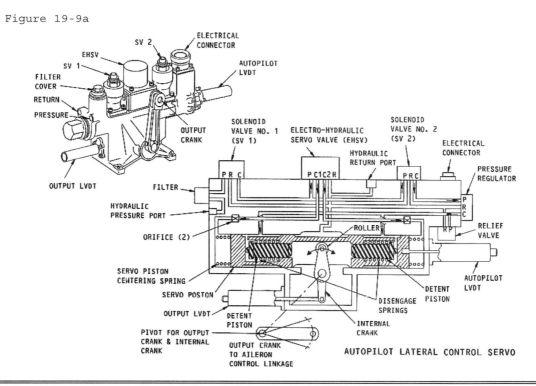

Autopilot Control Servos (cont'd.)

Linear Variable Differential Transducer

Each servo has two identical LVDTs. The actuator piston (servo position) LVDT functions as a linear follow-up transmitter for closing the loop around the EHSV. It is operated by the actuator piston. The output position (surface position) LVDT is operated by the intermediate crank, which is connected to the surface control linkage. Both LVDTs are variable reluctance transformers with an output that varies directly with linear motion. The LVDT uses 26vac excitation from the associated FCC. They require nulling adjustments to be completed which match actuator piston with piston position when the servo is on the bench.

Pressure Regulator and Relief Valves

The pressure regulator and relief valve regulates and limits hydraulic pressure applied to the actuator piston and detent pistons. The pressure relief function allows manual inputs from the control linkage to override autopilot control.

Actuator Piston Assembly

The actuator piston translates autopilot input commands through the EHSV into mechanical positioning of the control surface. The actuator piston is normally centered by two springs and is moved right or left by hydraulic pressure from the EHSV when the autopilot is armed or engaged. The springs center the actuator piston when pressurization is released.

Two detent pistons inside the actuator piston are normally retracted by disengage springs. When the autopilot is engaged, SV2 opens. Hydraulic pressure through SV2 overrides spring tension and locks the detent pistons against the roller of the internal crank. Detent pistons may be forced back if sufficient force to overcome hydraulic pressure is applied manually through the control Linkage (camout).

The actuator piston LVDT provides an electrical signal proportional to actuator piston position. This signal nulls the autopilot command signal from the FCC to stop movement of the actuator piston.

Cranks

The internal crank roller is clamped by the detent pistons when the autopilot is engaged. The crank moves with the actuator piston. Motion of the internal crank moves the output position LVDT. One end of the output crank is directly connected to the internal crank at a common pivot point. The other end of the output crank is connected to the control surface linkage.

Mechanical Control Sequence - Autopilot

Initially, the actuator piston is fixed by the centering springs. With the autopilot not engaged (armed), SV1 is open, SV2 is closed; detent pistons are disengaged, and the internal crank is free to move within the piston cavity. The output position LVDT provides internal crank position to the FCC for autopilot synchronization to surface position. The FCC commands through the EHSV cause the actuator piston to follow the internal crank so that the crank remains centered within the cavity. The actuator piston position LVDT nulls the command signal to stop the actuator piston.

With the autopilot engaged, SV1 and SV2 are open, the detent pistons are pressurized and the internal crank is clamped in the center of the actuator piston. When the EHSV receives a command from the FCC, hydraulic pressure is ported to one end of the actuator piston. The detent pistons carry the internal crank with the actuator piston to its commanded position. The output crank moves the linkage to the surface power control actuators and the output position LVDT sends position information back to the FCC to null the command signal and stop surface movement.

Camout

Camout occurs when the output crank position (surface position) does not correspond to the actuator piston position (servo position) as sensed by differing LVDT outputs.

Continued on next page ...

Autopilot Control Servos (cont'd.)

The autopilot control servos are typically centrally located. For example, for lateral aircraft control, three Lateral Central Control Actuators (LCCAs) are located in the main landing gear bay (Figure 19-9b). They translate electrical autopilot commands into a mechanical displacement which then is typically sent to the power control units (PCUs) on the ailerons via control cables or rods.

Figure 19-9b

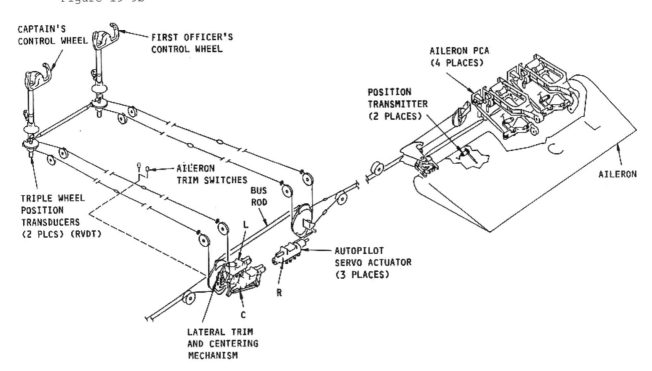

Rudder And Yaw Damper Actuator

The rudder and yaw damper actuator shown in Figure 19-10 differs from the previous actuator, used for ailerons or elevators. One important difference is that when the yaw damper operates the rudder, it does not move the rudder pedals, whereas the autopilot operations of the aileron and elevator actuators do move the cockpit controls.

Yaw damper is the name given to the rudder channel of the autopilot. It is called yaw damper because its principal function is to diminish the dutch roll of the airplane to an unobjectionable level. All subsonic jet aircraft perform dutch roll maneuvers while in flight.

The dutch roll maneuver is a continuous oscillation in which the airplane yaws first to the left while rolling to the right, and then yaws to the right while rolling to the left. The length of time required for one complete oscillation varies in different airplanes from about three to five seconds, but is constant for any particular airplane. It is an aerodynamic result of the airplane wings being swept back about 35° from straight out. This sweepback is necessary in order to prevent airflow over the top of the wing from reaching sonic speeds.

The yaw damper function of the rudder channel continuously operates the rudder in such a manner as to damp dutch roll oscillation to a minimum value

that is not noticeable to passengers. Rudder hydraulic actuators are constructed so that, when the yaw damper is operating, it does not move the rudder pedals, which would otherwise be in continuous motion.

The transfer valve, the on/off solenoid, and the LVDTs all correspond to the similar items in the previous actuator. The yaw damper actuator corresponds to the autopilot actuator in the previous drawing. The actuator follow-up system, which proportions the rudder movement to the amount of rudder pedal input or yaw damper control input, is omitted from this drawing for simplicity.

The lever system shown in the center of the actuator is extracted and shown in a detail view to the left of the actuator. Referring to both of these lever system presentations makes the operation easier to understand.

The yaw damp actuator and control valve are both shown with caging springs to the right. The white sections of the caging spring assembly are movable toward each other. But if no force is exerted, they remain against the stops in the body of the actuator, thus positioning the control valve or the yaw damp actuator in a neutral position.

Continued on next page ...

Figure 19-10

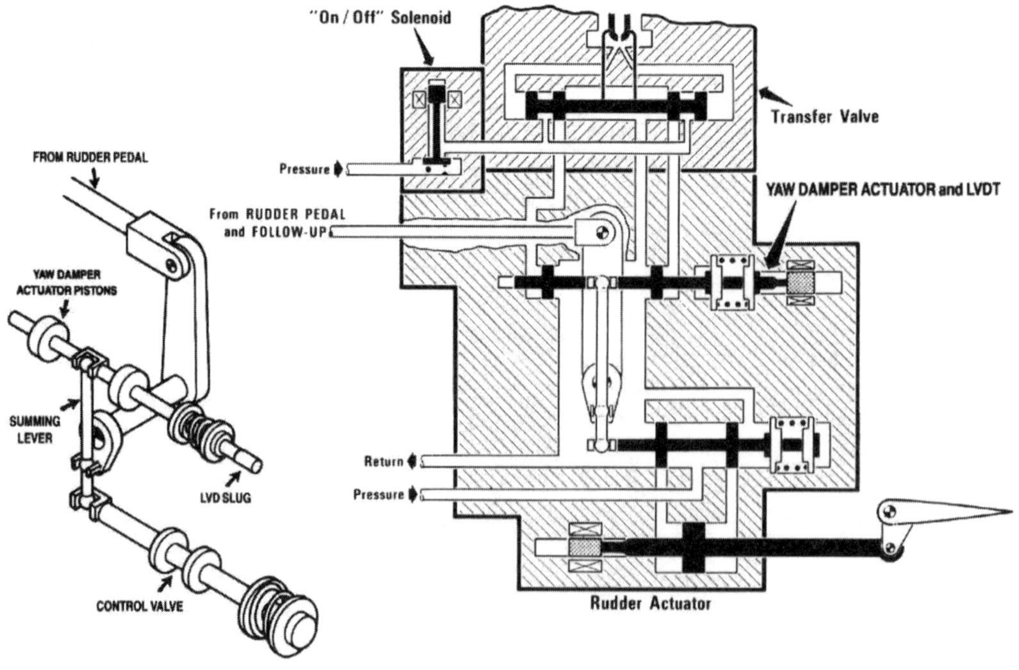

Rudder And Yaw Damper Actuator (cont'd.)

We will go through the rudder pedal operation first. Movement of the rudder pedals rotates the heavy long arm about its pivot shaft at the bottom. At the lower end of the shaft, a short lever moves the summing lever. The yaw damper actuator holds still because the caging springs maintain it in a neutral position. This results in the summing lever pivoting at its upper ball joint.

As it pivots around the upper ball joint, it pushes the control valve to one side or the other. When the control valve moves, pressure is ported to one side of the main actuator and return is ported to the other. This causes the rudder movement and follow-up action without movement of the yaw damper actuator.

The yaw damper actuator operation is as follows: A signal from the yaw damper to the transfer valve causes the upper black spool in the transfer valve to move to one side or the other. This ports pressure to one side of the yaw damper actuator piston assembly, and ports return to the other side. When the piston assembly moves, it moves the top of the summing lever which pivots at the ball joint in the middle (the rudder artificial feel system holds the rudder pedals still).

The summing lever then moves the main control valve. The control valve causes the rudder action. From this action it can be seen that the yaw damper moves the rudder without moving the rudder pedals.

This type of actuator, for some obscure reason, is called a "series" actuator. If the rudder actuator were to move the rudder pedals, it would be called a "parallel" actuator.

Figure 19-10

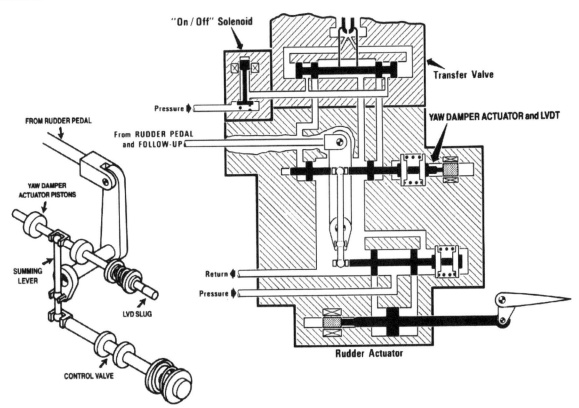

Yaw Damp Servo – Late Model Aircraft

The yaw damper servo provides left or right rudder inputs to the rudder power control actuators (PCAS).

Each yaw damper servo consists of an engage solenoid valve, an electro hydraulic servo valve and a hydraulically powered piston actuator assembly (Figure 19-10a). An electrical connector, hydraulic input/output ports, and a mechanical output shaft are utilized for airplane interface. An LVDT provides an electrical measure of output position.

28 volts DC controls the engage solenoid valve. Excitation voltage of 26 volts ac is provided from a Control Surface Electronic Unit (CSEU) power supply.

3000 psi hydraulic pressure is supplied to the EHSV via the solenoid valve. The Electro-Hydraulic Servo Valve (EHSV) directs the hydraulic fluid to move the output shaft in the appropriate direction of travel at a specified rate. Expended hydraulic fluid is output via the return port at a max pressure of 1000 psi.

Figure 19-10a

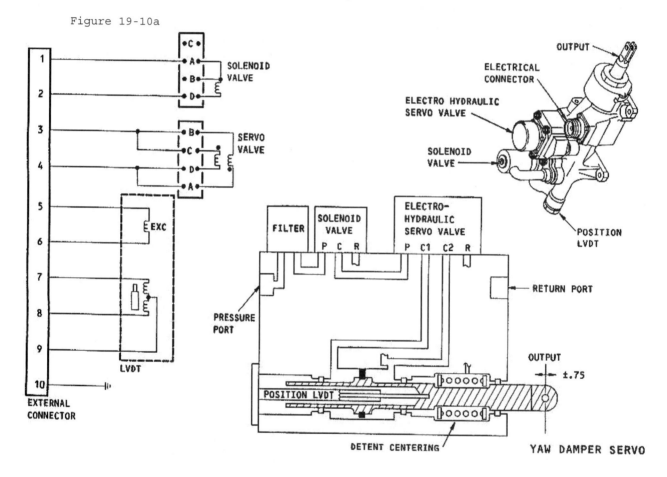

Error Signal No 1

Error signals are two wire signals which are developed back and forth across a null. They are commonly AC, but may be DC. The null has a particular significance, such as airplane wings level, airplane on a selected radio course, airplane on selected heading, and so forth.

The "not null" indications will have special directional significance of some sort. For example, left wing down, airplane to right of selected radio course, aircraft heading to the left of preselected heading, and so forth.

If the error signal is AC, the directional significance is given by the phase angle. If it is a DC signal, the directional significance is given by the polarity. In either case, the amplitude of the signal tells the amount of deviation from the null position.

E-pickoffs have many different forms and uses, but they all develop error signals. Figure 19-11 shows an E-pickoff with the armature in a balanced or nulled position. With the armature turned slightly counterclockwise, as in Figure 19-12, the output voltage is of one phase. With it turned slightly clockwise, as in Figure 19-13, the output voltage is of the opposite phase.

Figure 9-14 shows an autopilot turn knob potentiometer developing a null signal because it is at the center grounded point. When this knob is moved in one direction, as in Figure 19-15, the output is a voltage of one phase. With the knob moved in the other direction, as in Figure 19-16, the output is a voltage of the opposite phase.

Figure 19-11   **Figure 19-12** 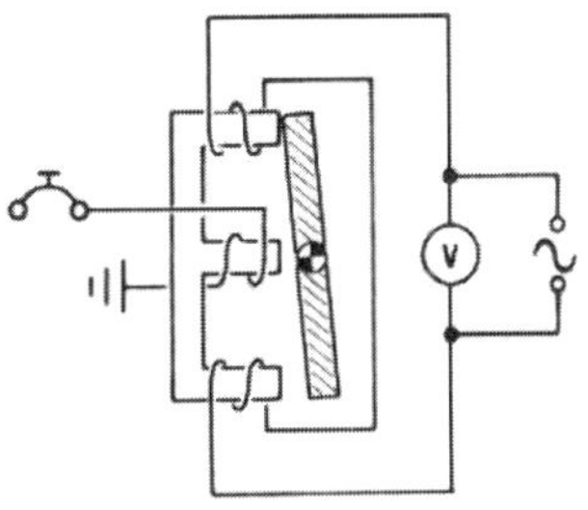   **Figure 19-13**

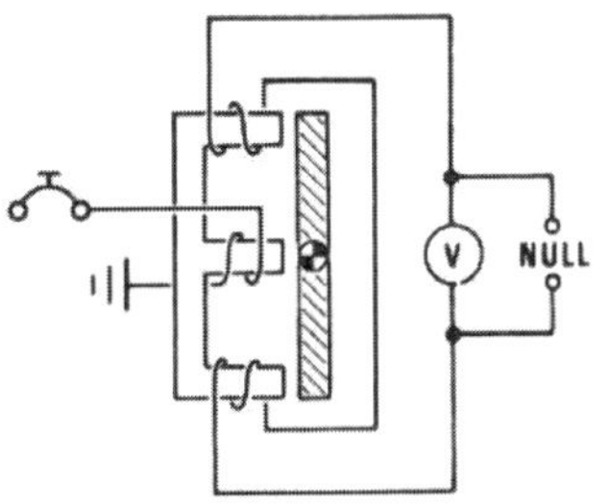

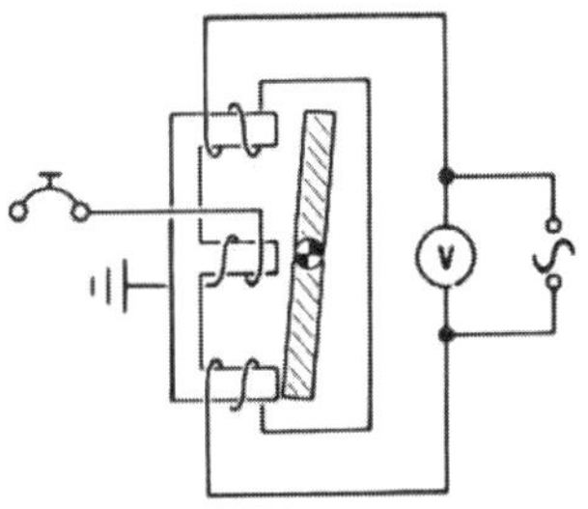

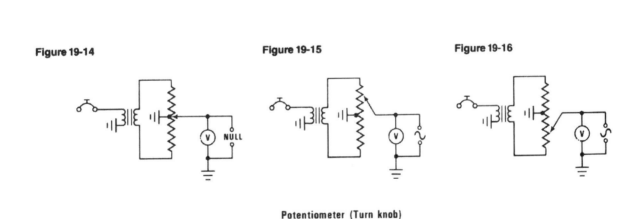

"E" Pickoff

Figure 19-14   **Figure 19-15**   **Figure 19-16**

Potentiometer (Turn knob)

Error Signal No 2

Figure 19-17 shows a vertical gyro attitude transmit synchro supplying attitude information to the stator of a servo-operated control synchro. The rotor positions signify that the output of the control synchro rotor is a null. If the transmit synchro is held steady while the control synchro rotor is moved in one direction, the output voltage is of one phase (Figure 19-18). Moving the control synchro rotor in the opposite direction provides an output voltage of the opposite phase (Figure 19-19).

Figure 19-20 is like Figure 19-17, with the output voltage a null. In Figure 19-21 the transmit synchro rotor has been moved and, therefore, the field on the control synchro is moved, producing an output voltage of one phase. In Figure 19-22 the transmit synchro rotor has been moved in the opposite direction, and the output voltage is of the opposite phase.

Figure 19-23 shows a navigation receiver providing a VOR or localizer signal to an indicator. In this case it is a null signal; the indicator needle is centered.

Figure 19-24 shows the airplane moved to the right of the radio beam; the signal is of one polarity and the deviation needle is to the left of the indicator.

Figure 19-25 shows the airplane moved to the left of the radio beam; a signal of the opposite polarity, and the needle moved to the right of the indicator.

In Figures 19-23, 19-24, and 19-25, the DC voltages have been modulated to AC voltages and the corresponding phases indicated.

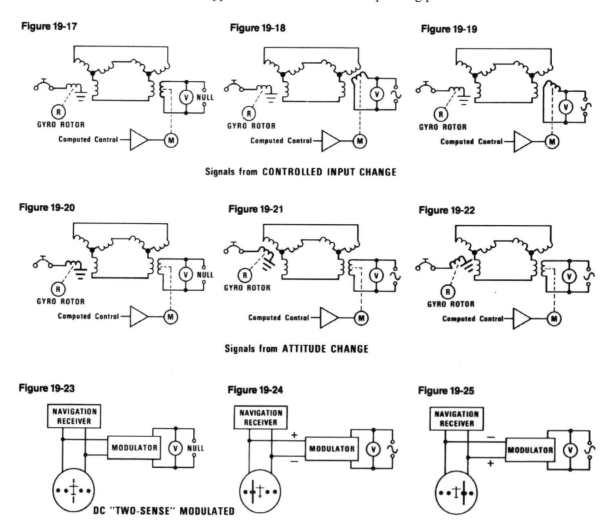

Figure 19-17   Figure 19-18   Figure 19-19

Signals from CONTROLLED INPUT CHANGE

Figure 19-20   Figure 19-21   Figure 19-22

Signals from ATTITUDE CHANGE

Figure 19-23   Figure 19-24   Figure 19-25

DC "TWO-SENSE" MODULATED

Accelerometers

Figure 19-26 illustrates the principle of an accelerometer. It is an E-pickoff device with its armature spring-loaded to a null position by two leaf springs. The output windings are connected in phase opposition to each other so that, when the armature is in a null position, the output signal is a null. Movement to one side causes the signal of one phase to predominate. Movement to the other side causes the signal of the other phase to predominate.

Figure 19-27 illustrates two accelerometers experiencing acceleration. Consider that you are holding the top one in your hand while seated in an automobile, with the armature pointed toward the front of the automobile, in the direction of the arrows. In this condition, the automobile is accelerating; that is, increasing in speed. The accelerometer armature is being pushed off to the right, just as you would be pushed back in your seat.

The lower accelerometer shows what happens if the driver puts on the brake. The speed is being decreased; you are being pushed toward the windshield, as is the armature of the accelerometer. If you are riding along at a steady 60 miles per hour, you are not being pushed forward nor backward; neither is the accelerometer armature being pushed away from its null.

Only changes of speed, in the direction of movement available to the armature, will be sensed by the accelerometer.

Figure 19-28 shows how paired accelerometers are often connected in aircraft. Two accelerometers are shown with their outputs connected together in such a manner that, if both accelerometers experience the same direction and amplitude of acceleration, their signals cancel.

In aircraft (such as the DC-8) which use accelerometers for inputs to the autopilot, they are mounted as shown in Figures 19-29, 19-30, and 19-31.

Figure 19-29 shows yaw accelerometers mounted so that their armatures can move in the direction indicated by the small arrows adjacent to the accelerometers. The aft accelerometer is located near the center of gravity and the forward accelerometer is located up front. If the airplane is yawing to the left, as indicated, the forward accelerometer sees the acceleration. The aft accelerometer does not see much movement because it is near the center of gravity and the airplane turns around the center of gravity.

Continued on next page ...

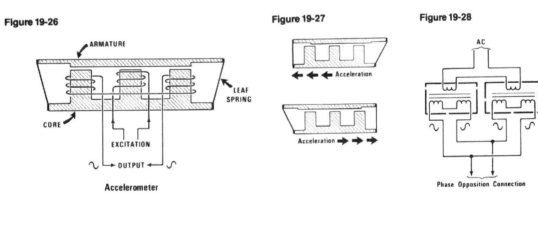

Figure 19-26

ARMATURE
LEAF SPRING
CORE
EXCITATION
OUTPUT

Accelerometer

Figure 19-27

Acceleration
Acceleration

Figure 19-28

AC

Phase Opposition Connection

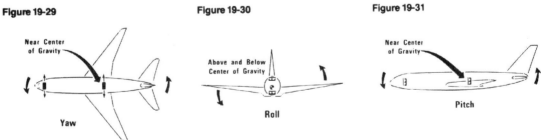

Figure 19-29

Near Center of Gravity

Yaw

Figure 19-30

Above and Below Center of Gravity

Roll

Figure 19-31

Near Center of Gravity

Pitch

Accelerometers (cont'd.)

These accelerometers are connected in phase opposition, so yawing of the airplane develops a net signal output from the two. If the airplane were to move straight sideways in either direction, both accelerometers would see the same movement and their outputs would cancel. The signal output from this combination can therefore be used to sense the yawing of the airplane for control of dutch roll.

Figure 19-30 shows a pair of roll accelerometers — one mounted near the skin of the airplane approximately above the center of gravity, and one mounted near the skin of the airplane directly below the other.

These accelerometers are also connected in phase opposition. If the airplane roll attitude changes, both accelerometers develop signals of the same phase. For example, when the airplane rolls as indicated by the arrows, the top accelerometer sees motion to the left, and the bottom accelerometer sees motion to the right. The sum of their signals can be used for inverse feedback to the roll channel for roll rate limiting and smoothing. If the airplane moves straight sideways, the accelerometer signals cancel each other.

Figure 19-31 shows a pair of pitch accelerometers. One is located near the center of gravity and the other is up front. If the airplane pitch attitude changes as indicated by the arrows, the aft accelerometer does not see much action since the airplane pitches around its center of gravity.

The forward accelerometer does feel movement, and develops a signal in accordance with the pitch action of the airplane. The combination of the signals from the two accelerometers can be used for inverse feedback in the pitch channel.

If the airplane moves straight up or down, both accelerometers develop signals which cancel each other. The signal from the aft accelerometer by itself can be used for vertical path smoothing, if desired in certain autopilot modes, because it will sense vertical motion of the airplane almost exclusive of pitching action.

CHAPTER 20 - AUTOPILOT YAW DAMPER "RUDDER CHANNEL"

Rate Gyro Dutch Roll Signals

Figure 20-1 illustrates an airplane flight path beginning at the left with a straight path. It soon changes to a constant rate of turn to the right. Near the end it resumes a straight path.

The rate gyro output directly below the flight path represents the 400 Hertz synchro signal developed by the rate gyro during this flight path. When the airplane is flying straight ahead there is no output from the synchro. During the time it is making a constant rate of turn, the output is of a particular phase with a constant amplitude.

The DC graph below the rate gyro output shows the demodulated and filtered output of a dutch roll filter. Only during the time that the rate of turn is changing is there an output from the dutch roll filter. While the rate of turn changes (at the left) from straight ahead and is building up to a constant rate of turn, the dutch roll filter output builds up, then falls off to nothing when the rate of turn becomes constant. While the rate of turn (at the right) is changing from constant right to straight ahead, the dutch roll filter output builds up with the opposite polarity, then falls off to nothing when the rate of turn becomes constant (straight ahead).

When the constant rate of turn changes to constant straight ahead flight, a condenser discharges. During the time the condenser is discharging, the DC input to the modulator changes from a null to negative.

Figure 20-2 shows the flight path and the changing turns that occur during a dutch roll. In a dutch roll

maneuver the rate of turn is constantly changing. It begins at the left with no rate of turn, builds up to a maximum rate of turn right (top of the first loop), changes to a zero rate of turn midway between the first top loop and the bottom loop, then develops into a maximum rate of turn left at the bottom of the bottom loop.

Since the rate of turn is constantly changing, the output of the rate gyro is constantly changing. The 400 Hz synchro signal graph begins at the left with no output because there is no rate of turn, builds up to a maximum rate of one phase (right turn), falls off to a null midway between turns, then builds up to a maximum rate of the other phase (left turn).

The DC graph at the bottom of Figure 20-2 is the dutch roll filter output resulting from the rate gyro input. The DC polarities are greatest when the rate of turn is greatest, and reverse when the direction of turn (phase of gyro signal) reverses.

Figure 20-3 is a super-simplified yaw damper, illustrating mainly the dutch roll filter. The dutch roll filter is a narrow band pass filter designed to pass only signals which change at the frequency of the dutch roll (reference Figure 20-6), which range from 1/5 Hz to 113 Hz. The rate gyro produces outputs for all turns, but only those related to dutch roll will appear at the input to the servo amplifier driving the rudder servo motor.

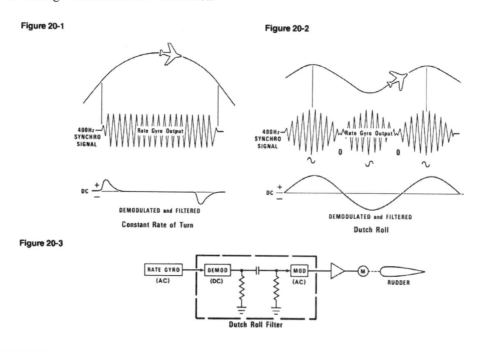

Figure 20-1

Figure 20-2

Figure 20-3

Conventional Yaw Damper Block Diagram

A yaw damper is so named because it dampens down the yawing of an airplane, which results from its tendency to perform the dutch roll maneuver. It is typically in the rudder channel of the autopilot, and may be referred to as such. But, whichever name it uses, its principal function is to dampen the yawing resulting from the dutch roll, and thereby also eliminate the rolling.

The name dutch roll has been given to a flight characteristic deficiency common to all subsonic jets because of their wing sweep back of about 35°. Dutch roll is sufficiently objectionable without dampening so that most subsonic aircraft are speed and altitude limited if their yaw dampers are inoperative.

Figure 20-4 is a block diagram of a conventional yaw damper. The complete rudder channel may or may not perform other functions, but this is its most important one. It would be possible to dampen the dutch roll by controlling the ailerons. But since there is a great deal more activity in the aileron channel, and since the whole dutch roll maneuver can be dampened by operating the rudder correctly, it is logical to put the dutch roll dampening function in the yaw channel.

Notice that the dutch roll is only dampened, it is not eliminated. The fundamental reason for its not being eliminated is that the signal to control it must come from the dutch roll itself.

In the signal source on the left of the block diagram, yaw rate or yaw acceleration is called out. This signal is typically supplied by a yaw rate gyro or yaw accelerometers.

The shaper/processor accomplishes whatever is necessary in the way of conversion, smoothing, dampening, limiting, and gain control. Its output goes to the dutch roll filter.

The dutch roll filter attenuates all signals which are not at the frequency of the dutch roll. Its output is a continuously changing command for left rudder, then right rudder, then back to left rudder and so on.

The servo amplifier amplifies this signal as required to control the servo and operate the rudder the correct amount to eliminate most of the dutch roll.

The small amount of dutch roll that is not eliminated is represented with a dashed line coupling the airplane to the signal source.

Figure 20-4

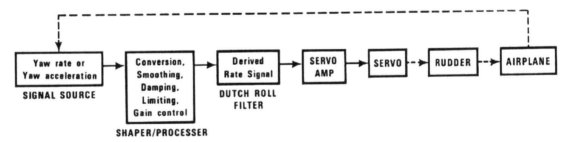

Simplified Yaw Damper Systems

Figure 20-5 shows a simplified yaw damper system using an accelerometer input to the computer, whose amplifier drives an electric motor clutched into the same cable system used by the cockpit rudder pedals to operate the rudder.

In this case, the servo motor operates not only the rudder, but also the rudder pedals since it is clutched into the rudder pedal cable system. The servo motor uses a sine synchro for follow-up to the computer. An airspeed signal from the central air data computer regulates the gain of the servo amplifier according to airspeed.

There may be an input from the roll channel of the autopilot for additional control during certain modes of operation of the autopilot.

Figure 20-6 shows a yaw damper system using a rate gyro input to the computer, whose output goes to the transfer valve of the hydraulic actuator controlling the rudder.

In this case, the actuator is of the type shown in Figure 19-7. This is a series yaw damper system; the yaw damper causes no motion of the rudder pedals. The follow-up signal is from the autopilot actuator LVDT. Gain control reduces the sensitivity at high airspeeds. Roll crossfeed may be present.

Figure 20-5

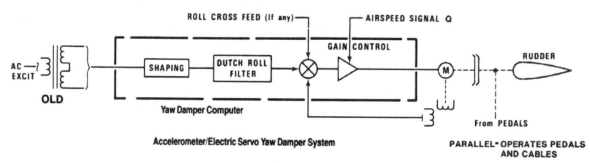

Accelerometer/Electric Servo Yaw Damper System

Figure 20-6

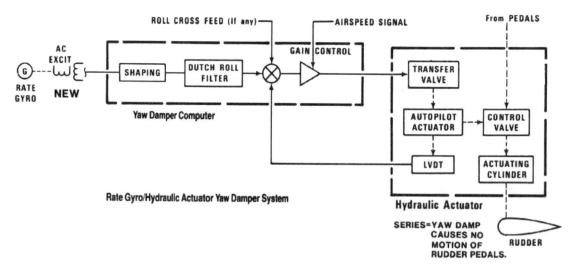

Rate Gyro/Hydraulic Actuator Yaw Damper System

Yaw Damper System – Late Model Aircraft

Using inputs from the air data computers (ADC's) and the inertial reference units (IRUs) the system commands the rudders to damp undesired yaw and to provide turn coordination.

The yaw damper modules use inputs from the air data computers and the inertial reference computers to derive rudder commands appropriate to flight conditions. These commands go to the yaw damper servos. Additionally, the modules monitor system operation and provide both manually initiated and automatic system testing (Figure 20-10).

The modal accelerometers supplement the yaw damp system by recognizing the frequency and phase of the body bending. Gust induced aft body lateral oscillations in long slender airplanes can affect comfort and ride quality in the rear section of the passenger cabin. The Boeing Company, used yaw damping to suppress lateral oscillations and control body bending modes. The U.S. Patent replaced the rate gyro with an accelerometer for detection of lateral motion, but does not fully compensate for body bending modes when airplane weight differs from the Design Gross Weight (DGW) due to fuel consumption. It is because of this deficiency that Boeing supplemented yaw damp accelerometers with modal accelerometers. Typically yaw damp accelerometers are located within the Inertial Reference Units and modal accelerometers are located in the aft fuselage area of the aircraft.

The yaw damper servos use electrical commands from the yaw damper modules to control hydraulic flow to an actuator piston. This motion is linked in series with manual and autopilot inputs to the rudder power control actuators. Maximum authority is typically + or - 6 with both channels active.

In the illustrated application, yaw damper and autopilot or manual rudder commands use different actuators and are mechanically mixed as is typical of the Boeing 757 and 767 aircraft. In other applications such as in the Boeing 737 and 747, the summing of the yaw damp and autopilot or manual inputs take place in a single Power Control Unit (PCU) package as is the case in Figure 19-10.

Figure 20-7

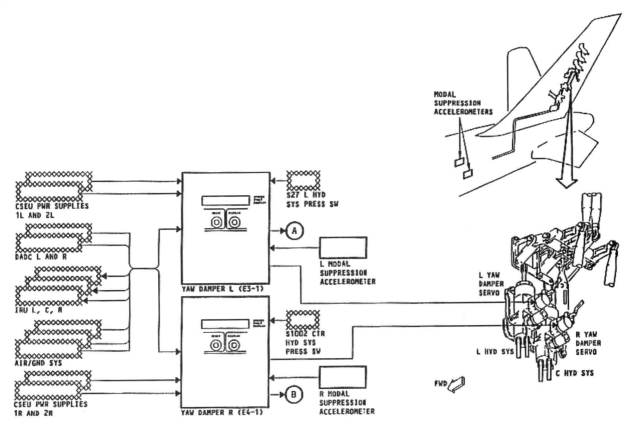

CHAPTER 21 - AUTOMATIC FLIGHT GUIDANCE SYSTEMS

Autopilot Control Panels No. 1

The number and types of autopilot modes of operation vary considerably among the different autopilot systems. All have certain basics in common, however. Before going into the autopilot systems, their inputs, outputs, controls, and signals it will be useful to look at some typical cockpit control panels and describe the general operation of autopilots.

Autopilot manufacturers are usually called upon to be able to supply, to different airline users, a certain amount of variety in modes of operation and types of controls. In recent years, the various airlines have saved themselves some money by agreeing beforehand to limit the number of variations that they impose on the manufacturer prior to the introduction of new airplanes.

In earlier airplanes, the control panels were typically mounted on a pedestal, whereas later air-craft have their autopilot control panels, flight director control panels, and VHF navigation frequency select panels just below the glare shield.

Figure 21-1 shows an autopilot control panel with a limited number of modes available. The pitch selector, on the left, is a rotary switch spring-loaded to the "Vertical Speed" position. For it to stay in the "pitch damp" position, a holding solenoid must be powered.

Other positions of the switch might have provided for "Indicated Airspeed Hold" or "Mach Hold". If the pitch channel of this autopilot is not in altitude hold, glide slope capture, or pitch damp, then it will be in vertical speed hold mode. In vertical speed hold mode, the autopilot maintains a constant rate of climb or rate of descent.

Pitch damp mode is intended for use during turbulence. In this mode, the pitch channel action is softened and the autopilot does not fight pitch attitude changes too hard.

The pitch wheel to the right of the pitch selector is detented in the "Altitude Hold" position (as shown).

If the autopilot, with the wheel in this position, has not captured glide path and is not in pitch damp mode, the pitch channel will maintain the altitude the airplane had at the time altitude hold mode was engaged.

Moving the pitch wheel forward by a small amount causes the autopilot to go to vertical speed mode and hold a small constant rate of descent. Moving the wheel farther forward increases the rate of descent. Moving the wheel aft causes the airplane to maintain a constant selected rate of climb.

The turn knob is detented in the position shown. Moving the turn knob all the way over to the right causes the airplane to take up a bank attitude of 25° or 30° right wing down. The turn knob is not spring-loaded to detent, and as long as it is left out of detent the airplane maintains a constant bank attitude.

If the turn knob were moved about half way toward a stop, the bank attitude would be 12° or 13°. If the autopilot is not in heading select mode and has not captured the localizer or VOR, then it is in "Heading Hold" mode if the turn knob is in detent. In that mode, it maintains whatever heading it had when the airplane leveled out with the turn knob in detent. The "Nav Selector" will be in "Turn Knob" position during heading hold mode, or when the turn knob is out of detent.

The modes during which the pitch wheel and/or turn knob are used are called "manual" modes, since the flight crew manually calls for certain pitch channel and/or roll channel operations.

The engage switch to the right of the turn knob is spring-loaded to the "off" position. It is locked in the off position with a locking pin which must be released with an electric solenoid. For the solenoid to be powered, much of the engage interlock circuit of the autopilot must be satisfied. If the engage interlock circuit is satisfied, the engage lever can be lifted.

Continued on next page ...

Autopilot Control Panels No. 1 (cont'd.)

If it is lifted to the "Yaw Damper" position and the yaw damper engage interlock circuit is satisfied, it will remain in that position, held there by the same solenoid, as long as the interlock circuit is complete. With the lever in the yaw damper position, only the yaw damper (rudder channel) is operating; the pitch and roll channels are not engaged.

Moving the lever to the autopilot position engages the yaw damper as well as the pitch and roll channel. For the lever to stay in the autopilot position, the autopilot interlock circuit must be complete. For the interlock circuit to be complete, the monitors of many peripheral systems must be satisfied.

For example, the vertical gyro monitoring must show a good vertical gyro; the directional gyro and compass system monitors must show a good compass system; servos must be clutched in; and the autopilot monitors must be satisfied.

As a rule of thumb, peripheral systems which are in use must be valid as indicated by their own

monitors, and autopilot functions must be valid as indicated by autopilot monitors. In general, a failure which would invalidate an autopilot in a particular mode will either disengage the autopilot or cause a warning to appear.

The rotary navigation selector switch on the right is spring-loaded to the "Turn Knob" position, and will return there whenever the turn knob is moved out of detent.

Moving it left to the "Heading Select" position causes the roll channel to maintain the heading selected with the heading select knob. With the switch in the "LOC/VOR" position, the roll channel can capture and follow either a localizer beam or VOR radial. With it in the "ILS" position, the roll channel can capture and follow the localizer beam, and the pitch channel can capture and follow the glide path.

Continued on next page ...

Figure 21-1

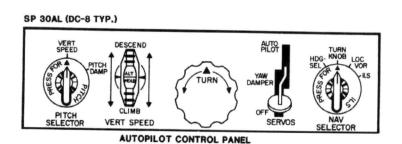

SP 30AL (DC-8 TYP.)

AUTOPILOT CONTROL PANEL

Figure 21-2

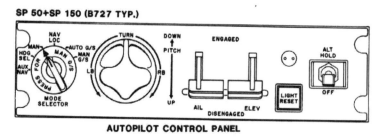

SP 50+SP 150 (B727 TYP.)

AUTOPILOT CONTROL PANEL

Figure 21-3

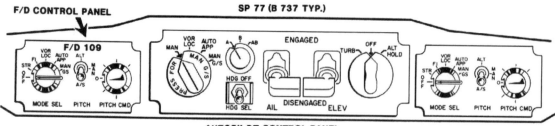

SP 77 (B 737 TYP.)

AUTOPILOT CONTROL PANEL

Autopilot Control Panels No 1 (cont'd.)

Figure 21-2 shows an autopilot control panel with a mode selector on the left. This mode selector is spring-loaded to the "Manual" position. With the mode selector in this position (autopilot engaged), the manual control is with the turn knob (on the right).

This turn knob is a combination turn knob and pitch knob. Moving the turn knob clockwise or counterclockwise causes the airplane to take up the desired bank attitude. The turn knob is spring-loaded to neutral for the pitch function. Moving the knob forward would cause the airplane attitude to change in the nose down direction. Moving the knob farther forward would cause the change to occur more rapidly.

As long as the knob is held out of pitch detent, the pitch attitude will continue to change, up to the limits imposed by the autopilot, typically 10° nose down and 15° nose up.

The "Heading Select" position of the mode selector functions as previously described. The "AUX NAV" position could be for INS control, Doppler control, or some other desired control. The NAV/LOC position would have been better named VOR/LOC, since it typically provides either capturing and holding a localizer beam, or capturing and holding a VOR beam.

The "Auto G/S" position is the same as the ILS position previously described. In that position, the roll channel can capture and hold the localizer beam, and the pitch channel can capture and hold the glide path. "Manual GIS" position probably has very little use, since all it does is to force capture of glide path.

This panel does not include the yaw damper engage control. The yaw dampener engage switches are separately located.

The engage levers are locked in the "off" position until interlock circuitry is satisfied, and the solenoid is held in the "engaged" position.

The square marked "Light Reset" is an autopilot disengage warning light. It will flash red if the autopilot is disengaged by a means other than with the disengage switches located on the control wheel. "Reset" means that the light can be turned off by pushing on the light itself.

The altitude hold switch is spring-loaded to "off", held in "Altitude Hold" position by a solenoid. If an autopilot is in altitude hold mode prior to capture of glide path, the altitude hold mode is automatically disengaged at the time of glide path capture. The same would apply for vertical speed hold, pitch attitude hold, indicated airspeed hold, or Mach hold.

Figure 21-3 shows autopilot and flight director control panels mounted below the glare shield. In this case, the two flight directors are separate systems, not involved with autopilot. The flight director systems are always used as monitors of autopilot performance during a Category II approach.

On the flight director mode selector switch "STR" stands for "Steering" mode, a special heading select mode during which the miniature runway of the attitude director indicator is used for flight director steering commands.

Continued on next page ...

Autopilot Control Panels No. 1 (cont'd.)

"FI" stands for "Flight Instruments", which is a fancy name for heading select. VOR/LOC is the same as the autopilot VOR/LOC mode. "Auto Approach" is "ILS" or "auto glide slope" mode described for the autopilots. "Manual Glide Slope" is another probably unused position forcing capture of the glide path.

The pitch switch is spring-loaded to the center "manual" position, solenoid held in "ALT" or "A/S" position. "ALT" mean altitude hold. "A/S" means airspeed hold.

The pitch command knob on the right would be used only during "Manual Pitch" mode and would select a desired pitch attitude within the limits of 15° nose up and 10° nose down.

The autopilot mode select switch presents nothing new.

The rotary switch marked "A", "B," and "AB" is used in two different ways. If the airplane has two autopilots, the switch will select either "A" autopilot,

or "B" autopilot, or both. Both would be selected for an automatic landing. If the air-plane has only one autopilot, the "AB" position is not used; position "A" selects a left elevator actuator for the pitch channel; position "B" selects a right elevator actuator for the pitch channel.

The switch marked "Heading Off" and "Heading Select" engages heading select mode for the roll channel, or eliminates heading control for the roll channel in manual mode.

The engage lever functions as previously described.

The pitch mode select switch is spring-loaded to "off". "Turbulence" compares with "Pitch Damp" in Figure 21-1. "Altitude Hold" mode operations are the same as previously described.

A separate pedestal-mounted panel holds the turn knob and pitch wheel, unless "control wheel steering" (discussed later) is used.

Figure 21-1

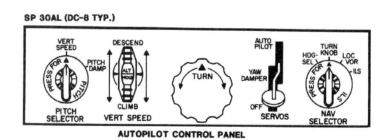

Figure 21-2

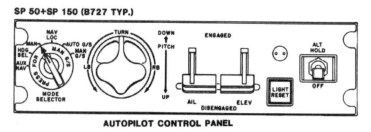

Figure 21-3

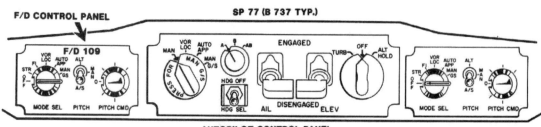

Autopilot Control Panels No. 2

Figure 21-4 shows a typical version of the B-747-100 autopilot panel (under the glare shield). Figure 21-5 is used on the pedestal. The flight directors could have had a "manual pitch" attitude select wheel near their on/off switches.

The autopilot computers are also the flight director computers, as is typical in late aircraft.

The auto-throttle system (A/T) is independent of the autopilot, but can be used in conjunction with it in an automatic landing. If the auto-throttle system is turned on, it will operate the throttles to maintain the airspeed selected with the knob and digital readout.

Only two autopilot engage levers are shown because only two of the three sets of computers are used as autopilots. Two are all that are required for automatic landing in Category II conditions, but there is space for a third engage lever for a third autopilot (to provide the airplane with Category III capability).

The engage levers are locked in the "off" position until considerable interlock circuitry is satisfied. If the lever is raised to the manual position, then only turn knob control and heading hold mode are available to the roll channel. Pitch wheel control for pitch attitude hold mode are used when the engage lever is in the "manual" position, unless in altitude hold mode.

If the autopilot has "Control Wheel Steering" (purchase option), the controller panel is eliminated, and the mid position of the engage lever is marked "CWS" for control wheel steering function.

With the autopilot in control wheel steering mode, applying a force in excess of four pounds or so to the control wheel causes the airplane to bank, and increases the bank, as long as the force remains on the control wheel. Releasing the control wheel then causes the autopilot to hold the bank angle achieved while the control wheel was being turned.

Moving the control column forward or aft causes the autopilot to change the pitch attitude of the airplane as long as there is a force applied to the control column. When the control column is released by the pilot, the autopilot maintains the pitch attitude achieved during the time the control column had a force applied to it.

The center of the panel shows two course select knobs and one heading select knob with readouts for each. Their compass operations are explained in

Figures 6-28 and 6-29. A course select knob also selects a pair of VOR radials if a VOR frequency has been selected. The left course select knob controls the inner mask and course select cursor on the pilot's HSI, and the right course select knob controls these items on the first officer's HSI. The heading select knob controls the heading select bugs in both instruments.

The switch marked "No. 1/Dual/No. 2" determines which navigation system and compass system will be used by the autopilots. If the switch is in "Dual" position, autopilot A will use No. 1 systems and autopilot B will use No. 2 systems. With the switch in either No. 1 or No. 2 position, both A and B autopilots will use the systems selected.

The mode select switch in the center of the panel controls the mode of both autopilots and all flight directors. It is not spring-loaded to any position. If an autopilot is in manual mode, however (only one autopilot can be used in manual mode at a time), then only the flight director mode is controlled by the mode select switch. Autopilots and flight directors can be used independently or simultaneously.

The switch marked "B/B" is a "Back Beam" switch. Back beam is a flight director function only, and can be used only with the mode select switch in the VOR/LOC position with an ILS frequency selected.

When this switch is moved to the "Back Beam" position with preconditions fulfilled, it disengages the autopilots, leaving only the flight directors effective. Switching to back beam mode makes it possible for the flight directors to command, following the localizer when coming in from the back side of an ILS runway.

The rotary pitch select switch on the right could have had more modes. It is spring-loaded to the "off" position.

The altitude select knob pre-selects a desired flight altitude, which shows in the digital readout window above it. If the select switch to the right is moved to "Altitude Select" position, the airplane can automatically capture the pre-selected altitude and hold it after capture. This three-position switch is spring-loaded to "off", solenoid held in "Altitude Hold" or "Altitude Select" position.

Continued on next page ...

Autopilot Control Panels No. 2 (cont'd.)

The two circles above and below the switch are dim green "on" indicating lights. The same type of light is above the "Back Beam" switch and the auto throttle "on" switch.

For the airplane to make an automatic landing or a Category II approach, both autopilots must be engaged and operative, and two flight directors must be operative.

The mode select panel shown in Figure 21-6 is from a DC-10. This autopilot has a mode annunciation panel (not illustrated) which indicates practically all of the modes and sub-modes of the autopilot.

The flight director on/off switches are on the extreme right and left ends of the panel. The two squares marked "Back Course" are illuminated push button switches. "Back Course" means the same thing as "Back Beam". The rectangle marked "Turbulence" is also an illuminated push button switch.

The VHF navigation frequency, including DME control, is incorporated into this panel, No. 1 on the left and No. 2 on the right.

The course select knobs perform the same function as they do in Figure 21-4.

The third section in from the left is labeled "ATS" for "Auto-Throttle System". Two are provided, one engage lever shown "on", and one "oil". "NI" button provides auto-throttle engine speed control, and the

knob with the triangle provides auto-throttle aircraft speed control.

The fourth panel in from the left has a heading select knob which functions the same as in Figure 21-4. It is also a push in/push out switch which controls heading hold and heading select modes. An outside concentric ring knob selects the bank angle limit for these heading modes. VOR/LOC, ILS, and land modes are push button selections.

The pitch wheel on the third panel from the right controls the vertical speed and altitude hold modes. An altitude select knob pre-selects a desired altitude, as in Figure 21-4.

Indicated airspeed hold, Mach hold, and CWS modes are push button selections. The push button for CWS would be used only with autopilot engage lever in the "Command" position, because if an autopilot engage lever is in the mid position, marked "CWS", then both the pitch and the roll channel are in control wheel steering mode only.

Two engage levers are shown, one for autopilot No. 1, one for autopilot No. 2. In the "ILS" and "Land" modes of operation, each autopilot has two channels, A and B, and two separate sets of hydraulic actuators which function separately, but simultaneously. The selection of one autopilot, which actually, for ILS, selects two (A and B), is sufficient for an automatic landing. A Category III landing could only be made with all four autopilots engaged and operative.

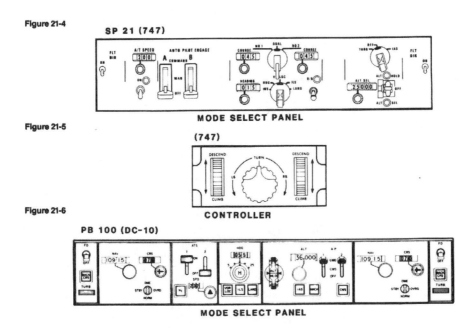

Figure 21-4

SP 21 (747)

MODE SELECT PANEL

Figure 21-5

(747)

CONTROLLER

Figure 21-6

PB 100 (DC-10)

MODE SELECT PANEL

Autopilot Control Panels No. 3

Figure 21-6a shows a typical B-757 / 767 autopilot panel. The selectable functions are:

F/D ON

Allows display of flight director command bars on the associated ADI. Command bars are removed from both ADI's when all 3 autopilots engage below 1,500 feet radio altitude in the approach mode. On the ground, turning one switch on with no autopilots engaged activates the associated flight director takeoff mode. In flight, the first switch turned on engages the associated flight director in the current autopilot mode. "OFF" Removes command bars from the associated ADI.

A/T ARM Switch

"ON" Arms the autothrottle for engagement. The autothrottle engages when EPR or SPD switch is pushed and when the VNAV, FLCH or GA modes are active. With flight directors off and no autopilots engaged, autothrottle reverts to basic SPD mode. "OFF" disarms autothrottle, preventing engagement.

EPR Switch

Engages autothrottle to hold reference EPR displayed on EICAS, subject to maximum speed limits. EPR displayed on each ADI. Changes thrust reference from T/O to CLB if above 400 feet radio altitude.

SPD Switch

Engages autothrottle to hold speed or mach indicated in IAS/MACH indicator. SPD displayed on each ADI. Changes thrust reference from T/O to CLB if above 400 feet radio altitude. Causes pitch mode to change to V/S if changing altitudes in FL-CH or VNAV.

IAS/MACH Indicator

Indicates current or selected speed if VNAV is not engaged. Command airspeed bugs are driven to this value. Indicator is blank when VNAV is engaged and command airspeed bugs are under control of the flight management computers. Changes from IAS to MACH automatically at mach .80 in a climb, and from MACH to IAS at 300 knots in descent.

SEL Switch

Changes the IAS / MACH indicator alternately from IAS to MACH.

Speed Knob

Changes the value in the IAS/MACH indicator when rotated. When pushed with VNAV engaged, transfers control of speed alternately from the flight management computers to the speed knob or from the knob to the computers. IAS/MACH indicator indicates existing speed when knob is pushed to transfer speed control to the knob, and blanks when knob is pushed to transfer speed control to the flight management computers. Selecting manual speed control does not disengage VNAV mode.

FL CH Switch

Engages flight level change mode to integrate AFDS pitch control and autothrottle for altitude change. FL CH displayed on each ADI. Changes thrust reference from T/O to CLB if above 400 feet radio altitude. When mode engages, the command airspeed bug moves to existing speed and the AFDS holds this speed, or speed set in the IAS/MACH indicator after mode is engaged. Autothrottle holds selected thrust for climb, or idle for descent. When altitude in MCP altitude indicator is reached, pitch control changes to altitude hold and autothrottle holds commanded speed. FL CH disengages VNAV mode.

L NAV Switch

Arms or engages lateral navigation mode of FMS transferring roll and heading control of AFDS to the FMS. With LNAV armed and airplane not on a heading which will intercept the active leg, the FMC MSG lights come on and NOT ON INTERCEPT HEADING appears in the scratch pad of both CDUs. If the active route leg is within the turning radius of the airplane, LNAV annunciates green (engaged) on each ADI. AFDS then follows active route. L NAV mode is disengaged by selecting HDG HOLD or HDG SEL mode, or when localizer capture occurs.

Continued on next page ...

Figure 21-6a

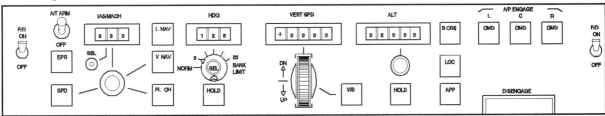

Autopilot Control Panels No. 3 (cont'd.)

V NAV Switch

Engages vertical navigation mode of FMS, transferring pitch and speed control of AFDS and autothrottle control to FMS if above 400 feet radio altitude. VNAV PTH or VNAV SPD displayed on each ADI. Changes thrust reference from T/O to CLB if above 400 feet radio altitude. Mode causes AFDS to fly vertical profile through climb, cruise, and descent as determined by FMS and modified by the pilots. V NAV mode can be terminated by any of the following:

- Selecting V/S or altitude hold
- Selecting SPD or EPR
- Selecting FL CH
- Capturing glide slope
- Reaching altitude in MCP altitude indicator before reaching FMS target altitude.
- Selecting GA mode

VNAV remains engaged when manual speed control is selected with the speed knob and FMS uses pilot selected speed for speed control. VNAV cannot be engaged after takeoff while the thrust mode is still TO below 400 feet radio altitude.

HDG Indicator

Indicates magnetic heading selected with heading select (SEL) knob if localizer has not been captured. Indicates selected localizer course after localizer capture. Heading bugs on HSI's are driven to the heading in indicator.

Heading SEL Knob

Changes the heading in the HDG indicator when rotated. Engages heading select mode when pushed. HDG SEL displayed on each ADI. Heading select mode causes AFDS to turn to and the hold heading in HDG indicator window.

BANK LIMIT Knob

AUTO - Limits bank angle for AFDS in heading select mode to 15 degrees at true airspeed above 250 knots, increasing progressively to 25 degrees, as true airspeed decreases below 250 knots to 200 knots. Limits bank angle to 25 degrees below 200 knots. The 5, 10, 15, 20 and 25 limits bank angle for AFDS to the selected value in heading select mode regardless of airspeed. It will not limit bank in LNAV.

Heading HOLD Switch

Engages heading hold mode manually. HDG HOLD displayed on each ADI. Heading hold mode causes AFDS to roll out and/or hold the magnetic heading existing at wings level. This mode engages automatically in flight when first flight director is turned on, or first autopilot is engaged in command. Heading hold is basic roll mode for flight directors and autopilot if no other roll mode is engaged.

VERT SPD Indicator

Indicates existing vertical speed when the V/S switch is pushed. Indicates vertical speed selected with vertical speed selector when mode is engaged. Range is -8000 to +6000 fpm in 100 fpm increments.

Vertical Speed Selector

Changes value in the VERT SPD indicator.

V/S Switch

Engages vertical speed hold mode. V/S displayed on each ADI. The AFDS maintains the vertical speed displayed in the VERT SPD window regardless of the resulting airspeed. Vertical speed is the basic pitch mode for AFDS if no other pitch mode is engaged. Engaging vertical speed mode from FL CH or VNAV mode causes autothrottle to engage in speed mode.

ALT Indicator

Indicates altitude selected with altitude knob. Establishes target and limit altitude for all AFDS, FMS and altitude alert functions. AFDS and FMS cannot fly through indicated altitude climbing or descending, and cannot depart indicated altitude except in vertical speed mode.

Continued on next page ...

Figure 21-6a

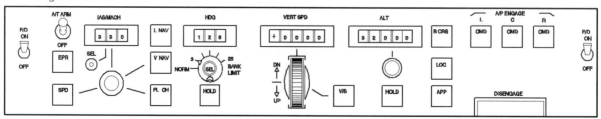

Autopilot Control Panels No. 3 (cont'd.)

Altitude Knob
Changes altitude set in the altitude window.

Altitude HOLD Switch
Engages altitude hold mode manually. ALT HOLD is displayed on each ADI. Causes the AFDS to capture and hold altitude existing at time switch is pushed.

B/CRS Switch
Arms the AFDS to capture and track inbound on the localizer back course and outbound on front course, if pushed before localizer capture. B/CRS displayed in white on each ADI prior to back course capture. It can be used only in conjunction with the LOC switch, and before localizer capture. It does not provide glideslope tracking (glideslope pointer is removed from ADI). If the LOC mode is captured before the B/CRS switch is pushed, the AFDS tracks the localizer front course and B/CRS cannot be selected. B/CRS mode is a single autopilot mode, therefore multiple autopilots cannot be engaged with this mode.

LOC Switch
Arms or engages the AFDS to capture and track the localizer front course. LOC displayed in white on each ADI prior to localizer capture. Capture does not occur if intercept angle is greater than 120 degrees. The previously engaged LNAV, heading select, or heading hold modes remain engaged until localizer capture. LOC mode is a single autopilot mode. Multiple autopilots cannot be engaged while in this mode.

APP Switch
Arms or engages flight directors and autopilots to capture and track localizer and glide slope. LOC and G/S displayed in white on each ADI prior to localizer and glide slope capture. Glide slope capture does not occur if intercept track angle is not within 120 degrees of the localizer course. Allows 2 or 3 autopilots to be engaged for dual or triple channel autoland and rollout.

CMD Switches
Engages the associated autopilot in vertical speed and heading hold modes if neither flight director is on, or if either flight director is in the takeoff or go-around mode. CMD displayed on each ADI. If either flight director is in any other mode, autopilot engages in the same mode. Engaging the autopilot removes the flight director command bars from ADI of the pilot who has selected the same FCC as the flight director source. Pushing the second autopilot CMD switch when in the APP mode arms the autoland system to engage the second autopilot below 1,500 feet radio altitude and arms the autoland functions of runway alignment, flare, landing and rollout. Pushing the third autopilot CMD switch when in the APP mode isolates the 3 electrical power source channels at any altitude, and arms the autoland system to engage the second and third autopilots and arm the autoland functions of runway alignment, flare, landing and rollout when below 1,500 feet radio altitude. LOC and G/S must be captured by each FCC for multi-channel engagement to occur.

DISENGAGE BAR
UP allows autopilots to be engaged. DOWN - Disconnects all 3 autopilots from The flight control servos preventing engagement of autopilots. The red strip is exposed when bar is down.

The Boeing 777 panel differs only very slightly from that of the Boeing 757 / 767 (Figure 21-6b).

One addition is the CLB CON feature which when two engines are in operation changes the autothrottle thrust reference to climb (CLB). When only one engine is in operation, it changes the thrust reference to maximum continuous (CON).

A second more significant difference the Autopilot engage switches The Boeing 777 places a single engage button at each end of the panel. Either switch will engage all available A/P channels.

Figure 21-6b

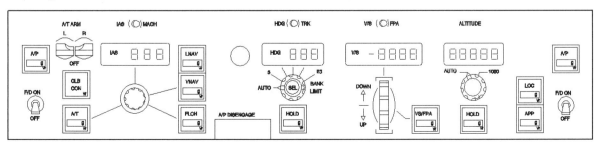

Operation Of Autopilot Force Limiting

Figures 21-8 and 21-9 illustrate why autopilots are always force limited. If an autopilot without force limiting were to fail in such a manner as to provide a hard over signal, the pilot might be hard pressed to manually overcome it, and the airplane might be put into a violent maneuver.

Figure 21-7 shows the possible difference in control column movement between a hard over signal with limiting and without limiting.

Force limiting is provided not only in the pitch channel, but also in the roll and yaw channels. The maximum output of an autopilot servo amplifier is limited with power source limiting resistors or other internal provisions.

In addition, the mechanical force is limited mechanically, downstream of the autopilot servo. Another safeguard is an autopilot disconnect switch located on the outboard horn of the control wheel, handy to the pilot's thumb.

Figure 21-7

Figure 21-8

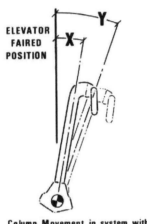

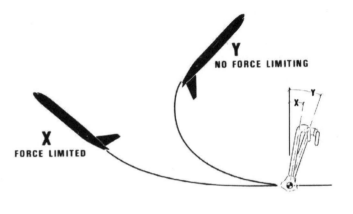

Figure 21-9

Typical Autopilot Modes

Figure 21-10 itemizes most of the typical autopilot modes. These also could be flight director modes, with the exception of control wheel steering and heading hold, in those aircraft which have common computers for flight director and autopilot.

Figure 21-10

PITCH CHANNEL

Pitch Attitude Hold

Turbulence

Vertical Speed Hold (Modern Default)

Glide Path Engaged (Captured)

Airspeed Hold (IAS or Mach)

Altitude Hold

Pre-Selected Altitude Captured

Control Wheel Steering

Auto Land

V NAV

Flight Level Change

ROLL CHANNEL

Roll Attitude Hold

Turbulence

Heading Hold (Modern Default)

Heading Select

Turn Knob

VOR Captured (On Course)

LOC Captured (On Course)

INS Captured (On Course)

L NAV

Control Wheel Steering

Auto Land

CHAPTER 22 - AUTOPILOT/FLIGHT DIRECTOR ROLL CHANNEL

Roll No.1: Attitude Stabilization

Figures 22-1 through 22-15 present a series of segmented autopilot roll channel operations. These illustrate one or two functions at a time in order to gradually build up to a more nearly complete roll channel.

Figure 22-1 shows the synchronizing action of a computer card servo motor loop prior to engaging the autopilot, and the autopilot maintaining the existing bank angle at the time of engage.

Remember that in all of these prints the indicated switch action will be to the triangle when the named condition exists. For example, when the autopilot is engaged, the two switches shown will move to the triangles.

While the autopilot is disengaged, the servo amp and servo motor are operative. Roll attitude information from the vertical gyro presents itself as a resultant field in the stator of the control synchro in the computer. The active servo motor loop maintains the control synchro rotor at a position perpendicular to the resultant field in the stator.

Therefore, at any time prior to engage, the position of the control synchro rotor is a function of the bank angle of the airplane. If the position shown represents wings level, then when the airplane banks to the right, the control synchro rotor will turn a corresponding number of degrees clockwise (prior to engage).

To visualize this on the schematic, suppose the airplane is banked 20° to the right. The vertical gyro transmit synchro rotor would have to move 20° clockwise with respect to its stator. The resultant field in the control synchro rotor is correspondingly moved 20° clockwise. The servo motor loop causes the control synchro rotor to follow the field, and the rotor also is turned 20° clockwise.

Prior to engage, the solenoid-operated valve in the aileron power unit (not indicated here) would be closed. The transfer valve would have no hydraulic power, and the autopilot actuator LVDT would give a null signal.

When the autopilot is engaged, the output of the servo amp is disconnected from the servo motor. The fixed field of the servo motor (not drawn) keeps the servo motor braked, and any output from the now fixed control synchro rotor is fed to the transfer valve amplifier. At engage, the solenoid operated valve in the aileron power unit is opened, the transfer valve has hydraulic power and the ability to move the ailerons. At engage, the computer control synchro rotor has a null output because it has been kept that way by the synchronizing operation of the servo loop.

Since the airplane has a 20° right wing down attitude at engage, the autopilot will maintain that roll attitude. If the right wing rises, the vertical gyro moves the field in the control synchro stator, developing a not null signal of one phase in the control synchro rotor. This causes the transfer valve amplifier to operate the transfer valve to move the ailerons until the bank angle is restored to 20°.

Any change in airplane bank attitude after autopilot engage causes the vertical gyro transmit synchro to move the field in the control synchro, developing a not null signal in the control synchro rotor. The phase of the signal developed will cause the ailerons to operate in the direction required to restore the roll attitude existing at the time of engage.

It is the position of the control synchro rotor which determines airplane roll attitude.

Continued on next page ...

Roll No. 1: Attitude Stabilization (cont'd.)

Figure 22-1 illustrates the principle involved in the roll attitude hold mode. This is used in conjunction with control wheel steering.

Figure 22-2 illustrates an arrangement whereby the autopilot, at the time of engage, causes the airplane to come to a wings level condition if it is not already holding that attitude. The condition prior \to engage is the same as in Figure 20-1 prior to engage, with the servo motor loop holding the control synchro rotor perpendicular to the field in the stators, and therefore at a position corresponding to the airplane bank angle.

An additional synchro utilizing a sine winding is driven by the servo so that, when the airplane wings are level, the output of the stator is a null. The two synchro rotors are on a common shaft, and turn degree for degree.

If, prior to engage, the airplane has a 20° right wing down attitude, a 20° signal of one phase is developed and waiting in the stator of the sine synchro. At engage, that 20° signal is switched into the servo amp and, since it is the only input signal, is of the phase which causes the servo motor to drive

both synchro rotors to the wings level position. That is to say that the servo motor drives until the sine synchro signal is null.

At the time of engage, when the right wing was down 20°, the control synchro rotor began to move toward the wings level position. As soon as it began to move, it developed an error signal because it was no longer perpendicular to its field. The error signal was of the phase which caused the ailerons to operate and roll the airplane toward wings level. As the airplane rolled toward wings level, the vertical gyro transmit synchro rotor moved toward wings level, causing the field in the control synchro in the computer to follow the motion of the rotor.

There is a constant lag between the motion of the control synchro rotor (driven by the servo motor) and the resulting motion of the field in the stator (caused by change in the airplane bank attitude). This lag is necessary for a command signal to the transfer valve amplifier to hold the ailerons displaced, causing airplane bank attitude to change.

Continued on next page ...

Figure 22-1

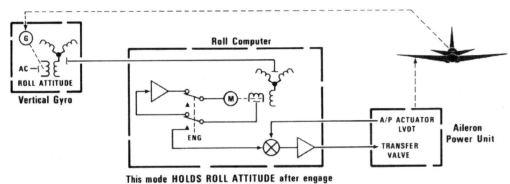

This mode HOLDS ROLL ATTITUDE after engage

Figure 22-2

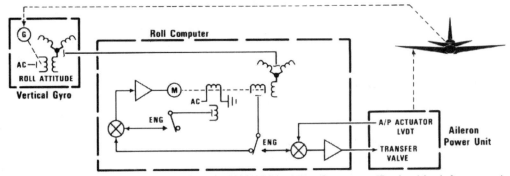

This mode levels airplane, if not level, and MAINTAINS WINGS LEVEL after engage (Synchronizing before engage)

Roll No. 1: Attitude Stabilization (cont'd.)

The control synchro rotor output continues until both it and the airplane attitude are in wings level position. The field is then perpendicular to the rotor. There is no input to the transfer valve amplifier, the ailerons are not displaced, and the airplane holds its wings level attitude.

If the airplane deviates from wings level attitude, the vertical gyro transmit synchro moves the control synchro field away from perpendicular to its rotor, developing an error signal which causes the transfer valve to return the airplane to wings level attitude. Again, it is the position of the control synchro rotor which determines airplane roll attitude.

Thousands of autopilots do use computer card mounted servo motor loops, driving control synchro rotors whose position does control airplane attitude in the manner described. Advances in solid state technology have made available miniaturized high density multiple unit components, which make their use in late model autopilots desirable. They are

sufficiently light so that large numbers can be used to replace almost all relays and mechanical devices such as the servo motor loop.

Nevertheless, the effect and operation of the solid state circuits is the same as described, so far as the line mechanic is concerned, because they perform the same functions as the mechanical devices they replace. The concepts, relatively easily grasped, of the operations of the servo motor loops, are very useful in understanding the late model autopilots which accomplish the same results with solid state devices.

It really does not matter to the line mechanic whether his roll computer has relays, vacuum tubes, and servo motors, or nothing but solid state devices. The important thing is for him to understand the general operation of the computer.

Following illustrations further illustrate the function of a roll computer servo motor loop.

Figure 22-1

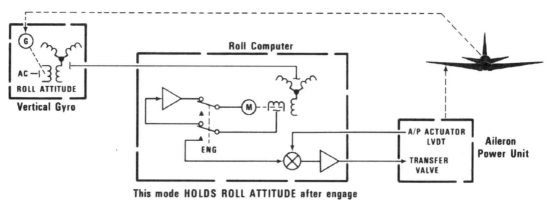

This mode HOLDS ROLL ATTITUDE after engage

Figure 22-2

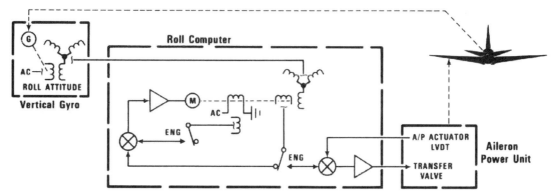

This mode levels airplane, if not level, and MAINTAINS WINGS LEVEL after engage (Synchronizing before engage)

Roll No 2: Turn Knob And Rate Gyro

Figure 22-3 introduces a rate gyro input and connects up the turn knob. Prior to engage, synchronizing of the control synchro rotor in the computer occurs just as in Figure 22-2. After engage, with the turn knob in detent supplying no signal from its potentiometer, the autopilot maintains the airplane in the wings level attitude.

The rate gyro supplies a signal only during the time the airplane roll attitude is changing. It is a roll rate gyro, which sees changes in roll attitude.

The more rapidly the roll attitude is changing, the larger the signal from the rate gyro. It is of one phase if the airplane is rolling to the right, and of the opposite phase if the airplane is rolling to the left. Its output is always connected into the computer servo amplifier, providing an inverse feedback signal. Its

signal always opposes whatever roll action is taking place. It thereby provides a tendency to limit the rate of roll and helps to smooth the roll maneuver.

If the turn knob is moved out of detent clockwise, the phase of the signal from its potentiometer calls for right wing down.

The farther it is moved, the greater the bank angle called for. The servo motor drives until the follow-up signal from the sine synchro equals and cancels the turn knob command signal. This establishes the amount of rotation of the rotor of the control synchro, which in turn establishes the bank angle of the airplane.

Subsequent figures illustrate this operation. Typically, the turn knob outputs are shop set so as not to call for more than 25° or 30° of bank angle.

Figure 22-3

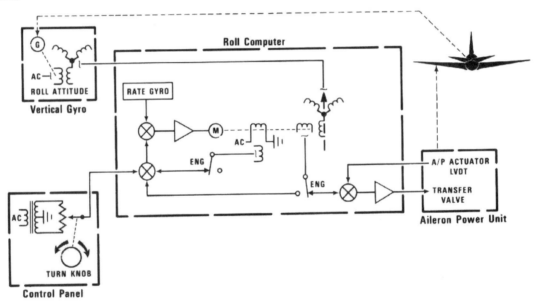

Without Turn Knob Command, this mode holds wings level

Roll No. 3: Turn Knob Banking Airplane

In Figure 22-4, we have moved the turn knob to its counterclockwise limit. The signal into the servo motor is causing the servo to drive its synchro rotors toward the 30° position. The sine winding of the resolver is providing some follow-up signal, which diminishes the turn knob command from the lower summing point.

The rate gyro is also providing an opposition signal, which further diminishes the input to the computer servo amplifier. There is still an input to the computer servo amplifier since it has not yet driven its rotors to the 30° position. The control synchro rotor is leading the movement of the field on its stator.

The movement of this field is caused by the changing bank angle of the airplane. The amount that the control synchro field lags the movement of the control synchro rotor is a function of aileron displacement, required to cause the airplane to follow the movement of the control synchro rotor.

The signal from the control synchro rotor into the summing point is being cancelled by the follow-up from the autopilot actuator LVDT, resulting in a null signal input to the transfer valve amplifier.

This means that the transfer valve has displaced the autopilot actuator far enough to displace the ailerons far enough to cause the airplane to roll at the same rate that the control synchro rotor is moving. Further displacement of the ailerons is not necessary (refer to Figure 17-6 for control surface actuator operation).

This condition illustrated in Figure 22-4 prevails until the computer servo motor has driven its synchro rotor 30° away from the wings level position. Following that, the airplane attitude continues to change until it achieves a 30° left wing down roll attitude, illustrated in Figure 22-5.

Figure 22-4

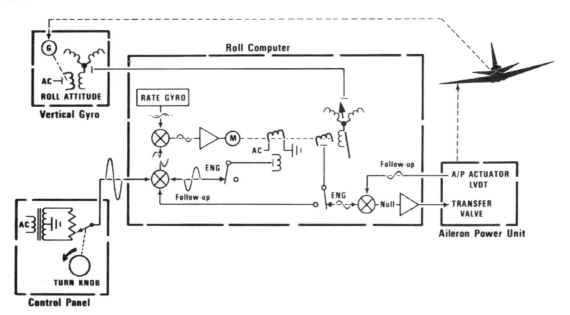

**With Turn Knob Command (MAXIMUM) computer servo motor turns
synchro rotors, forcing airplane to begin banking**

Roll No 4: Turn Knob Has Banked Airplane 30°

Figure 22-5 shows the airplane in a 30° left bank. The vertical gyro transmits this position, and has moved the field of the control synchro rotor 30° away from the wings level position, perpendicular to the control synchro rotor.

The follow-up signal from the computer resolver sine winding is a 30° signal canceling the turn knob input at the summing point (follow-up equals command). The output of that summing point is a null. The rate gyro sees a steady bank attitude and has no output, and there is no longer an input to the computer servo motor amplifier.

The synchro rotors have stopped in the 30° position. With no input to the transfer valve from its amplifier, the follow-up signal from the actuator LVDT has caused the transfer valve to drive its actuator to its neutral position, and the ailerons are therefore faired.

The last statement constitutes a small exaggeration of the fact. There is enough dihedral angle (wings moved upward from parallel to each other) to cause the airplane to slowly right itself if the ailerons are left in the faired position, but the amount of aileron displacement required to maintain a given bank angle is relatively small.

So there is, in fact, a small signal input to the transfer valve amplifier. This small signal is provided by a small difference between the airplane angle called for by the control synchro rotor and the actual airplane bank angle represented by the position of the field on the control synchro stator. Since it is such a small signal, we will ignore it in this and future illustrations.

Moving the turn knob back toward detent would result in the sine winding follow-up signal exceeding the turn knob input to the summing point. The output of that summing point to the computer servo amplifier would then be of the opposite phase from that illustrated in Figure 22-4.

This would cause the servo motor to run in the opposite direction, driving its synchro rotors back toward the wings level position. The airplane would then be forced to take up whatever bank angle is represented by the rotor of the control synchro.

The operation of the computer control synchro in controlling the transfer valve autopilot actuator and airplane bank angle, as described in Figures 22-1 through 22-5, are fundamental to the operation of the autopilot. Be sure you understand this operation before proceeding, because later autopilot operation explanations assume that you do understand it.

Figure 22-5

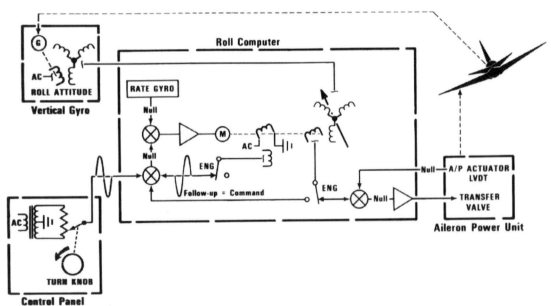

Airplane attitude has followed control synchro rotor (ailerons are not
deflected more than very slightly)

Roll No. 5: Limiters, Roll Accelerometers

Figure 22-6 introduces the bank angle limiter, the roll rate limiter, the function of the roll accelerometers, and a tachometer generator. Their operations are shown in the three illustrations following.

The function of the bank angle limiter is to limit any input signal to a value which can be cancelled by the resolver sine winding follow-up signal, as needed for that particular mode of operation. The maximum bank angle limit is usually 25° to 30°. The limiting action of the bank angle limiter will be changed automatically as modes change. For example, a typical a 20° bank angle limit is used in "VOR On Course" mode (discussed later).

When the output of the limiter is cancelled by the resolver sine winding, it means that the servo synchros are causing the maximum airplane bank angle allowed for that particular mode.

The maximum output of the roll rate limiter will cause the servo motor to drive its synchro rotors no faster than is desirable for a particular mode. The rate at which the control synchro rotor turns is the rate at which the airplane changes its bank angle.

Roll rate limiter outputs range from 1 1/2° per second to 7° per second.

These servo motors typically drive tachometer generators, whose output is fed back into the servo amplifier to effect roll rate limiting and smooth the roll maneuver.

The roll accelerometers, located near the upper and lower skin of the fuselage near the center of gravity, are mounted so that their armatures see only sideways movement. They are connected in phase opposition so that if the airplane moves straight sideways, without rolling, their outputs cancel. In that case, there is no roll attitude change. They do not cancel if there is a roll attitude change.

For example, if the airplane is banking to the left, the upper accelerometer sees a left movement and the lower accelerometer sees a right movement, and their outputs add. In the roll channel they provide the same function as the rate gyro illustrated in Figure 22-3. Since they provide an acceleration signal rather than a roll rate signal, that signal needs to be changed (by a filter) to a roll rate signal.

Figure 22-6

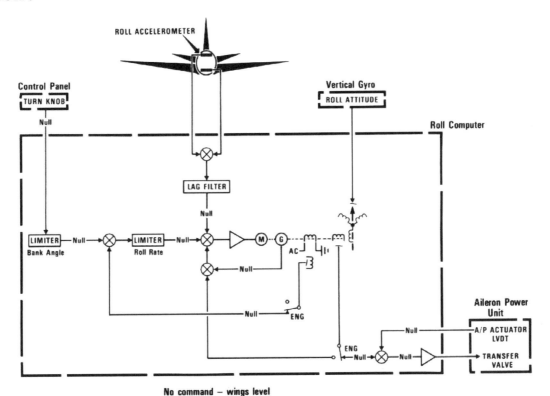

No command — wings level

Roll No. 6: Bank Angle And Rate Limiter Action (1)

Figures 22-7, 22-8, and 22-9 are used to illustrate the functions of the bank angle limiter, roll rate limiter, servo tachometer generator, and accelerometers. For simplification, we make an untrue assumption about the turn knob command signal. We assume that it can supply a signal greater than the maximum bank limiter output.

The turn knob signal is usually connected downstream of the bank angle limiter, since it is shop set to a desired maximum bank angle limit. For purposes of illustration, let's assume that the turn knob can call for a 45° bank angle.

In Figure 22-7, the condition represented is that the turn knob output has been suddenly introduced,

and the servo motor has not yet driven its synchro rotors away from the wings level position. The turn knob signal is reduced by the bank angle limiter to one calling for 30° left wing down. The roll rate limiter has further reduced the signal to one which will cause the servo motor to drive its synchro rotors no faster than (let us say) 5° per second.

The synchro rotors have not yet begun to move. Therefore, there is no signal into the transfer valve amplifier. The airplane attitude is not changing, so the accelerometers have no output. The servo motor tachometer generator also has no output. The airplane is wings level and ailerons are faired.

Figure 22-7

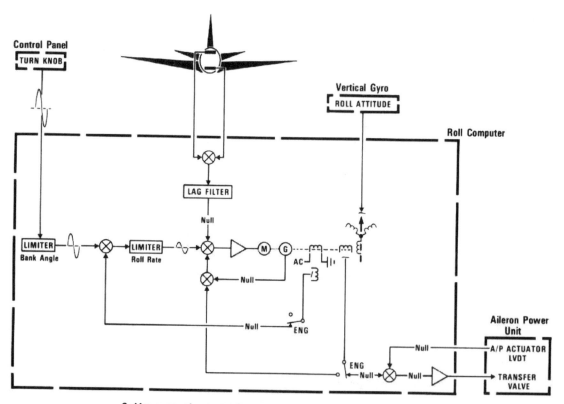

Sudden command — no synchro movement yet

Roll No 7: Bank Angle And Rate Limiter Action (2)

In Figure 22-8, the airplane is rolling at the desired rate of 5° per second. The accelerometers are providing an opposition signal. The servo motor tachometer generator is also providing an opposition signal, but nevertheless the synchro rotors are moving at 5° per second.

The airplane attitude, represented by the vector arrow on the stator of the control synchro, is lagging

the control synchro rotor by a couple of degrees, developing a signal into the summing point at the lower right of the roll computer.

The transfer valve amplifier has driven the autopilot actuator until its LVDT follow-up signal equals the control synchro rotor output, and cancels it. The input to bank angle limiter and its output remain constant.

Figure 22-8

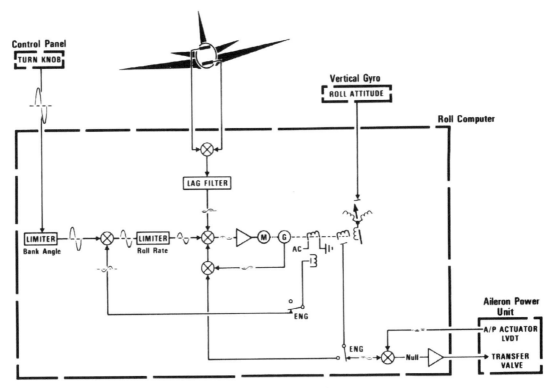

RATE LIMITER LIMITS SPEED of rotor movement. Airplane following rotor movement, but lagging, so ailerons are displaced.

Roll No 8: Bank Angle And Rate Limiter Action (3)

In Figure 22-9, the servo motor has driven the resolver rotor and the control synchro rotor to the 30° left wing down position. The output of the sine winding at this time cancels the output of the bank angle limiter.

There is no longer an input to the computer servo amplifier. There is no output from the accelerometers. The tachometer generator and motor have stopped running. The airplane bank angle equals the control synchro rotor displacement, and there is no output into the transfer valve amplifier.

The bank angle limiter has limited the maximum bank angle of the airplane.

Figure 22-9

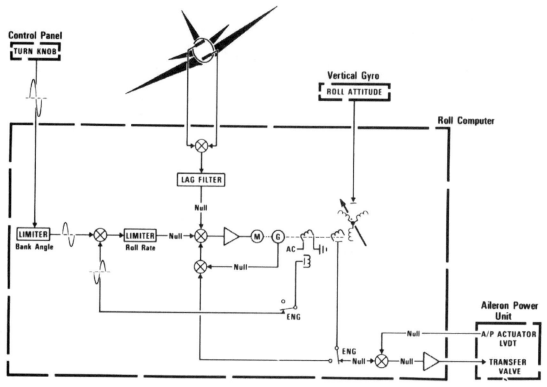

BANK ANGLE LIMITER determines Bank Angle

Lift Compensation For Bank Angle (Versine)

Figures 22-10 and 22-11 illustrate the need for "lift compensation" when the airplane is banked, and one way of developing this compensating signal.

The airplane in Figure 22-10 is holding its altitude. In such a case, the sum of the lift on both wings equals the weight of the airplane, so the airplane neither rises nor sinks. The lift is a force exerted perpendicular to the airplane wing surface.

Figure 22-11 shows the airplane in a left bank with the same forward speed, therefore, the same lift. Because the lift is not being exerted vertically, it is not capable of balancing the weight of the airplane. The airplane, without compensation, would begin to sink.

The vector triangle above the right wing of the airplane illustrates, with the dashed line, the amount of vertical lift that has been lost. The hypotenuse of the triangle represents the lift exerted perpendicular to the wings. The vertical side of the triangle represents the vertical component of lift. The angle on the bottom is the bank angle of 30°.

Since engineers are mathematically inclined, they don't hesitate to name the loss in vertical lift by its trigonometric function name, "versine". Unfortunately, the term has carried over into line mechanic operations also. The trigonometric function versine is equal to one (for unity) minus the cosine of the angle under consideration.

Referring to the vector triangle, you can see that the cosine of the angle under consideration is the adjacent side, namely, the vertical component of lift. That portion of lift which has been lost is a function of bank angle. It is unity minus the cosine, or versine. That value represents the amount of lift which needs to be added in some manner so that the airplane will not lose altitude.

The compensation is made by pitching the aircraft nose up to increase the angle of attack, and therefore the lift on the wings. If the autopilot is controlling the airplane, whenever the airplane banks, there will be an additional nose up signal to the pitch channel (versine) to make up for lost lift resulting from that particular bank angle. You will see it referred to both as "nose up lift compensation" and "versine". They mean the same thing as far as we are concerned.

Continued on next page ...

Lift Compensation For Bank Angle (Versine) (cont'd.)

The schematic in Figure 22-10 shows a common way of developing the required lift compensation voltage. The resolver is the one shown in the previous roll computer (but with its cosine windings added), driven by its servo motor (the sine winding develops the follow-up signal to the servo motor amplifier).

Illustrated is the wings level position of the resolver rotor. The sine winding signal is null and the cosine winding signal is maximum. The calibrated voltage fed into the summing point is a voltage equal but opposite in phase to the maximum output of the cosine winding. When the wings are level, the output of the summing point is a null, since no compensation is required when the wings are level.

Figure 22-11 illustrates the airplane in a 30° bank, and the resolver rotor rotated 30°. The voltage of the cosine winding has been diminished, so now the calibrated voltage predominates, producing output from the summing point. The amplitude of that voltage from the summing point is mathematically the versine of the bank angle, since it represents one minus the cosine. This voltage becomes an input to the pitch channel to raise the elevators the required amount to compensate for the loss of lift.

The amount of lift lost between wings level and a 5° bank angle is small, but the loss of lift between a 30° bank angle and a 35° bank angle is large. The versine signal always provides the required amount of nose up compensation, at least within the normal operating bank angles of commercial aircraft.

Figure 22-10

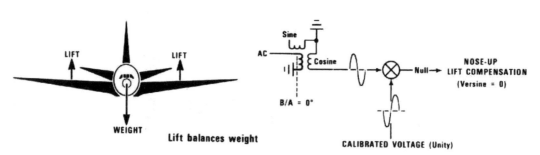

Figure 22-11

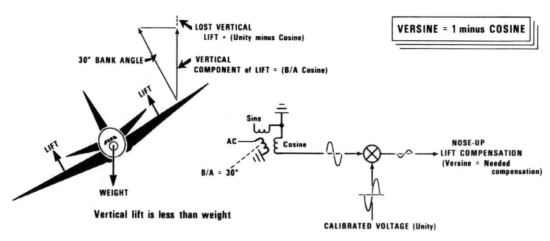

CWS Force Transducer

Figure 22-12 illustrates the principle of developing a "control wheel steering force" electrical signal. The control wheel steering "force transducer" is shown directly connected into the aileron control linkage so that it feels whatever force is applied to the control wheel.

The wavy outline of the transducer case indicates that its length can be changed. If it is pushed, the length will diminish. If it is pulled, the length will increase. The armature of the E-pickoff is attached to one end of the case and the core is attached to the other end.

When no force is applied to the control wheel, the armature is at its mid position, and the signal out is null. If the control wheel pushes on the case, the E-pickoff signal is one phase. If it pulls on the case, it is of the opposite phase.

The amount of shrinking or expanding of the length of the transducer is mechanically limited by stops which are not shown, so that if the transducer case fails, or even if it were completely removed, the only effect would be to add some slack in the linkage.

Transducers may be installed, as indicated, in the drive linkage itself. Others, more compact, are located at the upper part of the control column.

A similar force transducer is in the pitch channel to sense control column forces (in a fore and aft direction).

When autopilot is operating in "control wheel steering" mode, the transducers provide signals to the roll and pitch channels to operate the ailerons or elevators as desired by the pilot.

The advantage of control wheel steering is that the pilot applies relatively small forces to his control wheel or control column compared to the forces required in manual operation to overcome the forces of the artificial feel systems. It can be compared to power steering in an automobile.

Figure 22-12

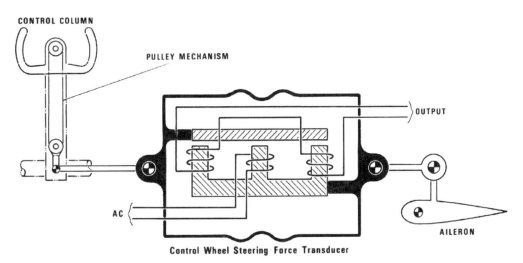

Control Wheel Steering Force Transducer

"E" Pickoff signal tells magnitude and direction of CONTROL WHEEL FORCE applied to aileron system

Roll No. 9: CWS And Versine

The roll computer in figure 22-13 is drawn in the attitude hold mode. The attitude hold switch in the center of the computer is actuated to its triangle, clamping the servo motor — a term used for a servo motor whose motion is prohibited. As long as the control synchro rotor cannot move, the airplane bank angle cannot change.

The lift loss compensation circuit has been added.

A level detector at the lower left always looks at the CWS transducer signal. When it sees more than about four pounds of force applied to the control wheel, it operates its switch, which connects the transducer output into the command circuit; and it cancels the attitude hold mode (attitude hold switch goes to its circle).

It also disconnects the sine winding follow-up signal, so that as long as the force is applied, the computer servo continues to run, calling for an increasing bank angle.

As soon as the pilot releases the control wheel, the roll channel reverts to attitude hold mode, and the autopilot maintains whatever bank angle has been achieved during the time the pilot was operating the control wheel.

If the pilot wishes to level the airplane, he moves the control wheel in the opposite direction. The level detector operates, canceling attitude hold mode, and the reversed phase of the signal from the transducer runs the servo motor in the opposite direction, causing the airplane to move towards wings level position. The bank angle limiter is bypassed because the pilot may wish to bank the airplane in excess of normal autopilot bank angle limits.

Figure 22-13

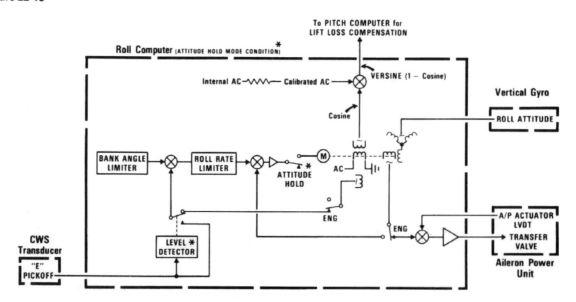

* A predetermined amount of force on control wheel triggers Level Detector and cancels "Attitude Hold" mode

Autopilot Heading Select Mode

Figure 22-14 illustrates the operation of the autopilot in heading select mode. The first airplane position shows the autopilot maintaining a heading of 90°, which is the selected heading shown by the heading select cursor (triangle) at 90° on the compass card. The cursor is also at the upper lubber line of the Horizontal Situation Indicator, so the signal from the heading select synchro is null.

The second position of the airplane shows, on the HSI, that the pilot has selected a new heading of 150° by moving the heading select bug to the 150° position on the compass card.

Since the heading select error signal into the autopilot is a direct function of the separation of the heading select bug from the upper lubber line, he has introduced a very large signal into the computer,

calling for maximum bank angle to the right. The autopilot banks the airplane to its maximum bank angle, and the airplane turns to the right.

The compass card begins to rotate counterclockwise, and the heading select bug with it. As the airplane approaches the new selected heading, the amplitude of the heading select signal diminishes, calling for less and less bank angle. When the heading select bug gets to the upper lubber line, the airplane is on its new selected heading, and the autopilot is holding wings level.

If the airplane deviates from the selected heading, the bug moves away from the lubber line. The heading select error synchro then develops a signal causing the airplane to bank until the selected heading is restored.

Figure 22-14

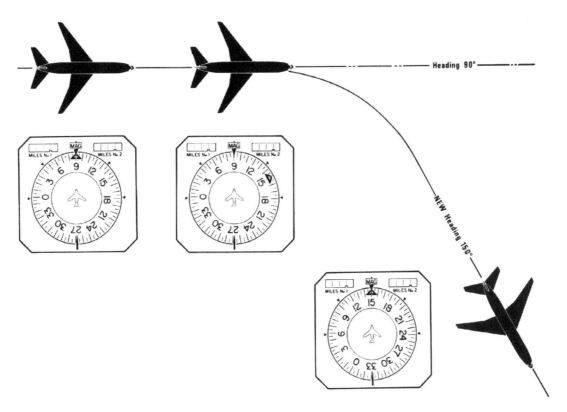

Roll No. 10: Heading Hold, Heading Select And Roll Error

Figure 22-15 illustrates "heading hold" and "heading select" modes.

This roll computer uses a roll error signal from the vertical gyro rather than the roll attitude signal used in the previous computers (refer to Figure 5-19 for vertical gyro signals).

The resultant operation is the same, but the layout required is different. The computer control synchro in this case is a resolver using its sine winding output for attitude control. When the airplane attitude is wings level, the roll error signal is null. When the computer is calling for wings level, the output of the sine winding is null. The sine winding of another resolver is used for follow-up to the computer servo amplifier.

If the computer calls for a 10° bank angle when airplane wings are level, the servo motor drives the control synchro rotor 10° away from the wings level position, originating a signal input to the transfer valve amplifier. As before, the transfer valve operates the autopilot actuator, which in turn, operates the ailerons until the airplane attitude is the commanded 10° of bank angle.

At that time, the roll error cancels the control synchro input to the summing point. Any airplane roll attitude different from the commanded bank angle results in a difference signal of one phase or the other out of that summing point to the transfer valve amplifier. That will operate the ailerons as required to restore the commanded 10° of bank angle.

Use of the extra servo motor system (upper left of the roll computer) is a common way of developing a "heading hold" signal. This servo motor system may be physically located in the yaw computer, but regardless of where it is located, its function is the same.

The directional gyro typically has two heading output synchros, one for the compass system and one for the autopilot. If the autopilot is not in heading hold mode, the servo loop in the upper left is driving its control synchro rotor to follow any heading change evidenced by movement of the resultant field in the stator of that control synchro. Thus, while not in heading hold mode, the control synchro rotor output is kept at a null.

When the autopilot is in heading hold mode, the servo motor in the upper left is clamped, and any output of the control synchro rotor is fed into the bank angle limiter. Any change in heading during heading hold mode causes the field in the control synchro to move, developing an output signal from the rotor.

The phase of this signal causes the airplane to bank in the direction required to bring the air-plane back to the desired heading, thus canceling the output of the synchro rotor and returning the airplane to wings level attitude.

If the autopilot is in heading select mode, the heading select error signal from the compass system controls the bank angle to maintain the selected heading.

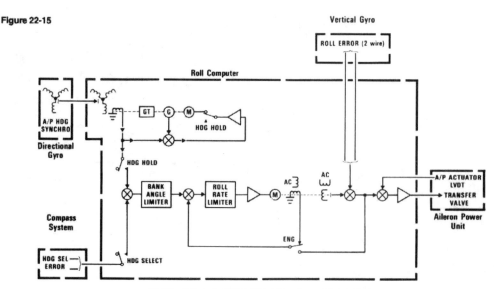

Figure 22-15

"Roll Error" instead of "Roll Attitude" requires a different layout

Beam Sensors

Radio deviation signals are sensed, but are not used in roll and pitch channels of the flight director and autopilot until "capture" or "engage" occurs. The term capture is used to describe the point at which radio deviation will be used in the autopilot and/or flight director.

The "capture" point is determined by the Vertical Beam Sensor (VBS) in the pitch channel and the Lateral Beam Sensor (LBS) in the roll channel. The VBS and LBS are voltage level sensing circuits that satisfy certain switching functions and apply radio deviation to the signal chain.

In Figure 22-16, the beam sensor is the NOT circuit. When the deviation signal is large, there is no output from the NOT circuit.

When the deviation decreases to a predetermined level, a Logic 1 is available from the NOT circuit to provide an input to the AND gate. The other input to the AND gate is available when the correct autopilot and/or flight director mode is selected.

Once the AND gate is satisfied, the output is used to "latch up" the AND gate, even if the deviation signal would exceed the level of LBS or VBS operations. The only way to unlatch the AND gate would be to deselect or change modes.

In some airplanes, the capture level may be variable. For example, it may use the angle of intercept as part of the criteria to determine capture. Capture could occur very close to a VOR radial, if the angle of intercept is shallow.

Figure 22-16

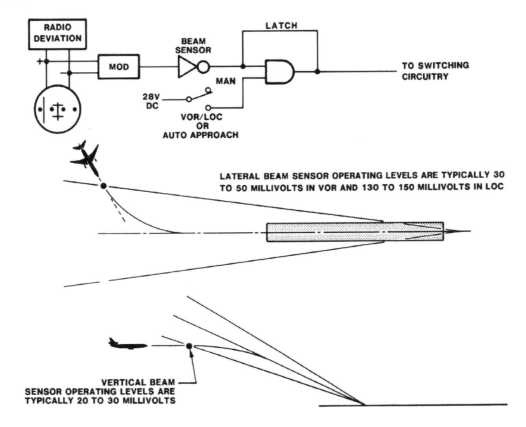

LATERAL BEAM SENSOR OPERATING LEVELS ARE TYPICALLY 30 TO 50 MILLIVOLTS IN VOR AND 130 TO 150 MILLIVOLTS IN LOC

VERTICAL BEAM SENSOR OPERATING LEVELS ARE TYPICALLY 20 TO 30 MILLIVOLTS

Autopilot/Flight Director Roll Channel System Interface

Figure 22-17 illustrates typical inputs and outputs of a roll computer. This more closely represents a 747 autopilot, but is not basically too much different from any late model airplane autopilot.

The roll computer develops outputs not only for the autopilot system, but also for the flight director system. Only one roll computer, aileron power unit and attitude director indicator are shown, but the other two computers (indicated in the flight director layout) would have the same kinds of inputs and outputs.

In practically all cases, there are two separate and equal systems providing inputs. For example, No. 1 and No. 2 inertial navigation systems, No. 1 and No. 2 compass systems, No. 1 and No. 2 VHF navigation systems, No. 1 and No. 2 central air data computers, and two separate windings on the control wheel force transducer. No. 1 systems' in-puts are typically used for computers No. 1; and No. 3 and No. 2 systems' inputs for computer No. 2.

The captain and the first officer both have flight director selector switches with which they can select a computer to provide commands to their flight director command bars and signals to their flight director mode annunciators. A relay with many contacts is controlled by the flight director selector relay.

The mode select panel has control of whether the flight director command bars will be in view or not. If the flight director switch for the Attitude Director Indicator (ADI) is in the "off" position, the command bars will be biased out of view. If a failure occurs which renders the flight director commands invalid that also would bias the command bars out of view, and at the same time the flight director warning flag would appear in the ADI.

"Cam out" is a term used to describe the condition prevailing when an autopilot is overpowered. It could be overpowered either by the flight crew or by another autopilot, when two or more autopilots are operating together.

The monitor and logic unit can detect a cam out condition by an excessive signal from the transfer valve amplifier, or by too great a difference signal between the aileron LVDT and the autopilot LVDT. The monitor and logic unit performs other monitoring functions, such as autopilot interlock validity and mode compatibility.

Extensive internal monitoring of autopilot and flight director operations is continuously taking place in all of the computers, as well as in the monitor and logic unit. In addition, each black box has its own self test function which can be performed by the mechanic to help him isolate a faulty unit.

Figure 22-17

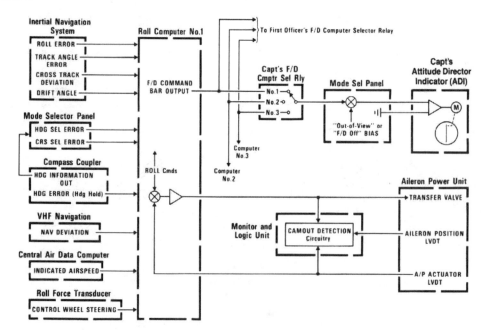

A/P F/D Roll Computer Base Schematic

Figure 22-18 shows a base print for a hypothetical but typical present generation roll computer. By "base print" is meant that this is the basic print which will be used to illustrate various modes of operation, in succeeding figures, by means of heavy lines showing signal paths in use.

At the upper left, an indicated airspeed input is shown controlling gain change. A Mach signal may also be used to control gain changes within the roll computer. Gain changes are made as needed in many different places throughout the computer to fit the requirements for various modes and airspeed conditions.

Near the center of the diagram are arrows labeled "F/D" (flight director) and "A/P" (autopilot). This is the point at which the flight director commands diverge from autopilot commands.

The lower part of the drawing is for the autopilot only, and the upper right portion develops the flight director commands to the ADI roll command bar (refer to Figure 19-2 for autopilot/flight director comparison).

Bear in mind that the computer symbology used is not intended to represent actual components — only the actions of the computer. For example, everything is solid state, whether analog or digital computer.

Figure 22-18

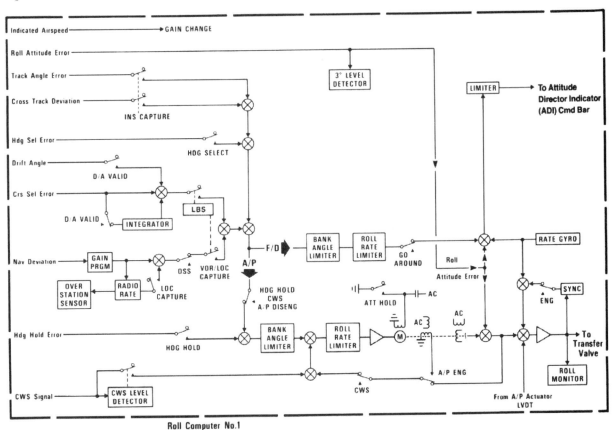

Roll Computer No.1

A/P F/D Roll Computer Base Schematic – Digital Computer (Typical Boeing 737)

The purpose of both FCC's roll control, is to provide lateral axis, flight directed control, using manual operated flight controls with the A/P not engaged or engaged in CWS (If equipped). The FCC also provides A/P commands to automatically operate the flight controls using the A/P in CMD (Figure 22-18a).

The FCC's have separate A/P command signal processing to operate their respective system aileron A/P actuator. Either provides the same mechanical input to both aileron PCU's. The FCC uses both input and output signals. The analog inputs to the FCC originate from the roll force transducer (is so equipped), VHF navigation receiver and Low Range Radio Altimeter (LRRA). These are connected to an analog to digital (A/D) converter inside the FCC for internal processing. The signals for all computation processing, gain control, A/P and F/D commands, from the IRU, FMC, DADC, MCP, are digital processed by ARINC 429 receivers, within the FCC, for internal use.

Roll attitude and roll acceleration are normally from the local IRU for F/D operation, and from the opposite IRU for A/P operation. The system uses both digital and analog inputs for operating mode control.

Digital outputs are sent to an EFIS Symbol Generator for EADI Flight Director (F/D) roll command bar movement and to visually inform the crew of the mode and status of the Digital Flight Control System (DFCS).

An interlock signal is sent to the Autoflight Status Annunciator, to visually display for AP warning, failed, or BITE conditions. A/P engage logic is processed both in the MCP and the FCC, to determine mode selection.

The aileron position sensor is used prior to engaging the A/P, to allow the FCC to synchronize the A/P actuator to the surface position.

Once engaged, the digital roll A/P command, computed to satisfy engaged mode requirements, is converted to an analog command. The signal is then amplified and applied to the aileron A/P actuator transfer / servo valve. The transfer valve converts the electrical signal to a hydraulic signal. This is used in the A/P actuator to drive an output arm, to mechanically move the input torque tube of both PCU's.

Aileron actuator position transducer (LVDT) feedback signal tells the FCC that the actuator answered the A/P command. The aileron surface position sensor signal tells the FCC when the ailerons are faired, and how much to move the ailerons to get them back to faired, once the mode required corrections are accomplished. Operation from this point is the same as manual flight control operation. The A/P roll command computer then determines the roll command and roll rate, using spoiler position signals at flaps greater than 27 units. An Aileron Force Limiter reduces the A/P aileron output when the flaps are full up, to prevent over control.

The F/D signal processing is very similar to the AP signal processing. A/P Command mode logic is used, but the F/D command processing is accomplished as a separate control. The digital output is transmitted to the EFIS Symbol Generator where it is processed to be displayed on the EADI.

Continued on next page ...

A/P F/D Roll Computer Base Schematic – Digital Computer (cont'd.)

Figure 22-18a

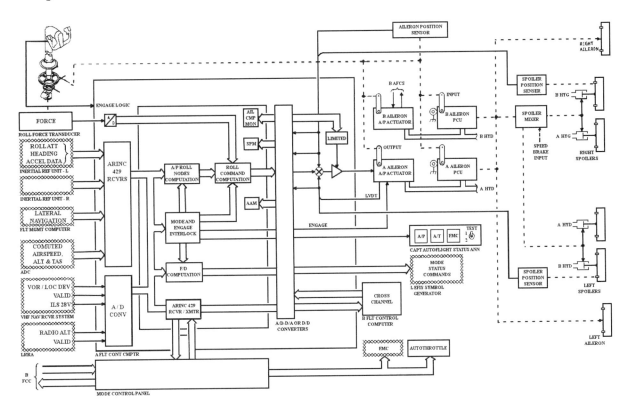

Roll Control Block Diagram

Roll Computer: A/P Disengaged Mode; F/D Heading Select Mode

Figure 22-19 shows the autopilot disengaged while the flight director is in heading select mode. The flight director could be in any selected mode while the autopilot is disengaged.

When the autopilot is disengaged, there are two synchronizing actions going on, keeping signals null at the two points called out at the lower right of the diagram.

One of these synchronizing actions is the little servo motor system driving its control synchro rotor to follow up on roll attitude changes. The roll attitude error signal (upper left) shows up at the summing point just to the right of the control synchro rotor.

If the roll attitude is not wings level, the servo motor loop drives the control synchro rotor to a position corresponding to the bank angle of the

airplane. When the control synchro rotor is at a position corresponding to the bank angle of the airplane, its signal cancels the roll attitude error signal. A null is thus maintained at the indicated point.

The synchronizing action involves the transfer valve amplifier at the lower right of the diagram. When the autopilot is not engaged, the synchronizer ("SYNC" in the little block just above the null point) is connected into the transfer valve amplifier through the summing point. When the autopilot is not engaged, the solenoid valve on the hydraulic actuator is not open and the transfer valve has no hydraulic power. Therefore the output from the autopilot actuator LVDT is null.

Continued on next page ...

Figure 22-19

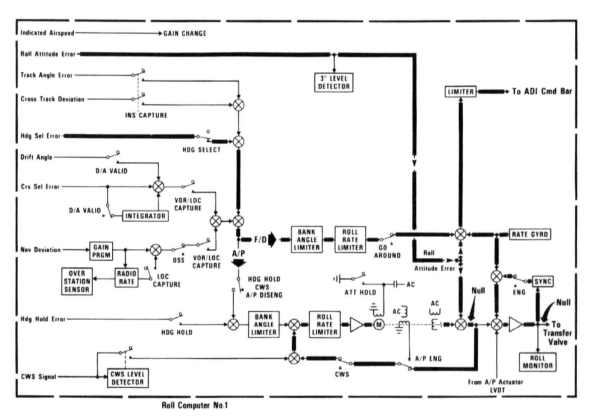

Roll Computer No.1

Roll Computer: A/P Disengaged Mode; F/D Heading Select Mode (cont'd.)

The rate gyro could develop temporary signals as the airplane roll attitude changes. Very small temporary signals could show up from the servo motor section during the time it is synchronizing. The synchronizer box will invert any output from the transfer valve amplifier and feed it back into the signal circuit in opposite phase so as to keep the output of the servo amplifier always at a null when the autopilot is disengaged.

These synchronizing actions make sure that, at the time the autopilot is engaged, there will be no standing signals to cause sudden aileron deflection.

With the flight director operating in heading select mode, the heading select error input is switched into the signal circuit (heading select switch activated to the triangle). If the airplane is not on the heading

selected, there will be a heading select error signal fed through the bank limiter, the roll rate limiter, the go-around switch, and a summing point to a limiter which limits the movement of the ADI command bar.

A heading select error signal commands the pilot to lower the right or left wing.

When he has lowered it far enough, the roll attitude error signal cancels the heading select error signal out of the roll rate limiter. Because the airplane is banked, it turns back toward the selected heading, diminishing the heading select error, and therefore diminishing the commanded bank angle. When the heading select error has been eliminated, the commanded bank angle is wings level.

Figure 22-19

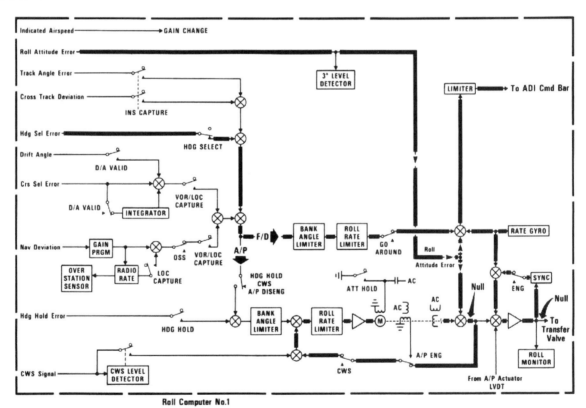

Roll Computer No.1

Roll Computer: A/P Attitude Hold Mode

In Figure 22-20, the flight director is ignored because it could be in any selected mode.

When the autopilot is in attitude hold mode, its servo motor is clamped by the "attitude hold" switch which grounds out the input to the fixed field of the servo motor. Since the servo motor cannot move, the synchro rotors cannot move, and the autopilot will force the airplane to maintain whatever bank attitude is represented by the position of the synchro rotors.

The signals into the transfer valve amplifier are shown dashed because they would exist only if the airplane bank attitude does not correspond to the commanded attitude. Attitude hold mode in the roll channel exists only when "control wheel steering" has been selected, and only if there is no force applied to the control wheel.

Figure 22-20

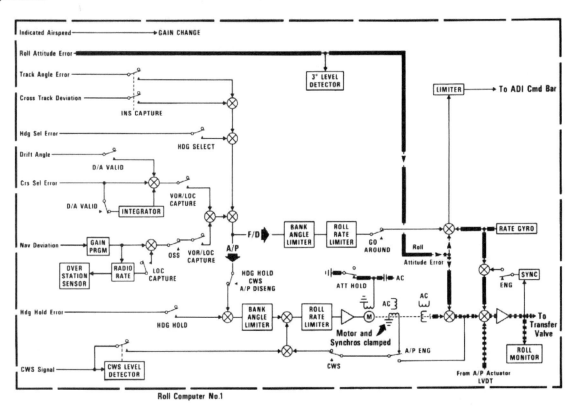

Roll Computer No.1

Roll Computer: A/P Control Wheel Steering Mode

Figure 22-21 shows the autopilot in control wheel steering mode; a force in excess of about four pounds is being applied to the control wheel. The output of the control wheel force transducer is switched into the servo motor circuit ahead of the roll rate limiter, through the switch operated by the control wheel steering level detector. As long as the pilot maintains a force on the control wheel, the servo motor continues to run, increasing the airplane bank angle.

When he releases the control wheel, the autopilot reverts to attitude hold mode.

If the pilot wishes to return the airplane to wings level or bank in the opposite direction, he applies a

force on the control wheel in the opposite direction. This reverses the phase of the control wheel steering signal, causing the servo motor to run its synchro rotors in the opposite direction.

When the autopilot is in control wheel steering mode, the rest of the computer is disconnected from the autopilot system by the switch above the bank angle limiter.

The signals into and out of the transfer valve amplifier are shown in solid black lines because, as long as the CWS level detector is operated, the ailerons are operated.

Figure 22-21

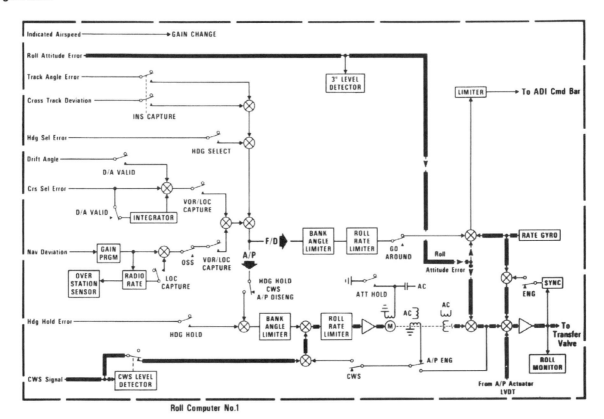

Roll Computer No.1

Roll Computer: A/P Heading Hold Mode

Figure 22-22 shows the autopilot in heading hold mode. This time the source of the heading error signal is different from that in Figure 22-15 (refer to Figure 6-26 for the explanation of this signal).

It originates in the compass coupler when the "external control" "clutches in" the spring-loaded rotor of the heading error signal. The "external control" is a voltage sent to the compass coupler when the autopilot is in heading hold mode. The

effect on the autopilot is the same as illustrated in Figure 22-15.

If the heading changes after heading hold mode is engaged, an error signal banks the airplane as required to return it to the desired heading.

The signals into and out of the transfer valve servo amplifier are shown dashed because they exist only if there is a heading hold error.

Figure 22-22

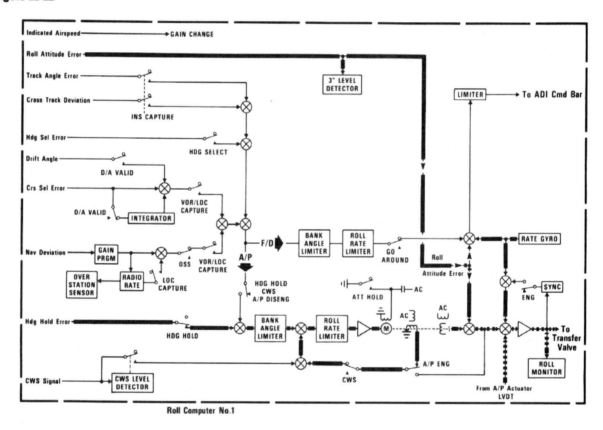

Roll Computer No.1

VOR (LOC) Capture

Figure 22-23 illustrates a typical capture of a VOR selected course. VOR or LOC capture is a mode switching function accomplished by a circuit called "lateral beam sensor".

The airplane is shown approaching the selected course at an angle of 45°. The autopilot could be in heading hold mode, heading select mode, or CWS mode. The autopilot mode select switch will have been moved to the VOR/LOC position. VOR capture in most autopilots occurs at about one dot (5°) of deviation from the selected radial.

In this illustration, the 270° radial is captured. Capture of the 90° radial while traveling the same direction on the other side of the VOR station would be the same operation.

At the time of capture, the intercept mode is automatically discontinued and VOR capture mode initiated. During VOR capture modes, the principal input signals are course select error and radio deviation. At the time of capture (from the right side of the beam), radio deviation calls for a left turn, course select error calls for a right turn. Unless the intercept angle is unusually small, the course select error signal predominates, causing the airplane to

turn toward the right, making a smooth approach to the radial.

As the airplane gets closer to the selected path, the deviation signal diminishes and the course select error signal diminishes. The course select error signal will always predominate, however, because if it did not, the deviation signal would diminish until the course select error did predominate.

During the capture mode, the bank angle limit is typically 25° or 30°, and the roll rate limit is on the order of 4° to 7° per second. When the airplane gets reasonably close to being established onto the selected course, "VOR ON COURSE" mode is automatically initiated.

The ON COURSE mode is automatically initiated when deviation is small, the course select error is small, and the bank angle is small. The "on course" mode causes the bank angle limiter to drop its limit to about 10°, and the roll rate limiter to drop its limit to perhaps as little as 1 1/2° per second.

It also arms for the operation of the "over station sensor", which will cut out the radio deviation signal when the rate of change of radio deviation exceeds eight millivolts per second.

Figure 22-23

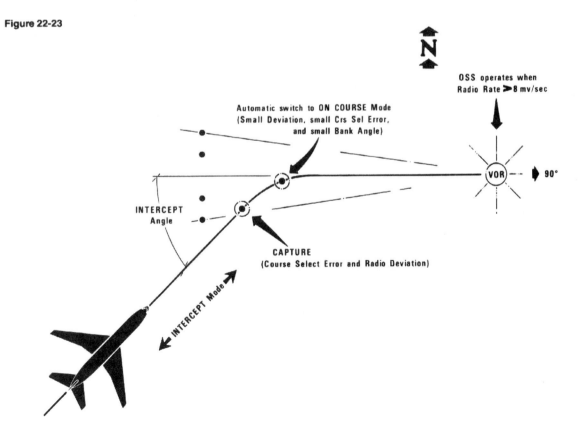

Beam Capture Maneuver

Figure 22-24 illustrates the conditions existing during the beam capture maneuver. The conditions illustrated pertain both to the capture of a VOR beam and to the capture of a localizer beam.

In airplane "A", capture is represented as occurring at two dots of deviation on the HSI. This would be 10° of deviation from the VOR beam, or 2° of deviation from the localizer beam. With an intercept angle of 45° and a selected course of 90°, the course select error (heading error) signal is 45°.

The heading error input to the roll computer calls for right wing down. The deviation signal calls for left wing down. Heading error predominates and the autopilot puts the right wing down. A smooth approach without overshoot results unless the intercept angle is too great, the aircraft speed is too great, or the airplane is too close to the station.

The farther away from the station the airplane is, the greater the geographic distance represented by 10° of error for VOR (or 2° of error for localizer).

So the farther away from the station the airplane is, the more room there is available for the capture maneuver.

Airplane "B" is inside of the first dot of deviation, with the heading error reduced to 25°. The combination of these signals still calls for right wing down, and, until the airplane comes on to course, it will be turning right.

If it overshoots, the beam capture maneuver will have to be reversed from the other side. Once the airplane crosses over the beam, the radio deviation and heading error both call for right wing down, and that condition persists until the heading error reverses itself and opposes the deviation.

At some time before the airplane gets onto the course from the left side, the heading error predominates, calling for left wing down, and the airplane comes onto course without overshoot.

Airplane "C" illustrates the airplane on course and on heading.

Figure 22-24

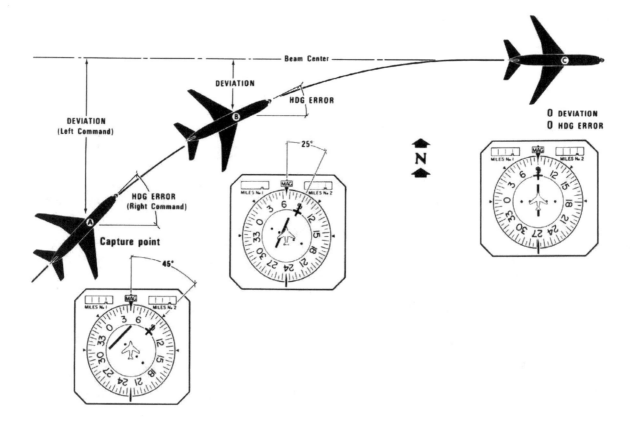

DC Integrators

Figure 22-25 illustrates the operation of a non-amplifier, limited type DC integrator. It is limited because the output voltage could not exceed the input voltage. The chart at the top shows an input from time zero to time X. In these drawings we are saying that it will take one-half time X, as indicated in the output chart, at which time it levels off until the input disappears.

When the input disappears (at time X), the condenser gradually discharges over one-half time X and returns the output to zero.

Condenser charging time, specified in this example as one-half time X, is the "time constant" of the integrator. Earlier technology used condensers; later technology uses such things as operational amplifiers.

Time constants range from 15 seconds to 200 seconds, depending upon their use in the circuit.

Figure 22-26 illustrates the operating principle of a DC integrator which is also an amplifier. In this case, the circuit is the same, except that an amplifier has been added in front of the RC network.

Time constants and amplification factors are adjusted to fit the needs of a particular operation. What is shown here is that, as long as the input persists, even though it is of a steady value, the output of the integrator circuit drops off at the buildup rate. Sometimes, in special applications, the buildup and drop-off rates are different.

Figure 22-25

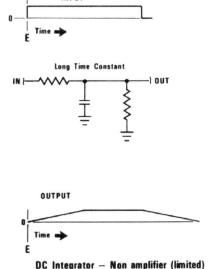

DC Integrator — Non amplifier (limited)

Figure 22-26

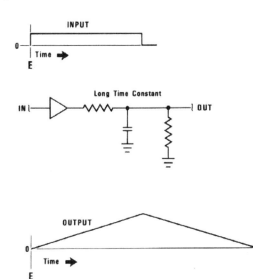

DC Integrator — Amplifier

AC Integrators

Figure 22-27 shows essentially the same integrators as in Figure 22-25, except that it employs a demodulator at the input and a modulator at the output so that it can handle AC signals.

The input this time is a steady amplitude AC signal persisting for time X. The integrator charges its condenser in time one-half X, then its output holds steady until time X, when the input disappears. The output of the integrator then falls off during the same time that it took to build it up.

Figure 22-28 is essentially the same integrator as Figure 22-26, except that it employs AC signals for the input and output, and uses a demodulator and a modulator.

A constant amplitude AC signal input is present for time X. All during time X, the output of the integrator continues to build up. When the input disappears at time X, the output of the integrator falls off at the same rate at which it built up.

Figure 22-27

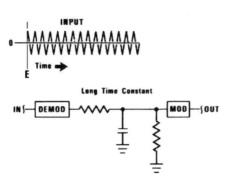

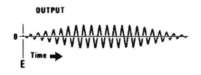

AC Integrator — Non amplifier (limited)

Figure 22-28

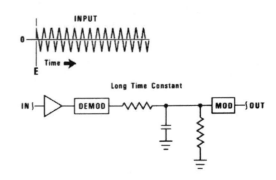

AC Integrator — Amplifier

Mechanical Integrators

An integrator gets its name because it performs a function similar to the calculus function of integrating. We have no need to understand that mathematically, but the concept of integrator operation is important to us.

To further illustrate the principle of integrators, Figure 22-29 shows a mechanical integrator. It is a non-amplifier and limited. It is limited by virtue of the fact that the sine winding output of a resolver will eventually cancel the input signal at the summing point ahead of the servo amplifier.

A low ratio gear train is in the output of the servo motor, causing it to very slowly drive the resolver rotors. When a fixed value input signal appears at the summing point, the servo motor begins to run, but moves the synchro rotors very slowly so that the sine winding outputs develop very slowly.

The speed at which they develop can be controlled by the gear ratio and by the authority of the tachometer generator signal. This type of integrator does not usually have a time constant in excess of about 20 seconds.

Figure 22-30 shows a mechanical integrator which is also an amplifier. It is an amplifier because there is no feedback signal to the input, so that as long as the input is present, the servo motor continues to drive the synchro rotor, increasing the output signal.

If the synchro rotor were driven beyond 90°, the output signal would diminish. Practical applications will not use it that long.

The output of either of these integrators could be limited by mechanical stops and slip clutches on the synchro rotor shaft. If the servo motor continued to run beyond the time when the stop was reached, the clutch would slip, causing no damage.

Whenever you see an integrator called out in a schematic, you will know that the output is being built up slowly at a predetermined rate.

Figure 22-29

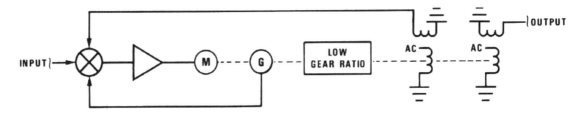

Mechanical Integrator – Non amplifier (limited)

Figure 22-30

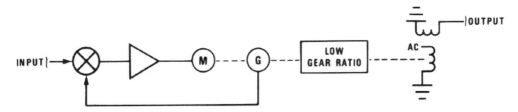

Mechanical Integrator – Amplifier

Integrator Symbols

Figure 22-31 shows a schematic symbol for an integrator which performs an amplifying function without signal inversion.

Figure 22-32 shows the same integrator with the signal inversion symbol. This particular in-version symbol is not an industry standard, but it is one that is typically used. It is pretty much self explanatory, representing that two wires have been transposed to invert the sense of the signal. If it were a DC signal, the polarity would be re-versed. If it were an AC signal, the phase would be reversed.

Figure 22-33 represents a non-amplifier integrator not inverting the signal.

Figure 22-34 represents the same integrator inverting the signal at its output. The inversion symbols could be presented ahead of the integrator instead of after.

You will not always find integrators so specifically defined in a schematic. Often you will find an amplifier indicated when, in fact, it is not an amplifier; or a rectangle indicated when it is, in fact, an amplifier.

Sometimes the signal inversion operation will not be indicated. These conditions would exist only in prints where you would be expected to be able to deduce the type of operation of the integrator from its function in the circuit.

Figure 22-31

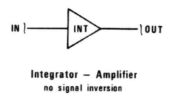

Integrator — Amplifier
no signal inversion

Figure 22-32

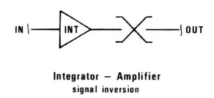

Integrator — Amplifier
signal inversion

Figure 22-33

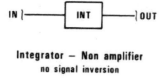

Integrator — Non amplifier
no signal inversion

Figure 22-34

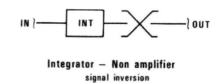

Integrator — Non amplifier
signal inversion

Crosswind Standoff — No Correction

Figure 22-35 shows what would happen to an airplane, controlled by an autopilot, while flying a VOR or localizer course if a crosswind were to come up, and no crosswind correction capability were included in the autopilot computer. (Autopilots always have a crosswind correction capability).

In position "A", the airplane is on the radio beam and on the selected heading (deviation needle centered and course select cursor at the lubber line).

In position "B", the airplane has been moved off the radio beam by the crosswind (dotted pattern). The heading has not yet changed.

In position "C", the autopilot has begun to change heading back toward the beam.

In position "D", the autopilot has achieved a stable "standoff" condition (defined later).

Under airplane "A" is an HSI showing the pilot's indication of the fact that the airplane is on the chosen heading of 90°, and centered on the radio beam.

Under airplane "B" the HSI shows the airplane heading has not changed yet, but the aircraft has been moved off the beam center. The two controlling signals into the autopilot roll channel are radio deviation and course select error. If the radio beam is to the left of the airplane, the deviation signal in the roll computer calls for a left turn. The vector chart below the HSI (airplane "B") shows that the deviation signal in the roll computer is calling for a left turn, and there is no course select error signal.

Under airplane "C" the HSI shows that a course select error of 5° has been developed as a result of the autopilot turning the airplane toward the beam center. The radio deviation signal has in-creased to one dot. The chart below the HSI shows that the radio deviation signal, calling for left turn, still exceeds the course select error signal, calling for a right turn.

Under airplane "D", where steady "standoff" conditions have been achieved with a steady crosswind, the two controlling signals are "standing each other off". The HSI shows a course select error of 10°. The deviation needle shows a deviation error of 11/2 dots. The chart below the HSI shows the deviation signal, calling for a left turn, balanced by the course select error signal, calling for a right turn.

Consequently, the roll channel does not have an output to the ailerons, and the airplane wings are level. This situation is described by saying that heading error is "standing off" radio deviation.

The heading error signal (course select error) is "required" by the crosswind. In order for the air-plane to follow the radio course, it must head into the wind; its heading must be different from its direction of travel, or ground path.

This condition could only develop in an autopilot without crosswind correction (typically accomplished with an integrator circuit). Understanding what would happen without the integrator correction for crosswind will make it easier to understand the function of an integrator circuit in providing crosswind correction.

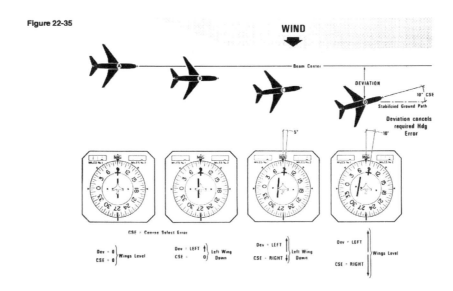

Figure 22-35

Crosswind Correction By Integrator

Figure 22-36 illustrates how an integrator would function to correct for the crosswind standoff condition developed in Figure 22-35. Assume that an integrator is switched into the roll computer after the stabilized standoff condition was built up in Figure 22-35.

The time of switching in the integrator is illustrated in Figure 22-36 with airplane "A". The vector chart below the airplane shows the deviation signal calling for left turn balanced by the course select error signal calling for right turn. Our integrator in this case is the one illustrated in Figure 22-34: inverting, non-amplifier.

The input to the integrator is the course select error signal. Over a period of time, usually on the order of 100 seconds, the output of the integrator will build up to equal its input. The sense (phase or polarity) of the input will be inverted by the integrator. In this case, the heading error input calling for a right turn will be inverted by the integrator whose output will call for a left turn.

Airplane "B" shows that the integrator output has begun to build up. Since the integrator output is now canceling part of the required heading error signal, a smaller deviation signal cancels the remainder for a stable condition. Therefore, the air-plane has moved closer to the beam center.

Airplane "C" shows a later time when the integrator output has further built up, canceling more of the required heading error signal, leaving less deviation necessary for a stable condition.

Airplane "D" shows a later time when the integrator output has been built up even more,

leaving only a small amount of deviation necessary to cancel the remainder of the required heading error.

Airplane "E" shows that the output of the integrator now equals its input, and therefore complete cancellation of the required heading error is being accomplished with the integrator output; no deviation error is needed to achieve a stable condition. The airplane is now following the radio beam center, with the required heading error cancelled by the output of the integrator.

As long as airplane speed and wind conditions remain constant, this situation will prevail.

In normal autopilot, the integrator action would begin as soon as any heading error develops, and the airplane would not have been moved as far off course as we have indicated.

The crosswind correction function could have been accomplished with an amplifier, non-inverting integrator (Figure 22-31), with the input signal being the deviation signal. In that case, some small amount of residual deviation (determined by amplification factor) would be required to hold the necessary output.

Integrators have other uses in autopilot, and their operation can often be observed in a parked airplane when a mechanic is checking the autopilot. If the mechanic does not understand what the autopilot is doing, he may think something is wrong. For example, the control wheel may mysteriously creep all the way to the right or left, or the control column may mysteriously creep forward or aft.

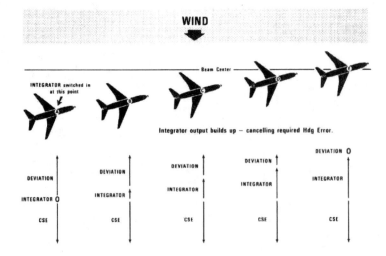

Figure 22-36

Radio Rate Concept

Figure 22-37 charts several different functions all related to the same progression from left to right, and time-sectioned by vertical lines. The purpose of the drawing is to help you see the relationships between airplane position relative to beam center, air-plane motion relative to beam center, deviation signals, rate signals, deviation indications, and circuit concept of how a rate signal is developed.

The term "rate" means "rate of change" of any referenced signal. Here it is "radio rate". All conditions pertain to the airplane flight path illustrated at the top. Arbitrary values and polarities have been assigned to the deviation signal.

During the first five time units, the deviation signal changes from - 4 to O. During the next five time units, the deviation signal changes from 0 to + 4. The next four time units show the deviation signal constant at + 4, and during the next ten time units it changes through 0 back to - 4 again.

The deviation indicator on the left illustrates the fact that the deviation needle in the HSI would move from all the way on the left to all the way on the right during the first ten time units. The deviation indicator on the right shows that the needle in the HSI would move from all the way on the right to all the way on the left during the last ten time units.

During the first ten time units, the deviation signal is becoming more and more positive, regard-less of its actual polarity. As long as the deviation signal is

becoming more positive, the sense of the rate signal is the same. During the last ten time units, the deviation signal is becoming more and more negative, regardless of polarity. Therefore, the sense of the rate signal is the same and reversed, regardless of the polarity of the deviation signal.

The rate circuit shown connected to the deviation output of the VHF navigation receiver will have electrons moving from the ground, through the resistor into the condenser, as long as the deviation signal becomes more positive.

Since we have charted a steady change in the deviation signal, the rate of electron movement through the resistor is constant during the first ten time units. The polarity of the rate signal is positive because of the direction of electron flow. During the last ten time units, the direction of electron flow through the resistor is reversed, because the deviation signal is changing in the negative direction, and the rate signal is negative.

At the end of the first ten time units, when the deviation signal becomes positive, there is a lag in the rate signal which will be a function of the RC time constant of the rate signal circuit. It will take a few seconds for the rate signal to diminish to zero. There was a similar lag in the buildup of the rate signal.

Continued on next page ...

Radio Rate Concept (cont'd.)

In the chart we have assumed an instantaneous rate signal which, in fact, does not exist. Virtually all of the rate signal circuits in autopilots and flight directors do have a time lag which varies according to design requirements of the particular mode in use. These time lags ordinarily range from about two seconds to about eight seconds.

The chart just above the deviation indicators shows the sense relationship between deviation and rate signals which exists if a radio rate signal is used during a radio mode of operation of an autopilot or flight director. The sense of these signals always follows the rule which says that, if the airplane is approaching the beam center, the deviation and rate signals are opposed; and, if the air-plane is leaving the beam center, the deviation and rate signals aid each other.

During the first five time units, when the airplane is approaching the beam center, the deviation signal is negative and the rate signal is positive. During the next five time units, when the airplane is leaving the beam center, both the deviation and rate signals are positive.

During time units 14 to 19, when the airplane is approaching the beam center, the deviation is positive and the rate is negative; and, during the last five time units when the airplane is leaving the beam center, both the deviation and rate signals are negative.

Rate of change signals are sometimes developed from other computer signals and used in the computer; for example, heading error, radio altitude, roll attitude, or from any other signal whose rate the design engineer may find useful.

Figure 22-37

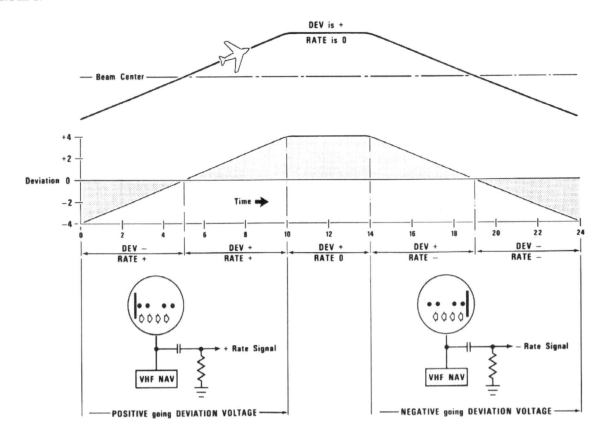

Radio Rate in AC Computer

Figure 22-38 shows the principle of the development of AC signals from DC radio deviation signals, to be used in an AC computer. The DC radio deviation signal into the modulator at the left is changed to an AC deviation signal of phase and amplitude corresponding to the polarity and amplitude of the DC deviation signal. A demodulator then develops a DC signal to be used in generating a DC rate signal which could be used as such, or modulated into an AC rate signal.

Whether the computer is an AC computer, as was the case in most earlier autopilots, or whether it is a DC computer, as is the case in most autopilots in airplanes being produced today, does not really matter to us. We are interested only in the effect of input signals and the operations performed on them in the computer so as to be able to understand the computer outputs.

Figure 22-38

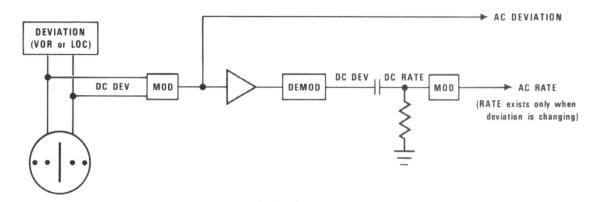

Correlation Of Deviation, Rate, And Heading Error

Figure 22-39 illustrates the relationships between radio deviation, radio rate, and heading error signals as used by an autopilot or flight director during a radio captured mode.

This chart shows that the sense of the course select error signals and the sense of the radio rate signals always correspond.

The radio rate and course select error signals will always oppose the deviation signal if the airplane is approaching the beam center, and will always aid the deviation signal if the airplane is leaving the beam

center. The purpose and result of this rule of operation is that the airplane, when approaching the beam center, will always be turned toward the direction of intended flight on the beam center as the airplane approaches it.

This results in a smooth curved approach to the beam, with the airplane coming on to the beam without overshoot, in most cases. It also results in more rapid correction back toward the beam if the airplane is leaving the beam.

Figure 22-39

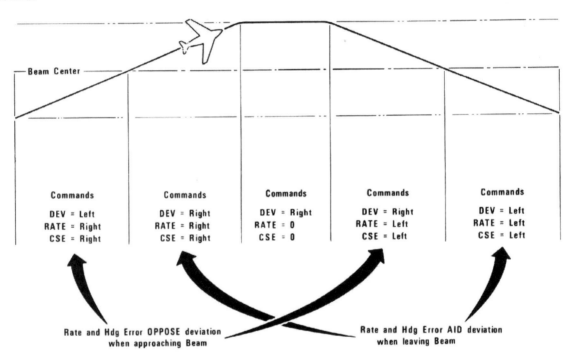

Need For Heading Error Or Radio Rate

Figure 22-40 shows that an autopilot or flight director would never keep the airplane on the radio course if deviation only were used as the controlling signal. Using only deviation would cause the airplane to continue turning toward the course as long as it is off course. For example, if it is off the course to the left, the deviation signal would always call the airplane to turn toward the right until it reached the course center, and by that time it would be headed in such a direction as to cross the course center.

After it crossed the course center, the deviation signal would cause the airplane to turn toward the left and to continue turning left until it came to the course center again. At that time, it is once more headed in the direction which would cause it to cross the course, and would not be turned toward the course again until after it had crossed.

This operation is called "wandering" and will result if, for some reason such as broken wires, the course select error signal is lost as an input to the roll computer (unless the computer is also using a radio rate signal). Once the airplane is on or near the course, a heading error signal is not necessary if a

deviation rate signal is also employed, since heading error and radio rate accomplish the same function in the roll computer.

Figure 22-41 illustrates the action of the radio rate or heading error signal in smoothly putting the airplane back on the course if, for any reason such as encountering a crosswind, it moves off.

If the airplane moves off the course, the first signal into the computer is deviation error. This signal (first illustration to the left) would cause the airplane to turn towards the left. The radio rate signal and the heading error signal later developed both call for turning the airplane to the right.

At some time before the airplane reaches the course, the deviation signal would be completely balanced out by either heading error, rate, or both, at which time the airplane wings would be level.

Further progress of the airplane toward the course center would diminish the deviation signal, leaving the heading error and/or radio rate to predominate, causing the airplane to turn toward the right. From that time on, the heading error and/or rate signal predominate over the deviation signal, causing the airplane to come smoothly on to the course.

Figure 22-40

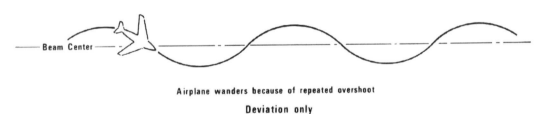

Airplane wanders because of repeated overshoot

Deviation only

Figure 22-41

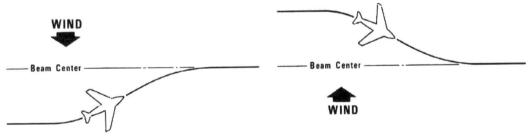

RATE or HDG ERROR OPPOSITION predominates before course is reached — this prevents wandering

Deviation + Hdg Error and/or Rate

Roll Computer: A/P-F/D VOR-LOC Captured Mode; "D/A Not Valid"

Figure 22-42 shows, in heavy outline, the signals in use if the autopilot and flight director are in the VOR or localizer captured mode, with the "drift angle not valid". "Drift angle" and its use is covered later.

The drift angle valid switches are shown in their "not valid" position. The course select error signal is therefore switched into the integrator. This represents the condition demonstrated in Figure 22-36. If there is a crosswind, there must be a "crab angle" (required heading error) into the crosswind; and therefore, there must be a course select error signal, resulting from the airplane not being on the heading selected.

Over a period of 100 seconds or so, the integrator develops an output signal equal to its input. It reverses the sense of that signal so that the integrator

output cancels the course select error signal at the summing point, during steady on-course conditions.

If it is a VOR captured mode, the "over station sensor" will be armed to switch out the radio deviation signal during the time the airplane may pass over the VOR station. If it does, the over station sensor operates the switch labeled "OSS" to the right of the radio rate circuit.

The gain programmer in the deviation circuit functions only in localizer captured mode, and only during the time of final approach. Its time of operation is usually determined by the radio altitude of the airplane.

The radio rate signal is not usually used during VOR modes, but is almost universally used during localizer captured modes.

Continued on next page ...

Figure 22-42

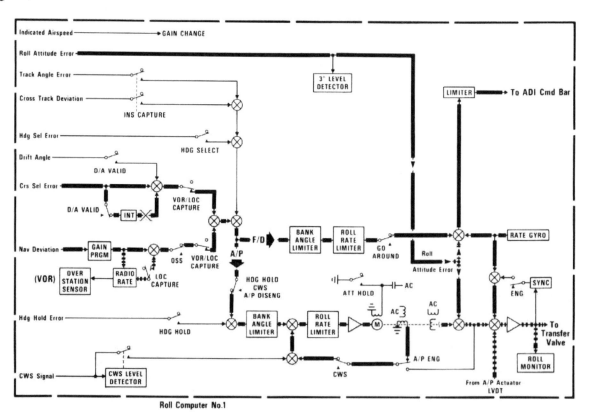

Roll Computer No.1

Roll Computer: A/P F/D/VOR-LOC Captured Mode; "D/A Not Valid" (cont'd.)

The course select error signal, through its VOR/LOC capture switch, and the deviation signal, through its VOR/LOC capture switch, are summed at the summing point to the right of the switches. These are the two controlling signals during a VOR/LOC captured mode. From that point, the computed command signal is split up so that one part goes to the flight director and the other part to the autopilot.

The flight director signal is bank angle limited and roll rate limited before it is summed with its follow-up signal, which is roll attitude error. From there it goes to the ADI command bar limiter, where the command presentation in the ADI is limited to a predetermined maximum.

If the pilot is following the commands of the ADI command bar, he keeps it centered by banking the airplane enough to cause enough roll attitude error to balance any computer command out of the roll rate limiter, which keeps the ADI command bar centered.

The rate gyro signal is an inverse feedback signal used in all modes.

The autopilot signal is bank angle limited and summed with the follow-up signal from the sine winding of the first resolver in the servo motor output. The amplitude of the command out of the bank angle limiter determines how far the servo motor will drive its synchro rotors. It will drive until the follow-up signal equals the command signal out of the bank angle limiter.

The sine winding of the resolver to the right determines the bank angle of the airplane, since, if the roll attitude error signal does not equal and cancel the output of this sine winding, there is an output to the transfer valve to operate the ailerons until the roll attitude error does cancel the command from the control resolver.

The output of the summing point to the right of the servo motor system is shown dashed, since that signal is present only during the time that aileron operation is necessary.

Figure 22-42

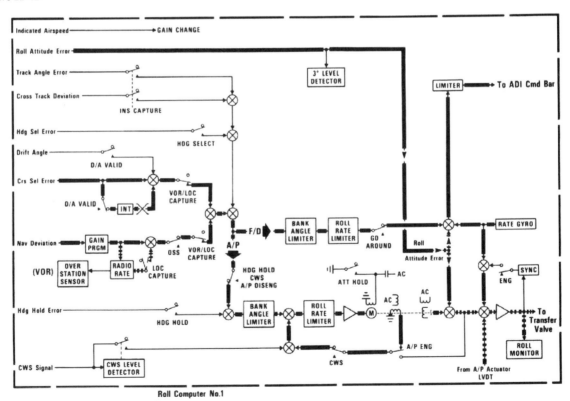

Roll Computer No.1

Some Inertial Navigation Terms

Figure 22-43 illustrates some inertial navigation terms necessary for understanding the autopilot operation during inertial navigation modes.

An Inertial Navigation System computes its own great circle track from "waypoint" to "waypoint". A series of waypoints are set up by the pilot prior to takeoff. These represent his intended flight path from departure to destination and are inserted into the computer in terms of latitude and longitude.

In between any two waypoints, the inertial navigation computer constantly supplies information relative to the desired track between these waypoints. The desired track, so far as an autopilot or flight director is concerned, corresponds to a radio course.

The cross track deviation signal developed by the INS corresponds to the radio deviation signal developed by a radio navigation receiver.

The track angle error signal is the difference between the ground path of the airplane (actual track) and the desired track between waypoints.

Notice that track angle error is not concerned with airplane heading. It is only the actual ground path track of the airplane relative to the desired track.

The diagram in Figure 22-43 shows an airplane in a crosswind condition with the required crab angle; but the actual path of the airplane along the ground corresponds with the desired track between waypoints. There is, therefore, no track angle error.

There is, however, a "drift angle", which corresponds to the course select error signal necessary for an airplane following a radio course under similar conditions. The Inertial Navigation System will compute the drift angle regardless of whether the airplane is following an INS desired track, following a radio course, or just being flown manually and not following any particular path.

We can say that the drift angle equals crab angle, which equals the course select error signal, if the autopilot or flight director is following a radio course.

Figure 22-43

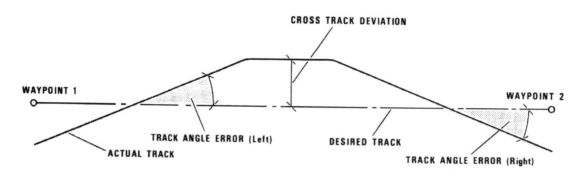

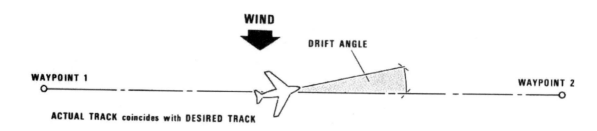

Roll Computer: A/P-F/D Captured Mode; "D/A Valid"

Figure 22-44 shows the same sort of situation as Figure 22-42, except that this time the drift angle signal is valid.

The two drift angle valid switches are operated to their triangles. The lower drift angle valid switch has disconnected the integrator, and the upper drift angle valid switch has switched in the drift angle signal from the INS. The drift angle signal is used to cancel

the required course select error signal when there is a crosswind. This eliminates the need for an integrator.

Use of the drift angle signal from the INS is a very handy way of getting immediate, constant, and accurate cancellation of the required crab angle for crosswind correction.

Figure 22-44

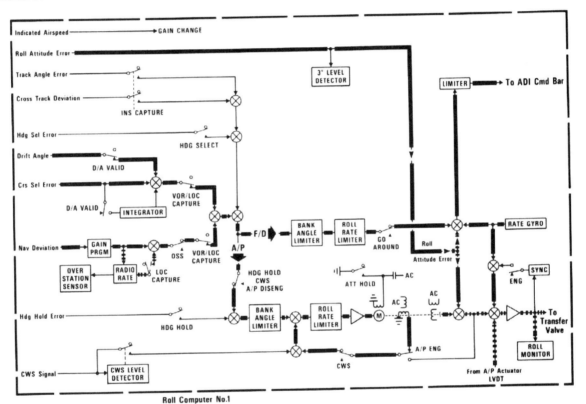

Roll Computer No.1

Roll Computer: A/P-F/D INS Captured Mode

Figure 22-45 shows the signals in use in the roll computer during the INS captured mode. Cross track deviation is used in a manner similar to radio deviation.

In the INS captured mode, track angle error is used in a manner similar to the use of course select error in radio mode. There is a considerable advantage here, however, in that track angle error is independent of aircraft heading; it is only the angular separation of the ground track from the desired track. No correction for crosswind is required, since actual aircraft heading is not involved.

Figure 22-45

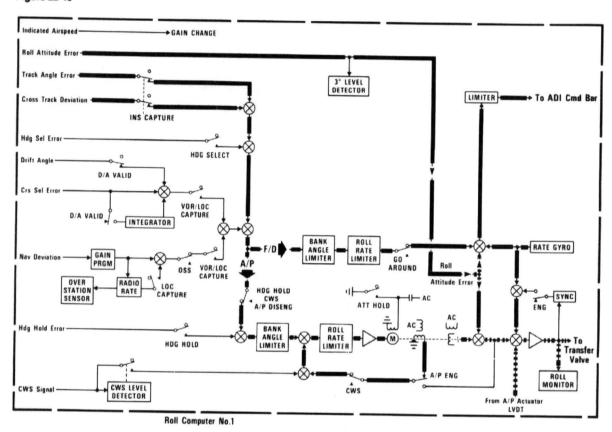

Roll Computer No.1

Roll Computer: F/D Go-Around Mode

Figure 22-46 shows the roll computer and flight director in "go-around" mode. The term "go-around" tends to be a little misleading since it does not involve turning the airplane, in autopilot or flight director usage. The full meaning of the expression go-around is to give up an attempted ILS approach and go around to try again.

The pilot would give up his approach and go around for another try if, at any time during the approach, the ground station tells him that the visibility has become too low for him to make his landing; or if the traffic has become too heavy and the control tower has decided that he should move out of the approach pattern; or if the conditions of approach indicate that he is too far off course and he would have to make too much correction to get back on in time for landing.

During a Category II approach when the autopilot is controlling the airplane, the pilot must be able to see to land the airplane manually when he reaches an altitude above the runway of 100 feet. If he is not able to see to land by that time, he will go around. At that time, he would apply power to the engine and climb out at the maximum safe rate.

So far as the autopilot or flight director is concerned, the go-around maneuver consists only of climbing out at safe rate, not turning. Although autopilots have the capability, particularly in combination with auto-throttle systems, of

performing the climb-out maneuver, typically it is a manually controlled maneuver using the flight director.

The go-around mode is ordinarily initiated in a flight director (or autopilot) by operation of palm controlled switches on the throttle levers, just be-low the knobs. The pilot can operate these switches at the same time that he moves the throttle levers forward.

In most autopilot/flight director systems, initiation of the go-around mode by the pilot disconnects the autopilot and puts the flight director into the go-around mode.

In the roll channel of the flight director, the only command to the pilot is to keep the airplane wings level. This schematic, therefore, shows the autopilot in the disengaged mode. The only signal input to the flight director channel is roll attitude error. As long as the airplane wings are level, the ADI command bar will be centered. If the airplane wings are not level, the ADI command bar will command the pilot to bring the wings to level condition.

With a Category III equipped aircraft on a triple channel approach, go-around mode will cause the autothrottle to increase thrust to go-around limit. The roll computer remains engaged and will continue to track the localizer if captured or maintain aircraft track if it is not. The computer will remain in go-around until another roll mode is selected

Figure 22-46

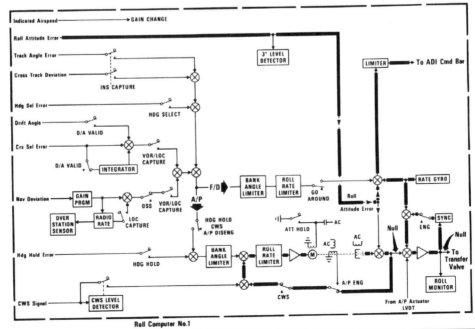

Roll Computer No.1

CHAPTER 23 - AUTOPILOT/FLIGHT DIRECTOR PITCH CHANNEL

Basic Attitude And Altitude Hold

Figure 23-1 shows a super simplified autopilot pitch channel which maintains the pitch attitude existing at the time of engage. This is called "pitch attitude hold" mode. The schematic is drawn in the autopilot disengaged mode.

The pitch attitude signal from the vertical gyro goes to the stator of the control synchro of the servo motor loop in the pitch computer. The servo motor loop, at all times prior to engagement, maintains its synchro rotor perpendicular to the field in the stator, so that the synchro rotor position always corresponds to the pitch attitude of the airplane.

At the time of engage, the control synchro rotor output is switched from the servo motor loop and fed into the transfer valve servo amplifier. The servo motor is simultaneously clamped. From then on, if the pitch attitude deviates from what it was at the time of engage, the field of the control synchro is not perpendicular to the rotor and a signal is developed of one phase or the other.

That signal goes to the servo amplifier, the transfer valve, and autopilot actuator system in the elevator power unit to operate the elevators as required to restore the airplane attitude to what it was at the time of engage, and return the synchro field to a position perpendicular to the rotor.

Continued on next page ...

Figure 23-1

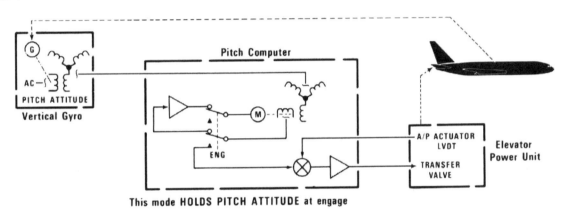

This mode HOLDS PITCH ATTITUDE at engage

Figure 23-2

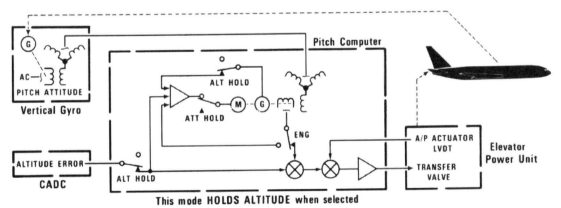

This mode HOLDS ALTITUDE when selected

Basic Attitude And Altitude Hold (cont'd.)

Figure 23-2 shows the super simplified pitch channel in altitude hold mode. In this mode the speed of operation of the servo motor system is considerably slowed. Prior to engage, the servo motor operates quickly to cause its control synchro rotor to follow up on pitch attitude changes.

At the time of engage, the servo motor is slowed by switching in a tachometer generator signal given sufficient authority so that any command into the servo motor amplifier meets from the tachometer generator. It could also be slowed by a gear train gear shift, accomplished with a solenoid changing gears to a low gear ratio (the servo motor operation can be looked upon as that of a mechanical integrator).

When a pilot operates a switch to altitude hold mode, altitude error from the CADC is switched into the computer. The altitude error in the CADC is developed from a synchro which is spring-loaded to null when its clutch is disengaged. When altitude hold mode is energized, the clutch is engaged, and any subsequent changes of altitude are reflected in the output of the CADC error synchro (refer to Figures 16-10 and 11).

Deviation by the airplane from the altitude it had at the time of engaging altitude hold mode develops an altitude error signal, which is fed immediately to the transfer valve servo amplifier, causing the elevators to operate as required to return the airplane to its desired altitude.

The servo motor loop in the pitch computer functions to maintain the control synchro rotor at a position corresponding to the necessary pitch attitude to hold, over the long term, the desired altitude. For example, if the pilot changes the airplane speed, a different pitch attitude would be required. At the time the pilot changes the aircraft speed, the airplane would deviate from the desired altitude by dropping below, if the speed is diminished, or rising above, if the speed is increased. A temporary standoff condition would result.

If the airplane speed is diminished, the airplane drops below the desired altitude, an altitude error signal goes to the elevators to raise the elevators, moving the airplane back toward the desired altitude (changing the pitch angle to compensate for the loss of lift which resulted from diminished speed).

Change in aircraft attitude results in a signal from the rotor of the pitch computer control synchro. As the altitude error increases, the change in pitch attitude increases, increasing the signal from the servo loop synchro. Within a few seconds a temporary standoff condition is set up. In this condition, the altitude error cancels the servo loop attitude error signal at the point where the two signals join ahead of the servo amplifier.

Over the long term in this mode, about 20 or 30 seconds, the altitude error signal to the pitch channel servo loop drives the control synchro rotor toward the position corresponding to the new pitch attitude required for maintaining altitude.

As the pitch attitude error signal from the control synchro rotor diminishes as a result of control synchro rotor movement, the altitude error signal gradually brings the aircraft back to the desired altitude. By the time the servo motor loop drives the control synchro rotor to the new required pitch attitude, the altitude error has diminished to a null.

The function of the servo motor loop in the slow mode of operation just described is also used in other pitch channel modes, such as vertical speed hold, indicated airspeed hold, glideslope capture, and so on.

Autopilot/Flight Director Pitch Channel System Interface

At this point we begin to consider a simplified hypothetical autopilot/flight director pitch channel system (a series of illustrations) which would be typical of one of the wide-bodied aircraft now in production.

On the left are typical input signals to the pitch computer. In the airplane we would expect to find two each of these items, such as two central air data computers, two VHF navigation receivers, and so on. If the airplane is not using inertial navigation systems, it will use remote directional and vertical gyros for pitch and roll error inputs.

Only pitch computer No. 1 is shown. Pitch computers No. 2 and No. 3 outputs to the flight director selector relays are indicated. The inputs to No. 1 and No. 3 computers are almost all from No. 1 systems. The inputs to No. 2 computer are almost all from No. 2 systems.

In order for these wide-bodied aircraft to be certificated for Category II approaches, they must be capable of making an automatic landing. In order to make an automatic landing under Category II conditions, the aircraft must be using at least two autopilots simultaneously.

In order for the aircraft to be certificated for Category III landing, it must be using at least three autopilots simultaneously. The 747 makes use of three separately boxed autopilots/flight director computers; three pitch channels and three roll channels. The DC-10 uses four autopilots on approach and they are boxed in two separate sets of boxes. Two of the autopilots, one in each box, are incomplete in that they are designed to function only in the approach mode. In every case, each operating autopilot has its own set of hydraulic actuators.

Continued on next page ...

Figure 23-3

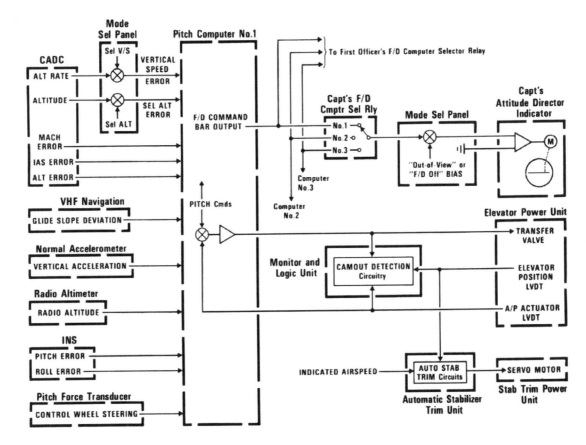

Autopilot/Flight Director Pitch Channel System Interface (cont'd.)

Much of the circuitry in a single pitch computer is common to the flight director command bar output and the autopilot transfer valve output. Downstream in the operation of the computer, however, these outputs are separated and treated differently as necessary for the differences between flight director and autopilot operation (Figure 23-3 illustrates this principle).

The flight director command bar output goes to the captain's and first officer's flight director computer selector relays. These relays are controlled by a flight director selector switch located on the flight instrument panel.

Either the captain or the first officer can select any one of the three channels to give him his flight director commands on his attitude director indicator. The selected computer output goes through the mode select panel. There it is subject to being overridden by "out-of-view" or "FID off" bias if the flight director is turned off, or if a monitored failure occurs. The out-of-view bias is generated by failure monitor operation.

In general, a failure of any input signal in use, or a failure of the pitch computer itself causes the flight director command bar to be biased out of view and a warning flag to appear.

If the autopilot and the flight director are both operating in the same mode, the same computed commands go to the autopilot and flight director. As long as the autopilot is functioning correctly, its response to its commands is correct, and therefore the flight director command bar should remain centered.

Among other things, the monitor and logic box has "cam out" detection circuitry. The term "cam out" means that the autopilot has been overpowered. In this case cam out detection occurs only if one autopilot is overpowered by another autopilot. The cam out detector considers an autopilot to be

overpowered if the output of its transfer valve amplifier is too large or too long, or if the elevator position LVDT signal is too much different from the autopilot actuator LVDT signal.

The automatic stabilizer trim circuits will function to trim the stabilizer if they see that the elevator is being held too far out of position for too long. The sensitivity of the automatic stabilizer trim unit is usually controlled by an airspeed function.

The servo motor used for trimming the stabilizer could be either electric or hydraulic. Dual circuits, dual motors, dual control and monitoring systems are always provided in automatic stabilizer trim operations for reliability and safety.

Indicated in the upper left mode select panel box are "Selected Vertical Speed" and "Selected Altitude" functions. "Selected Altitude" is short for "Pre-selected Altitude" and its operation is explained in the next diagram.

In the middle of Figure 21-1 is shown a vertical speed wheel. If the airplane is operating in vertical speed hold mode, the vertical speed wheel will be at a position such that a potentiometer connected to it will provide a signal calling for a particular vertical speed (altitude rate) signal. The altitude rate signal from the central air data computer opposes and is summed with the selected vertical speed signal.

If there is a difference, it shows up at the output of the summing point as "vertical speed error". It will be of a phase or polarity which will cause the autopilot to operate the elevators, or the flight director to move the command bar, as required to achieve the selected vertical speed.

The entire autopilot system, including all of the pitch and roll computers and associated boxes, is complicated by multiplication and cross monitoring. The basic principles of interface illustrated here apply, however, and make the functioning of all the interrelated parts comprehensible.

Altitude Hold Mode

Figure 23-4 shows an airplane being held at 40,000 feet altitude by the autopilot in altitude hold mode. There are two ways in which this operation is initiated. They are customarily distinguished by calling one of them "altitude hold" mode and the other "pre-select altitude capture" mode.

In the altitude hold mode, the autopilot or flight director latches onto and holds whatever altitude the airplane has at the time the pilot switches to altitude hold mode. In the other, the pilot pre-selects a desired altitude and captures it in a maneuver which can be compared to the capture of a radio course.

Figure 21-4 shows on the right of the mode select panel an altitude select knob which the pilot uses to pre-select a desired altitude. It also shows the switch would be moved down to "altitude select" if he wishes to pre-select, or up to the "altitude hold" position if he merely wishes to maintain the altitude which he has.

If he pre-selects a desired altitude, he approaches that desired altitude either in pitch attitude hold,

indicated airspeed hold, Mach hold, vertical speed hold, or control wheel steering mode.

As long as he sets up the autopilot to approach the desired altitude either from above or below, the autopilot automatically switches into "pre-select altitude capture" mode at a time determined by the autopilot. The capture point depends upon the rate at which the airplane is approaching the desired altitude, but is limited by a predetermined maximum distance from that pre-selected altitude.

At the time the autopilot or flight director reaches the capture point, the previous mode, whatever it may have been, is cancelled and the pitch channel goes into pre-select altitude capture mode.

From that time on, a "selected altitude error" signal and altitude rate control the autopilot and/or flight director. In a manner similar to a radio course capture operation, the altitude rate signal opposes the "selected altitude error" signal if the airplane is approaching the selected altitude, and aids the "selected altitude error" signal if the air-plane is leaving the selected altitude.

Figure 23-4

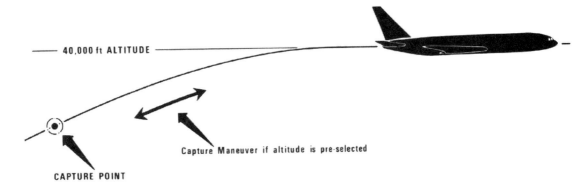

Altitude Pre-select Circuit

Figure 23-5 uses the principle of coarse and fine synchros, explained in Figure 4-14. The overall function of this system is to provide, to the pitch computer, error signals which result from the difference between a pre-selected altitude and the actual airplane altitude as seen by the CADC and modified by the barometric setting on the captain's altimeter (the central air data computer might itself be corrected by a setting of the altimeter baro knob, Figure 16-5, in which case the coarse and fine synchros in the captain's altimeter would not be needed for this system).

All of the synchros and their rotors at the time of installation in their respective boxes must be carefully positioned according to a set plan. The plan is as follows: If the airplane is at the pre-selected altitude, the signals from the coarse and fine altitude synchros in the CADC, after having passed through the differential synchros in the mode select panel, will show up on the stators of the synchros in the captain's altimeter so that the rotors of the synchros in the captain's altimeter will be perpendicular to their fields, giving null signals.

If the airplane altitude changes, the rotors of the coarse and fine synchros move, which causes movement of their respective fields in the stators of the captain's altimeter synchros, generating not-null signals in the rotors of these synchros.

Returning to the original null signal output condition, if the captain moves the knob on his altimeter, he will move the rotors of the coarse and fine synchros and their outputs will no longer be null.

Returning again to the null output condition, if the captain moves his altitude pre-select knob on the mode select panel, he will move the rotors of the differential synchros in the panel and produce not-null signals to the pitch computer.

Any one of these three operations is seen to have an effect on the error signal outputs from the captain's altimeter.

To demonstrate the pre-select altitude operation, assume that the airplane is flying at 10,000 feet and the captain decides he wants to fly at 25,000 feet. With the knob on the mode select panel, he selects in the digital readout window 25,000 feet. That introduces a large error signal from the altimeter coarse synchro to the pitch computer. The altimeter fine synchro output might be anything, depending upon the gear ratios of the coarse and fine synchro system.

The gear ratio is probably such that the output of the fine synchro rotor would pass through several nulls on the way from 10,000 feet to 25,000 feet. The captain will select an autopilot mode of operation, suitable to his flight plan, by which the autopilot will fly the airplane to the pre-selected altitude — for example, indicated airspeed hold, vertical speed hold, pitch attitude hold, or control wheel steering.

The coarse altitude error signal is used by the pitch computer to limit the maximum number of feet from the pre-selected altitude at which capture will occur. The capture will occur at a point, mainly determined by the airplane altitude rate, such that a smooth capture maneuver can be performed (reference Figure 23-4). At the time of capture, whatever mode the autopilot was in will be automatically dropped, and the fine altitude error and altitude rate signals will control the pitch channel.

Figure 23-5

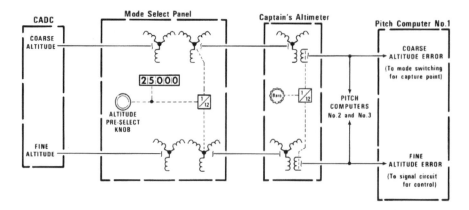

A/P-F/D Pitch Computer Base Schematic

Figure 23-6 is the pitch channel computer. It is the base print upon which is drawn heavy lines to illustrate signals and functions in use during particular modes, as was done for the roll channel. Looking first at some of the inputs, the indicated airspeed signal input at the top left accomplishes gain changes. It is required throughout the computer, to control the sensitivity of the autopilot in its various modes as a function of indicated airspeed. A Mach signal input might also be used for the same purpose.

The pitch attitude error input (next down) has an asterisk to refer it to, and carries the signal to the pitch attitude error input indicated in the lower middle part of the drawing.

Roll attitude error is used to generate a versine signal. Since these pitch and roll computers are virtually 100% solid state circuitry, the versine signal is generated with a solid state circuit using roll attitude error input. The versine signal is never disconnected, except perhaps during flare mode.

The altitude rate signal comes directly from the central air data computer, without passing through the mode select panel.

Skip the next four. The altitude hold error is the signal shown in the CADC of Figure 23-3 as altitude error.

Continued on next page ...

Figure 23-6

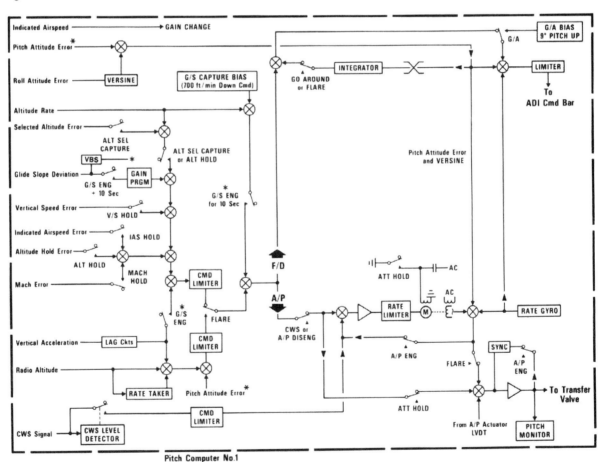

Pitch Computer No.1

A/P-F/D Pitch Computer Base Schematic (cont'd.)

The vertical acceleration signal comes from an accelerometer often called "normal" accelerometer. That does not imply that other accelerometers are abnormal. It is merely one of those engineering terms that we accommodate ourselves to. It is located near the center of gravity and oriented so that it senses vertical motion.

Three command limiters are indicated, one above the other. They function to limit any signal they receive to a value which will cause no more than a predetermined amount of pitch attitude away from horizontal. The amount of limiting varies from one mode to another, but will never be in excess of 15° nose up and 10° nose down.

In the upper right corner is a limiter in the signal path to the ADI command bar. That limiter does not limit pitch attitude, it merely limits the amount of displacement of the pitch command bar from its center position.

The servo motor loop shown in the lower right corner does not actually exist in an autopilot for a wide-bodied airplane, but the treatment and generation of signals is most easily represented by a servo motor loop.

The solid state computer accomplishes identical functions without the liabilities inherent in electromechanical devices. We do not care whether they use vacuum tubes and relays and electromechanical devices, or whether it is all solid state circuitry, as long as we understand the operations performed.

Near the center of the diagram is shown the point at which the flight director signals are separated from the autopilot signals. The flight director signals go up and the autopilot signals go down. To the left of the separation point, the flight director and autopilot have computer circuitry in common.

The servo motor fixed field input is used for convenience in clamping the servo motor by switching it to ground during attitude hold mode.

A/P F/D Pitch Computer Base Schematic – Digital Computer (Typical Boeing 737)

The purpose of both Flight Control Computers (FCC's) pitch control, is to provide longitudinal axis, flight directed control, using manual operated flight controls with the A/P not engaged or engaged in CWS (if so equipped). The FCC also provides A/P commands to automatically operate the flight controls using the A/P in CMD. The FCC's have separate A/P command signal processing to operate their respective system elevator AP actuator. Either actuator provides the same mechanical input to both elevator PCU's (Figure 23-6a).

The FCC uses both input and output signals. The analog inputs to the FCC originate from the pitch force transducers, alpha vane sensor (angle of airflow), flap position sensor, VHF NAV receiver, MCP, and LRRA. In addition, inputs from the elevator position sensor, and neutral shift sensor provide command feedback for A/P elevator control.

An input from the stabilizer position sensor provides stabilizer information for stab trim commands. These are connected to an analog to digital (A/D) converter inside the FCC for internal processing. The signals for all computation processing, gain control, A/P and F/D commands, from the IRU, FMC, DADC, MCP, are digital processed by ARINC 429 receivers, within the FCC, for internal use. Pitch attitude and pitch acceleration are normally from the local IRU, for A/P and F/D operation.

Digital outputs are sent to an EFIS Symbol Generator for EADI F/D pitch command bar movement, and to visually inform the crew of the mode and status of the DFCS system. An interlock signal is sent to the Autoflight Status Annunciator, for A/P warning, failed, or BITE conditions, to be visually displayed. Analog outputs are sent to the A/P actuator, the stabilizer trim servo, and the Mach trim actuator.

A/P engage logic is processed both in the MCP and the FCC, to determine mode selection. Hydraulic solenoids in the A/P actuator are energized to activate A/P control. The elevator position sensor is used prior to engaging the A/P, to allow the FCC to synchronize the A/P actuator to the surface position. Once engaged, the digital pitch A/P command, computed to satisfy engaged mode requirements, is converted to an analog command. The signal is then amplified and applied to the elevator A/P actuator transfer / servo valve. A/P elevator authority is 6 degrees max at low speeds, 2.5 degrees at high speeds, based on the FCC inputs of elevator feel, using airspeed vs. stabilizer position, and DADC airspeed.

The DADC provides total and static pressure inputs which are combined to generate differential pressure. This signal is combined with tabular feel pressure calculated as a result of stabilizer position and used to calculate total elevator authority.

The transfer valve converts the electrical signal to a hydraulic signal. This is used in the A/P actuator to drive an output arm, to mechanically move the input torque tube of both PCU's. Elevator actuator position transducer (LVDT) feedback signal tells the FCC, whether the actuator answered the A/P command.

The elevator surface position sensor signal tells the FCC when the elevators are at neutral, and how much to move the elevators to get them back to neutral, once the mode required corrections are accomplished. Either actuator, as selected, may be used for single channel operation or both are used for dual channel operation. Operation from this point is the same as manual flight control operation.

The FCC also provides outputs to the stabilizer for maintaining proper pitch attitude. A combination of stabilizer position and elevator output command is used to determine when the stabilizer should be repositioned. The output to the Mach trim actuator is used to reposition the elevators as a function of Mach number. This is accomplished by mechanical connections to the elevator feel and centering unit. The neutral shift sensor provides feedback to determine the elevator neutral position, based on the sum of stabilizer and Mach trim positions. The F/D signal processing for the EADI command bars is very similar to the A/P signal processing. A/P command mode logic is used, but the F/D command processing is accomplished as a separate control.

Continued on next page ...

A/P F/D Pitch Computer Base Schematic – Digital Computer (cont'd.)

Figure 23-6a

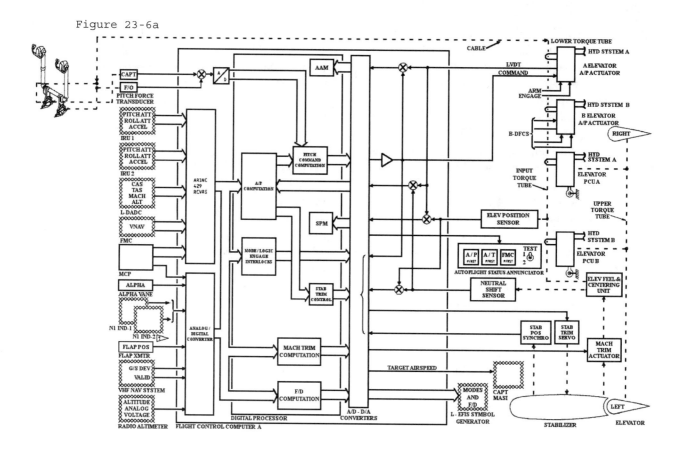

Pitch Computer: A/P Disengaged Mode; F/D Vertical Speed Hold Mode

Figure 23-7 shows the autopilot in the disengage mode and the flight director in the vertical speed hold mode.

The servo motor loop is in a synchronizing mode similar to the synchronizing mode of the servo motor loop in the roll channel. The only input to the servo motor amplifier is the difference signal that might temporarily exist between the control synchro and the pitch attitude error signal at their summing point. From this summing point, a null is maintained by summing the servo motor operating in a high gear mode, quickly following up on pitch attitude or versine changes, or rate gyro signals.

The synchro rotor position is a direct function of airplane pitch attitude, while the autopilot is disengaged.

The transfer valve amplifier (lower right) output is kept at a null by the operation of its synchronizer (sync). The synchronizer inverts any temporary signal from the amplifier and feeds it back into the amplifier to null its output.

The pitch monitor box at the lower right represents all of the computer and interlock monitoring functions of the pitch channel.

Continued on next page ...

Figure 23-7

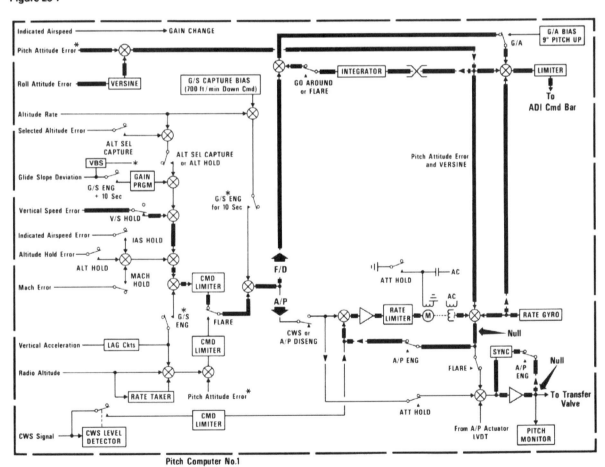

Pitch Computer No.1

Pitch Computer: A/P Disengaged Mode;
F/D Vertical Speed Hold Mode (cont'd.)

With the flight director in the vertical speed mode, if a vertical speed error signal develops, it goes through the flight director command limiter, through the flare switch below it, into the separation point between flight director and autopilot, to the top of the page and through the go-around switch to a summing point, and from there through a limiter to the command bar.

At the summing point it meets the pitch attitude error signal. Pitch attitude error goes also to the left through a signal inversion circuit into an integrator (time constant on the order of 15 to 30 seconds). Any pitch attitude error signal which persists for 15 to 30 seconds is cancelled by the integrator output.

To illustrate how the vertical speed mode operates, assume the airplane is climbing at a pre-selected vertical speed of 4,000 feet per minute, with a pitch attitude of 10° nose up. The 10° attitude error signal will have been cancelled by the integrator output. Since it is climbing steadily at the pre-selected speed of 4,000 feet per minute, there is no vertical speed error.

Suppose the captain advances the throttles. The airplane begins to climb too fast, developing a vertical speed error signal. This signal immediately shows up at the input to the command bar limiter, moving the command bar down, telling the captain he should put his airplane nose down.

As soon as the captain starts to move the airplane nose down, the pitch attitude error signal diminishes. When pitch attitude error diminishes, the integrator output is larger than the pitch attitude error, so they no longer cancel each other. The difference between these two signals then opposes the vertical speed error signal.

When the difference between pitch attitude error and integrator output becomes great enough to cancel the vertical speed error signal, there is no longer a signal to the ADI command bar and it will center. As the integrator, over its time constant period, adjusts itself to cancel the new pitch attitude error, the captain keeps the command bar centered.

As long as the captain keeps the command bar centered, the attitude of the airplane will be such as to cause the airplane vertical speed to equal the selected vertical speed. After a half minute or so, a new steady pitch attitude results with no vertical speed error.

Pitch Computer: A/P-F/D Vertical Speed Hold Mode

Figure 23-8 shows both the autopilot and the flight director in the vertical speed hold mode. Since the flight director vertical speed hold mode was previously discussed, we will take up only the autopilot portion here.

While the autopilot was disengaged, the servo loop had been keeping its control synchro rotor at a position corresponding to pitch attitude. When the autopilot was engaged, the two synchronizing operations previously described were disconnected with the autopilot engage switches.

If the airplane vertical speed equals the selected vertical speed, there is no vertical speed error signal into the servo motor amplifier or into the transfer valve amplifier. In addition, there is a null signal out of the summing point which is to the right of the control synchro.

Continued on next page ...

Figure 23-8

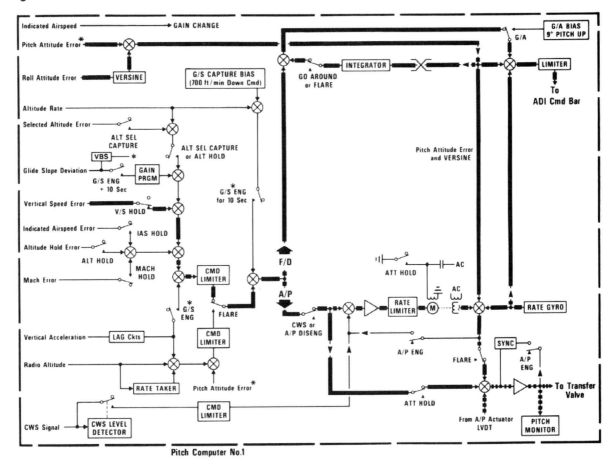

Pitch Computer No.1

Pitch Computer: A/P-F/D Vertical Speed Hold Mode (cont'd.)

Assume that at the time of entering the vertical speed hold mode, a vertical speed error does exist (it may be a large selected error). In going through the command limiter, it is limited to a valve which causes no more than 15° nose up or 10° nose down. That signal out of the command limiter goes immediately to the transfer valve amplifier. The transfer valve moves the autopilot actuator enough so that the autopilot actuator LVDT signal cancels the vertical command signal at the summing point ahead of the transfer valve amplifier. This would have displaced the elevators, which continually change the pitch attitude as long as they are displaced.

As the pitch attitude changes, a difference signal develops out of the summing point on the right of the control synchro rotor, which opposes the command signal. At the same time this difference signal is developing (pitch attitude changing), the vertical speed error is diminishing because airplane vertical speed is approaching selected vertical speed.

Also, at the same time, the servo motor is slowly rotating its control synchro rotor toward the new required pitch attitude. Within the first couple of seconds, the greatest change takes place: The vertical speed error is diminished; the resulting diminished command is partly cancelled by the difference signal; and the elevators are not too far from faired (A/P LVDT canceling residual command).

However, the pitch attitude is still changing slowly; there is still a vertical speed error (diminished); and the servo is still running. This condition continues until the vertical speed error is eliminated. Then the servo stops running and synchro rotor position corresponds to pitch attitude. So there is no difference signal, and the elevators are faired because there is no signal into the transfer valve amplifier. (The stabilizer trim may have operated, but more on that later.)

Pitch Computer: A/P Attitude Hold Mode

Figure 23-9 shows the autopilot in attitude hold mode. The mode of the flight director is not indicated since it could be in any selected mode. The servo motor has been clamped by the switch above it, grounding out its fixed field voltage. This also clamps the position of the synchro rotor.

Subsequently, whenever the pitch attitude error signal does not correspond to the airplane pitch attitude command by the position of the control synchro rotor, there is a difference signal into the transfer valve amplifier. This signal operates the elevators to restore the pitch attitude the airplane had when the autopilot went into the attitude hold mode.

Other signals are disconnected from the transfer valve amplifier by the switch ahead of it.

Figure 23-9

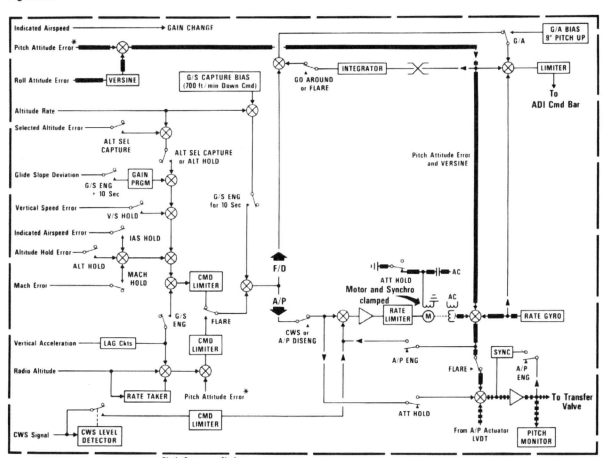

Pitch Computer No.1

Pitch Computer: A/P-F/D Altitude Hold Mode

In Figure 23-10, both autopilot and flight director are in altitude hold mode. Their operations are the same as in the vertical speed hold mode (Figure 23-8), except that the controlling input signal is altitude hold error instead of vertical speed error, and that there is a rate signal added. The rate signal is altitude rate, which exists only when the altitude is changing.

The rate signal opposes the altitude hold error signal as the airplane approaches the desired altitude and aids as it leaves the desired altitude. Its function corresponds to the radio rate signal function shown in Figures 21-38, 21-39 and 21-40.

The servo motor section can be thought of either as a mechanical integrator or as setting up the "basic required pitch attitude", over the long term, from which the airplane deviates as short term altitude hold error signals develop.

As the airplane fuel gradually burns out, the servo motor slowly drives its control synchro rotor in the nose down direction, decreasing the lift of the airplane as its weight decreases.

If the pilot changes the power setting of the engines, the servo motor will run slowly, over 15 or 20 seconds, to move its control synchro rotor to the new position corresponding to the new required pitch attitude.

Figure 23-10

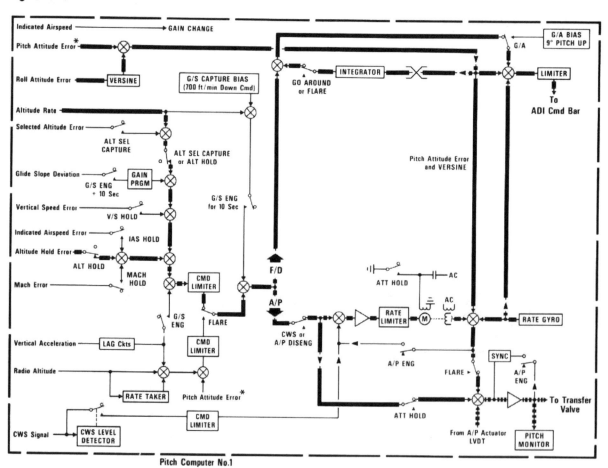

Pitch Computer No.1

Pitch Computer: A/P-F/D Selected Altitude Capture Mode

The "selected altitude capture" mode of flight director and autopilot is heavy lined in Figure 23-11. At the time of altitude select capture, the airplane will have been climbing or descending to-ward the selected altitude. There is an altitude rate signal corresponding to the rate of change of altitude, and a selected altitude error signal corresponding to the deviation from the selected altitude.

These two signals oppose each other because the airplane is approaching the selected altitude. It is the opposition of the rate signal to the altitude error signal which accomplishes the smoothly flaring capture maneuver, a curved path toward the selected altitude.

Figure 23-11

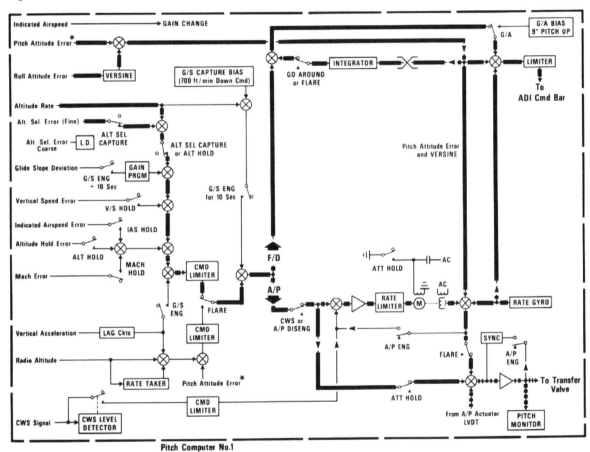

Pitch Computer No.1

Pitch Computer: A/P-F/D Mach Hold Mode

The operations of the autopilot and flight director in Mach hold mode (Figure 23-12) is the same as its operation in vertical speed hold mode, with the exception that the controlling signal is a Mach error signal.

Figure 23-12

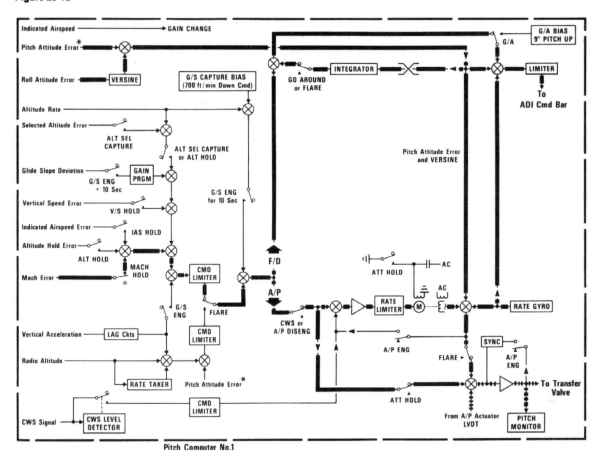

Pitch Computer No.1

Pitch Computer: A/P FID Indicated Airspeed Hold Mode

Indicated airspeed hold mode (Figure 23-13) is the same as vertical speed hold mode with the exception of the difference in the controlling signals.

Figure 23-13

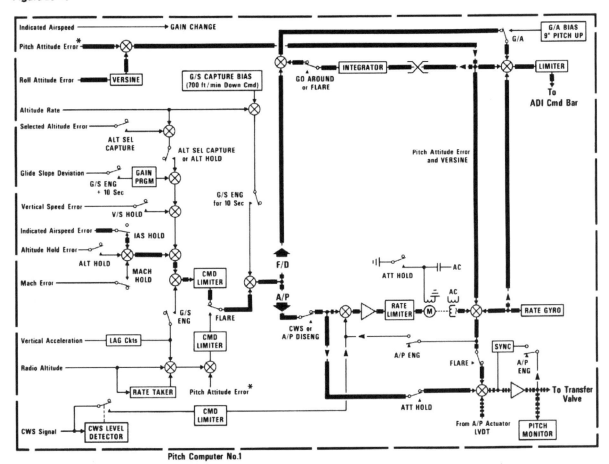

Pitch Computer No.1

Pitch Computer: Control Wheel Steering Mode

While the autopilot is in control wheel steering mode (Figure 23-14), the flight director could be in any mode selected by the pilot. He can utilize his manual control of the autopilot to follow the commands of the command bars in the ADI.

The control wheel steering mode ordinarily operates in conjunction with attitude hold mode. A control wheel steering maneuver will be initiated only if the pilot applies sufficient force to the control column, on the order of ten pounds, to cause the control wheel steering level detector to operate its switch.

In that case, it switches in the control wheel steering signal from the force transducer. That signal goes to the command limiter, where it is limited so as not to cause excessive airplane attitude changes. From there it goes to the servo motor amplifier, is rate limited, and runs the servo motor.

As long as the control wheel steering signal is present, the servo motor continues to run, turning the control synchro rotor, which changes aircraft pitch attitude. As long as the control synchro rotor is turning, there is a difference between its signal and the pitch attitude error signal.

This difference signal causes the transfer valve amplifier to keep the autopilot actuator moved far enough from neutral so that it develops a follow-up signal to cancel the difference signal. The elevators are therefore displaced far enough to cause the pitch attitude to change fast enough to follow up on the changes in position of the control synchro rotor.

As soon as the pilot stops applying force to the control column, the control wheel steering level detector opens its switch, and the autopilot reverts to attitude hold mode.

Figure 23-14

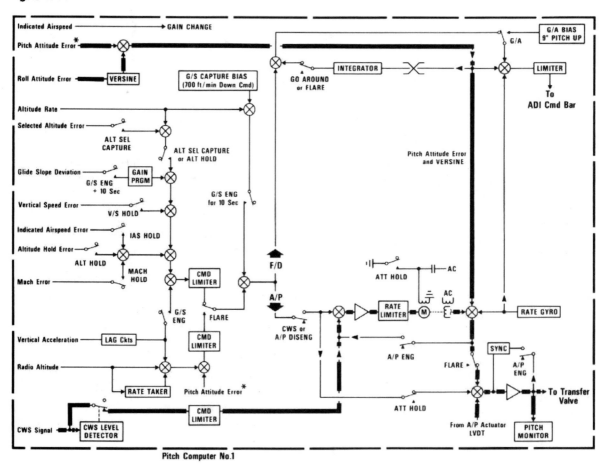

Pitch Computer No.1

Glideslope Capture Illustration

Figure 23-15 illustrates the method of capturing a glidepath from above, and Figure 23-16 shows a capture from below. The vertical "dot pattern" area represents a ten second period after glidepath capture. During the ten second period after capture, the autopilot causes the vertical speed to change to a 700 feet per minute rate of descent.

In Figure 23-15, the autopilot may have been descending at 1,500 feet per minute. When the airplane comes to within about one-third of a dot of glideslope deviation, capture occurs and the autopilot begins to move the airplane nose up, and continues to do so until the rate of descent is slowed to 700 feet per minute.

The mode of the autopilot prior to a capture could have been attitude hold, indicated airspeed hold or vertical speed hold. Whatever the mode of the autopilot prior to capture, it is automatically cancelled when glideslope capture occurs. The sensing circuit which initiates glideslope capture mode is called the "vertical beam sensor".

In Figure 23-16, the mode of the autopilot prior to capture, might have been altitude hold, indicated airspeed hold or whatever other mode the pilot wished to use. At the time of capture the mode in use is disengaged, and glideslope capture mode initiated. The vertical speed prior to capture was probably zero, so the airplane is nosed down until its vertical speed equals 700 feet per minute.

In capturing from above, the airplane probably undershoots the glidepath. In capturing from below, it probably overshoots. In either event, at the end of the ten second period following capture, the submode calling for 700 feet per minute down is disengaged, and the glideslope deviation signal is switched into the command circuit as the primary control. Although this pitch channel does not use it, a radio rate signal and/or an average rate of descent signal might be added. A vertical acceleration signal smoothes the descent and helps to hold close to the glidepath.

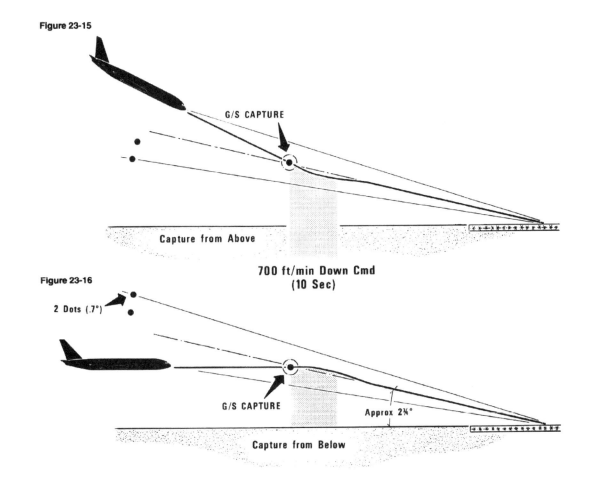

Figure 23-15

G/S CAPTURE

Capture from Above

**700 ft/min Down Cmd
(10 Sec)**

Figure 23-16

2 Dots (.7°)

G/S CAPTURE

Approx 2¾°

Capture from Below

Pitch Computer: A/P F/D Glideslope Capture For Ten Seconds Mode

Figure 23-17 shows the pitch computer mode which exists for ten seconds after glideslope capture. The switch below the "Glideslope Capture Bias" box marked "G/S ENG for 10 Sec", closes at the time of glidepath capture, and remains closed for ten seconds. Glideslope engage is synonymous with glideslope capture.

The "Glideslope Capture Bias" is a calibrated signal which equals and cancels (opposite phase) an altitude rate signal of 700 feet per minute down.

If the capture is from above, the altitude rate signal is greater than the glideslope capture bias. Their difference out of the summing point calls for elevator up until the altitude rate diminishes to 700 feet per minute. At that time, there is no longer a difference, and no longer a command to the elevators.

The vertical acceleration signal is modified to provide the desired smoothing effect. It opposes changes in vertical speed.

Figure 23-17

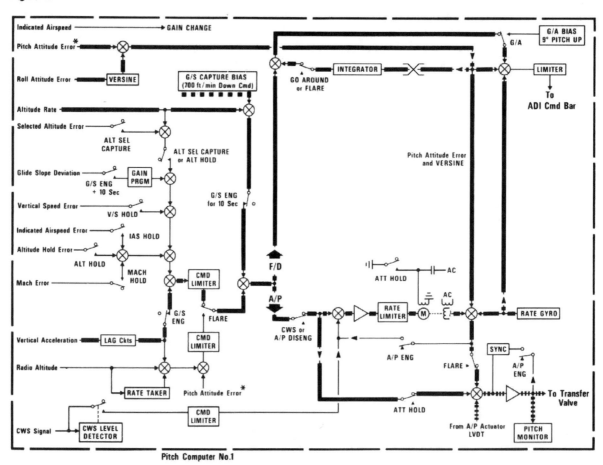

Pitch Computer No.1

Pitch Computer: A/P-F/D Glideslope Capture Plus Ten Seconds

Figure 23-18 shows the signals in use ten seconds after glideslope engage. The glideslope capture bias signal is switched out and the glideslope deviation signal is switched into the command chain.

The gain programmer operates, in late model autopilots, in accordance with information from the radio altimeter. In different autopilots the point of the beginning of the gain programming operation ranges from 15,000 feet above the run-way to 1,000 feet above the runway.

When gain programming begins, the glideslope deviation signal strength is diminished by the gain programmer as a function of radio altitude, compensating for the glideslope beam convergence.

For example, if the gain programmer begins its operation at 1,500 feet, when the airplane reaches 750 feet the gain programmer will be attenuating any glideslope deviation signal by 50%. By the time the airplane descends to approximately 50 feet above the runway, the gain programmer will be attenuating the glideslope deviation signal by 100%.

Gain programmers in earlier autopilots operated on a time base so that, over a period of perhaps 100 seconds, the glidepath signal would be attenuated by 75% and a second gain programmer would start to operate at the middle marker, attenuating the glidepath signal an additional 75% over a period of perhaps 25 seconds.

Figure 23-18

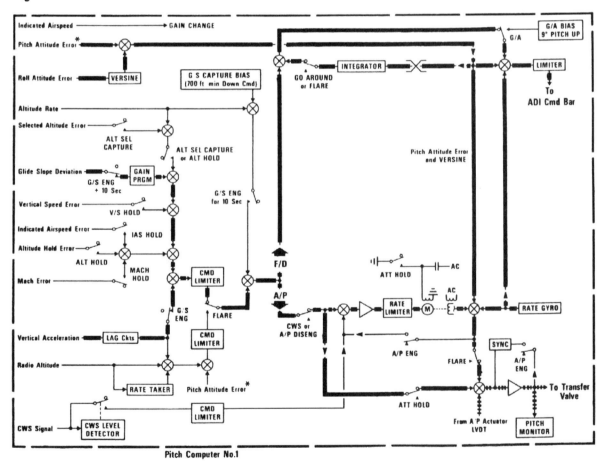

Pitch Computer No.1

Pitch Computer: A/P-F/D Flare Mode

Figure 23-19 illustrates the "flare" mode. "Flare" is the term used to describe the touchdown maneuver. The radio altimeter reading at the time of flare will typically be about 50 feet.

When flare mode is initiated, the switches marked "flare" are operated to their triangles. One in the lower right corner disconnects the servo motor synchro output, and another in the lower left quadrant switches in vertical acceleration, radio altitude, and unmodified pitch attitude error. The gain programmer will be attenuating glidepath signals 100%. (In order to avoid having many line crossovers in this diagram we show the pitch attitude error signal coming into the signal chain between the two lower command limiters.)

The flare mode of operation can be compared to a radio capture mode in the roll channel. The radio altitude signal corresponds to the radio deviation signal. The altitude rate signal ("rate taker" box) corresponds to the radio rate signal.

The pitch attitude error signal corresponds to the heading error signal. The vertical acceleration signal opposes sudden changes in the rate of descent.

At the time of flare, the pitch attitude error signal calls for nose on the horizon. The radio altitude signal calls for nose down. The radio altitude rate signal calls for nose up and predominates. We can say that the pitch channel is performing a capture maneuver which is a capture of the runway.

If the auto-throttle system is also in use, it retards the throttles to their aft stops when the radio altitude is about 30 feet.

On a category III equipped aircraft performing a triple channel approach, the A/P will not only support the flare function, but will go on to auto-land the aircraft. Once the aircraft's main gear makes contact with the runway the "rollout" mode becomes active and will correct for any approach crab angle and will track the runway centerline. The A/P will remain engaged until disconnected.

Figure 23-19

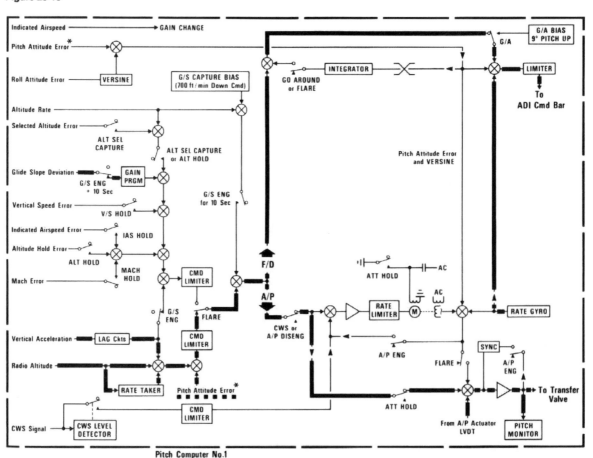

Pitch Computer No.1

Pitch Computer: F/D Go-Around

The flight director go-around mode is illustrated in Figure 23-20. Autopilots are capable of performing the go-around maneuver, but typically are not permitted to do so. Initiation of the go-around mode is usually accomplished with palm switches, located on the aft sides of the throttle levers, so that when the throttles are advanced, it is easy to press those switches with the heel of the hand at the same time. When the go-around mode is initiated, the autopilot is disconnected.

In the upper right corner is a box designated "go-around bias" which, in this flight director, calls for 9° nose up. That means that if the airplane attitude is 9° nose up, the pitch attitude error will equal and cancel the go-around bias signal. At the time of go-around initiation, the pitch attitude would be approximately nose on the horizon or perhaps 2° down.

The go-around bias signal would move the pitch command bar up to its limited position. The pilot would pull the control column back and hold it back until the pitch attitude equaled 9° nose up. At that time, the pitch attitude error and go-around bias cancel each other and the command bar is centered.

If the pilot deviates from the 9° nose up attitude, the command bar will move up or down, telling him to raise or lower the nose of the airplane (the roll channel at this time commands the pilot to maintain the wings level).

The go-around mode is used until the aircraft has achieved sufficient altitude so that it can decrease its rate of ascent and follow the control tower instructions relative to making another approach.

On a Category III equipped aircraft with the pitch computers engaged in approach, go-around mode will cause the autothrottle to increase thrust to go-around limit.

When 2,000 fpm climb is reached, autothrottle reduces thrust to maintain 2,000 fpm. When MCP altitude is reached, AFDS reverts to altitude hold and autothrottle adjusts thrust to hold selected airspeed.

Above 400 feet with the airplane level or climbing, the AFDS can be taken out of go-around by selecting another pitch mode.

Figure 23-20

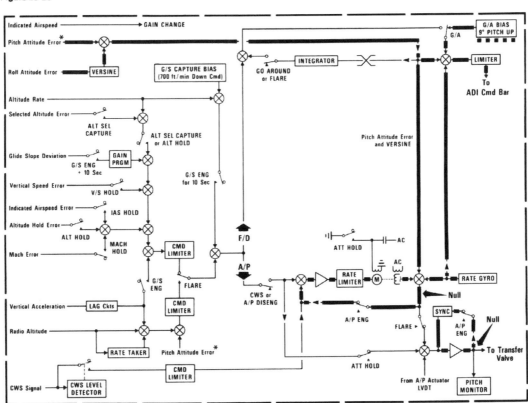

Pitch Computer No.1

Automatic Stabilizer Trim System

Figure 23-21 shows a highly abbreviated stabilizer trim system. Practically everything in a stabilizer trim system is duplicated for reliability and safety. The motor on the right drives a jack-screw whose nut is attached to the forward spar of the stabilizer. The stabilizer is pivoted on the aft spar. Turning the jackscrew one way or the other raises or lowers the nose of the stabilizer.

The motor itself could be hydraulically operated and electrically controlled, or it could be a three-phase electric motor. We have shown a DC electric motor for convenience.

There are always at least two stabilizer motors, and usually, both can drive into the same differential gear box. If both are driving, the rate of operation is greater than that of only one motor. If there are two electric motors, one is typically smaller and drives through a lower ratio gear train for slow speed operation. There might be one hydraulic motor used for the fast operation, and an electric motor used for

the slow operation. Fast is used during takeoff and approach, and slow is used for cruise. The slow operation is used by the autopilot system.

On the outboard horns of the cockpit control wheels are two switches mounted close together which can be operated by one thumb. They are labeled "Captain's and First Officer's Trim Sw's".

The control wheel trim switches operate the high speed system of the stabilizer, and levers on the pilot's side of the pedestal usually operate the same system. An additional low speed manual trim switch (usually located somewhere on the pedestal) can be used for cruise trimming of the stabilizer.

At cruise speeds, it is generally not desirable to operate the stabilizer in the high speed mode, so the pilot will use the pedestal switch for slower operation. At the lower speeds used for takeoff or approach, he would use the control wheel horn switches.

Continued on next page ...

Figure 23-21

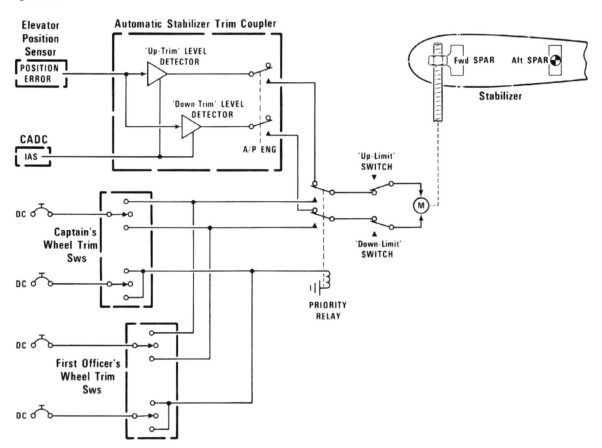

Automatic Stabilizer Trim System (cont'd.)

Generally, the manual operation of the stabilizer trim system by the pilot disconnects his autopilot. The reasoning behind this arrangement is that, if the pilot needs to trim the stabilizer, the autopilot is not doing its job correctly. Operation of either the captain's or first officer's wheel trim switches operates a priority relay which disconnects from the motor any signals that might be generated by the automatic stabilizer trim coupler.

The automatic stabilizer trim coupler has another set of level detectors, in addition to the ones shown, which perform the same job (duplication for reliability and monitoring). A monitoring system looks to see that both sets of level detectors are operating in agreement. A disagreement generates a warning in the cockpit.

The basic control signal for automatic stabilizer trim is the elevator position. Elevator position could be sensed directly by the position sensor, or

indirectly by the control surface LVDT shown in Figure 18-6.

If the level detectors see an elevator too far away from the faired position for several seconds, one of them will operate the servo motor to trim the stabilizer up or down as needed. As the stabilizer is trimming, the need for the autopilot pitch channel to hold the elevators out of the faired position diminishes. When the elevators get close enough to the faired position, the level detectors stop their operation of the servo motor.

An airspeed function usually controls the sensitivity of the level detectors. At cruise speeds, if the elevator is held perhaps as little as 1/4° away from the faired position, the stabilizer trim system operates. At approach or takeoff speeds, automatic stabilizer trimming is not initiated unless the elevator is held much farther away from the faired position.

Figure 23-21

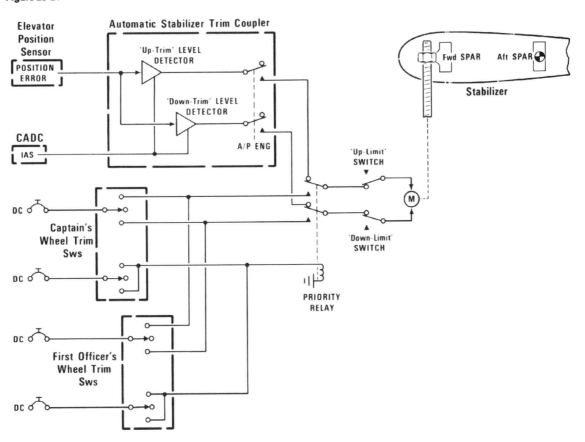

CHAPTER 24 - AREA NAVIGATION (RNAV)

Area Navigation

Area navigation (RNAV) is a navigation and guidance system which uses VOR bearing, DME slant ranging, and barometric altitude as its basic signal inputs to compute course and distance to a waypoint. Since the system can only function within the service area of a VOR/DME station, it cannot be used for overseas navigation.

Figure 24-1

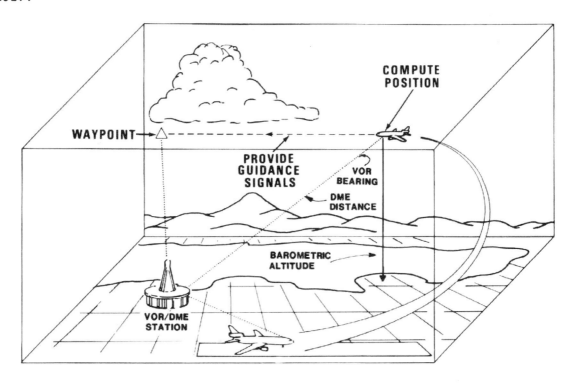

Area Navigation Block Diagram

A block diagram of a typical system is shown in Figure 24-2. The navigation computer receives a VOR bearing from the VOR receiver, DME distance from the DME interrogator, and altitude from the central air data computer.

A navigational data base is stored either within the navigation computer or in an external storage unit. The navigational data base contains all information needed regarding the routes between cities, the navigation aids (VOR/DME stations) and waypoints.

The control display unit is used to enter information into the computer and to display navigation information. In a typical commercial airplane installation, the computer may also send course deviation signals to the course deviation indicator, and lateral steering commands to the autopilot.

Figure 24-2

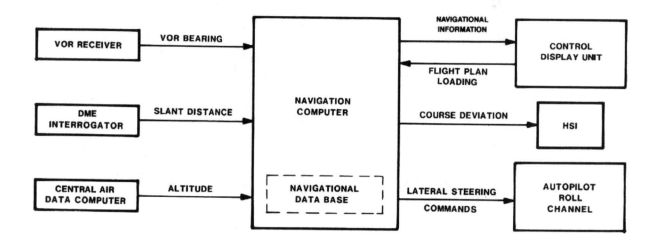

TYPICAL AREA NAVIGATION SYSTEM

Using Area Navigation

If we were to take a look at a flight plan between Chicago's O'Hare airport and Newark, we'd see something like the one in Figure 24-3. The flight would take us from one VOR station to another until, by a round about path, we arrive at Newark.

A more desirable flight plan would be a direct route.

Unfortunately, there can not be an ideal line-up of VOR stations between all stations. The area

navigation concept provides direct routes between airports.

Along each route there are waypoints towards which the airplane flies. The waypoint locations are established when the route is designed.

Each waypoint is associated with a specific NAV aid or VOR/DME station, such as in Figure 24-4.

Figure 24-3

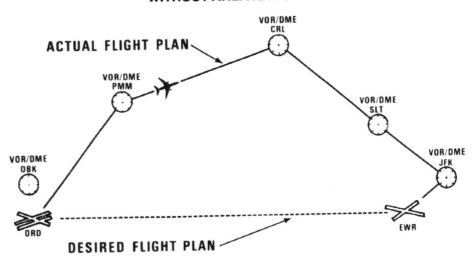

WITHOUT AREA NAVIGATION

Figure 24-4

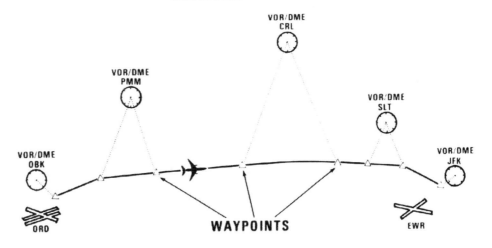

USING AREA NAVIATION

WAYPOINTS

Waypoint Characteristics

The navigational data base stored in the computer contains the following characteristics of each waypoint, as shown in Figure 24-5: latitude and longitude, altitude, frequency of its NAV aid, distance from the NAV aid, and magnetic bearing from the NAV aid.

If the VHF navigation system is tuned to the proper NAV aid, the area navigation computer will receive the information regarding the position of the aircraft in respect to the NAV aid as shown in Figure 24-6.

Figure 24-5

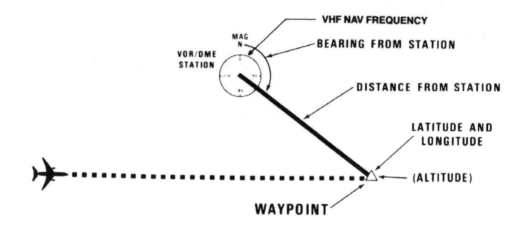

Figure 24-6

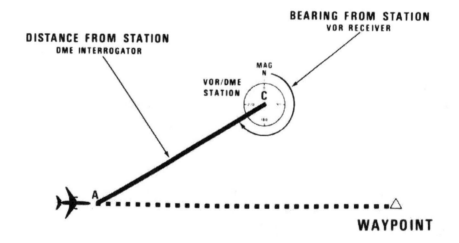

Rho-Theta Mode

Knowing the length of side A (DME distance), the length of side B (from the data base), and the angle of A (difference between the bearing of the aircraft and the bearing of the waypoint), we can compute the length of side A-B, which gives distance to the waypoint; and angle B, which is the course or track angle to the waypoint.

This combination is referred to as the Rho-Theta mode of area navigation, where Rho is the DME distance, and Theta is the VOR angle.

Figure 24-7

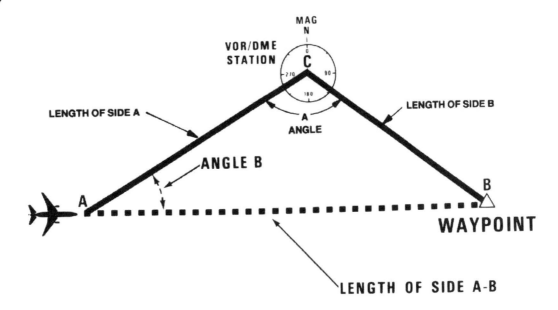

Rho-Rho Method

An improvement over Rho-Theta is possible using two DME distances. The navigation data base would be expanded to provide each waypoint with two NAV aid reference.

Improved position accuracy is achieved along with improved navigational accuracy. Rho-Rho is the preferred method of area navigation, as shown in Figure 24-8.

Figure 24-8

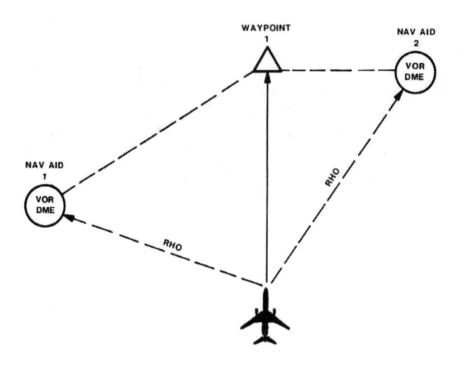

Rho-Rho RNAV Computers

To implement Rho-Rho, two RNAV computers are installed in the aircraft. The left computer receives VOR/DME from the left or number one system, while the right RNAV computer receives VOR/DME from the right or number two system.

The inertial sensor units (sometimes referred to as IRUs), provide an inertially corrected altitude, and acceleration data. An intersystem data base provides sharing of data between the two computers so that they may both use the same input signals; and

compare computer data to further improve navigational accuracy.

The VOR bearing is available to each computer, but not normally used. In the event of a loss of one of the DME signals, the computer will automatically revert to Rho-Theta mode.

In a modern system with dual computers, as the flight progresses from one waypoint to another, the computers automatically tune the VHF navigation frequencies for each side. If coupled to the automatic pilot, the airplane will track the flight plan.

Figure 24-9

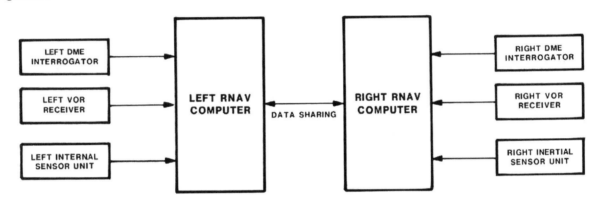

Airspace Routes

Area navigation provides more airspace between cities. In Figure 24-10, we show three additional routes being established, each with its own series of waypoints, referenced to the same NAV aids as the first route. This allows four times as much traffic in the same route area.

Figure 24-10

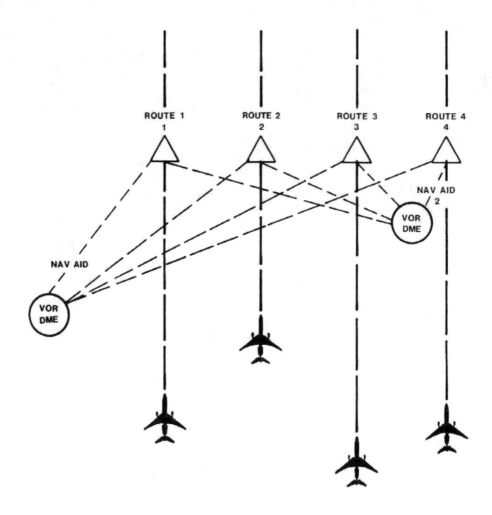

Frequency Scanning DME

If we are to add parallel routes and more flights to a geographic area, it is desirable to achieve the best possible navigational accuracy. This can be accomplished through use of frequency scanning DME interrogators. The interrogators can report distances to as many as five NAV aids, providing good reception.

The RNAV computer will use the ones that provide the best angles for mathematical computation, resulting in excellent course, distance and present position computation.

Figure 24-11

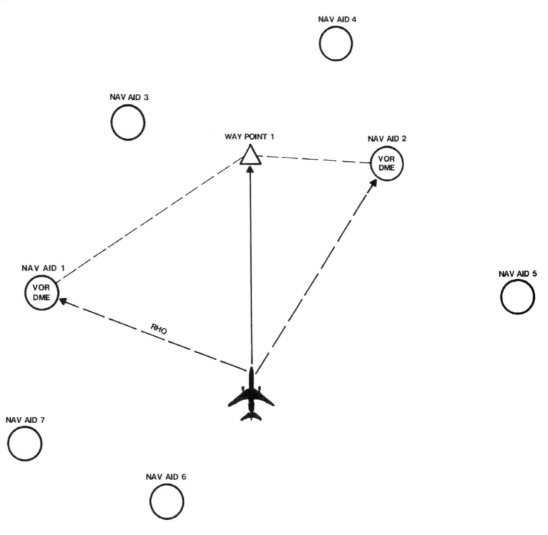

CNS / ATM

CNS / ATM is the acronym for Communication, Navigation. Surveillance / Air Traffic Management.

"FANS", "CNS / ATM or "'Free flight", no matter what you call it, the opportunity to save money and lower costs through satellite-based communication and navigation procedures is here today

What is FANS, CNS/ATM?

FANS - The Future Air Navigation System is the internationally agreed next-generation plan for more efficient communication, navigation, surveillance and air traffic management (CNS/ATM), based heavily on satellite technology (Figure 24-11a).

The FANS modification package typically consists of the following systems:

- Flight Management System software upgrade package
- GPS (Global Positioning System)
- SATCOM (satellite communications)
- AGARS/Data Link (Aircraft Communication And Reporting System)
- EFIS (Electronic Flight Instrument System)

These systems not only help to optimize the flight route but they improve communications between the aircraft and the ground, while the improved cockpit layout appreciably lessens the crew's workload.

Currently, United Airlines utilizes FANS on the 767-300, 747-400 and 777 fleets. The Airbus fleet (A319, A320) does not have FANS installed, but does use GPS as a navigational input.

Continued on next page ...

Figure 24-11a

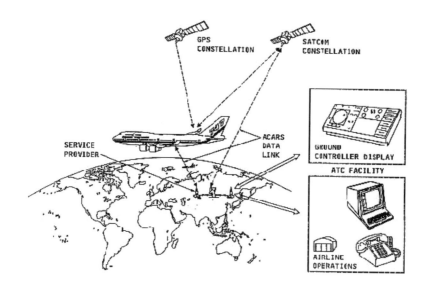

CNS / ATM (cont'd.)

Fans Block Diagram

Figure 24-11b shows the signal path of the Future Air Navigation System. Although the FMC has additional navigation inputs, the accuracy of the GPS is a main component of FANS. Flight plan updates, winds reporting, and ATC communication (data mode), is accomplished through the ARINC Communication Addressing and Reporting System (ACARS) over the VHF radio (center channel), when within range of a VHF station.

If the aircraft is out of VHF range, or if the VHF circuit is not operational, the SATCOM system would provide the same functionality.

For voice communications with Air Traffic Control, VHF and HF radio are the primary mode, with SATCOM as an authorized backup (secondary) system. For flight crew communications with System Aircraft Maintenance Control Center (SAMC), Airline Operations, or for passenger communications, the SATCOM system can be used as available.

Continued on next page ...

Figure 24-11b

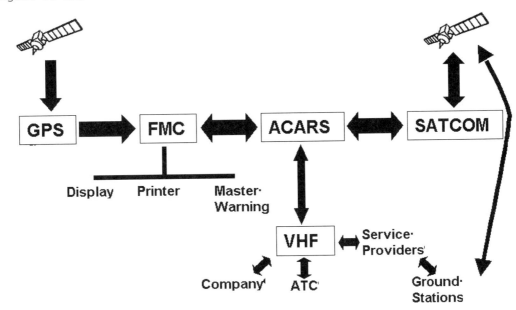

CNS / ATM (cont'd.)

Global Positioning System

The global positioning system (GPS) is a satellite-based radio navigation system which uses navigation satellites to calculate accurate airplane position and time (Figure 24-11c).

The first GPS satellite was launched in the summer of 1983. Today there are 24 GPS satellites in orbit, 21 primary and 3 spares. The life span of a GPS satellite is approximately 12 years. Currently, 2 replacement satellites are launched into orbit each year.

The 24 satellites orbit the earth at approximately 10,900 nm, effectively forming a constellation of

satellites. There are at least eight satellites visible to a GPS receiver at any time, anywhere on (or above) the earth.

For accurate positioning information, a GPS receiver must have a minimum of 4 satellites locked on. Three, will provide precise latitude and longitude identification, the fourth will add altitude information. In the event that only 3 satellite are locked on, the IRS system can act as the fourth satellite.

Continued on next page ...

Figure 24-11c

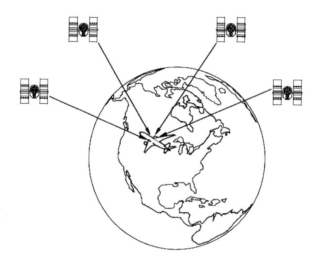

CNS / ATM (cont'd.)

Aircraft GPS System

There are two GPS systems installed on FANS equipped aircraft. Each system is identical, except for using on-side systems. For example, the left GPS System would be connected to the left FMC system and the left CDU.

The GPS system (Figure 24-11d) consists of the following:

- GPSSUs – GPS Sensor Unit. The GPSSU calculates the airplane position and updates the GPS clock. Some airplanes use a Multi-mode Receiver (MMR) in place of the GPSSU. The same functions are performed, however the GPS receiver is contained in the same Line Replaceable Unit as the ILS receiver.

- GPS Antenna – The antenna receives an extremely small L-band frequency signal from the satellite. Antenna cable impedance is 50 ohms, and the length of the antenna cable is limited to 15 feet due to signal loss.

- FMCs – The Flight Management Computer uses the GPSSU input and other navigation sensors to calculate airplane position and uses this information for flight plan management.

- CDUs – One of the Control Display Unit (Figure 24-11e) functions is to display the airplane's latitudinal and longitudinal position, and the source of this information. For example, if the GPS system was not operational, then the radio update mode might be DME-DME, or VOR-DME.

- DMU – The Data Management Unit monitors and stores airplane data, in this case, GPSSU data. Reports can be produced from the DMU for troubleshooting and performance monitoring.

- IRUs – The Inertial Reference Units send latitude and longitude to the GPSSUs for initialization. This allows the first satellite position fix to take place within 10 minutes from power-up. Short periods of adverse satellite coverage can occur. The GPSSUs use IRU data to aid in continued calculation of airplane position when not enough satellites are in view. This input also lets the GPSSUs require the satellites to re-enter the navigate mode quickly.

Continued on next page ...

Figure 24-11d

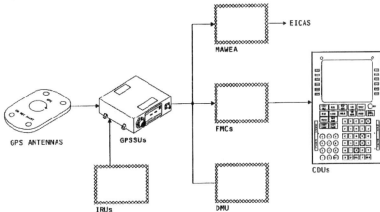

CNS / ATM (cont'd.)

Navigation Display

The radio position update mode in the lower right corner shows what navigation radio is currently being used by the FMC for position updating. Other possible modes would be D/D (DME-DME), V/D (VOR-DME) and L/V (LOC-VOR) to name a few. Below the radio position update mode is the IRS position update mode.

The current indication (Figure 24-11e) shows that the averages of all 3 IRUs are being used to indicate the IRS position. In the event of an IRU failure, the indication would degrade to the on-side IRU indication.

For example, if the center IRU failed, the left Navigation Display would indicate IRU (L), the right Navigation Display would indicate IRU (R), as the IRS position update mode.

NOTE: The GPS symbols are shown here side by side to illustrate the 2 GPS systems. In reality, the 2 symbols would normally be superimposed on top of each other, centered on the aircraft heading symbol (triangle).

Continued on next page ...

Figure 24-11e

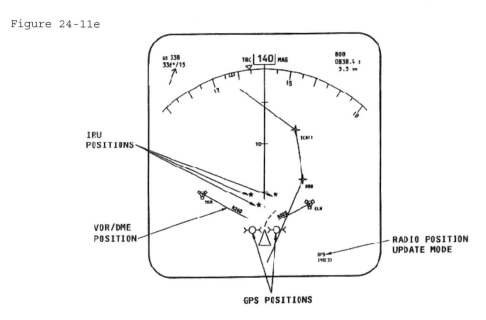

CNS / ATM (cont'd.)

Satellite Communications System

The satellite communications (SATCOM) system is a long range data and voice communications system. SATCOM is much less susceptible to atmospheric interference than VHF communications systems.

SATCOM uses 4 satellites in a geo-synchronous (stationary) orbit as relay stations between the airplane earth station and ground earth stations (Figure 24-11f). The earth stations use telephone lines or microwave links to complete the communication link between the airplane and the selected ground station.

All of the satellites are positioned above the equator at an altitude of 23,000 nm. They operate in a full duplex mode allowing simultaneous reception and transmission of a voice or data signal.

Continued on next page ...

Figure 24-11f

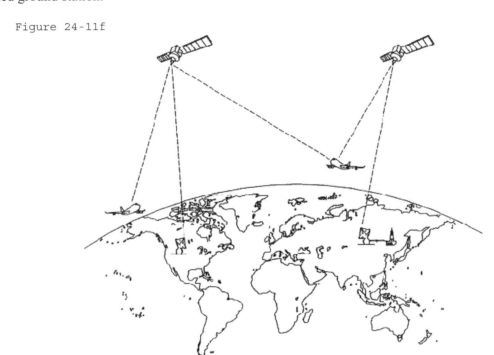

CNS / ATM (cont'd.)

Aircraft SATCOM System

On each of United Airline's FANS equipped aircraft, there is only one SATCOM system installed. There are two antenna systems installed, a Low Gain system and a High Gain system. Normal operation utilizes the High Gain system, for both voice and data communication. As a backup, the Low Gain system provides data communication only (Figure 24-11g).

The following component descriptions cover the transmit mode of operation. During the receive mode, the process is reversed.

- SDU – The Satellite Data Unit receives voice audio from the flight interphone system and digital data from the ACARS system. The SDU formats and converts these inputs to an intermediate frequency (IF) and sends them to the RFU. The SDU also selects the active antenna system (High Gain or Low Gain), and sends beam steering commands to the BSU.
- RFU – The Radio Frequency Unit converts the IF signal into an RF signal. It utilizes a very stable oscillator to generate carrier signals in the L-band frequency range (1530 – 1660.5 MHz).
- RF Splitter – The Radio Frequency Splitter routes the RF signal to both the High Gain and Low Gain, High Power Amplifiers (HPA).
- HPA (High and Low Gain) – The High Power Amplifiers are microprocessor controlled linear amplifiers. The microprocessor controls the output RF to

keep the reflected power level below 5 watts. Typical output power is approximately 40 watts.
- LNA/DIP (High and Low Gain) – The Low Noise Amplifier/Diplexer couples transmit and receive signals to the associated Antenna. The Diplexer portion allowing simultaneous transmit and receive operation. The LNA amplifies the very low-level received signal to usable levels for the RFU.
- BSU – The Beam steering unit receives beam steering commands from the SDU, and electrically positions the antenna elements for the best RF link.
- High Gain Antenna – A 32 element array that is 'steered' towards the satellite electrically. The 747-400's has one HGA mounted on top of the fuselage. The 777's and 767-300's have two HGA, one on either side of the fuselage. Aircraft routes (projected) were the driving force in determining the number and placement of the HGA.
- Low Gain Antenna – The Low Gain Antenna is omni-directional, and due to the low received signal strength, is for data communications only.
- RF Combiner – The RF Combiner routes either the Low Gain or High Gain received signal, whichever one is present.

Continued on next page ...

Figure 24-11g

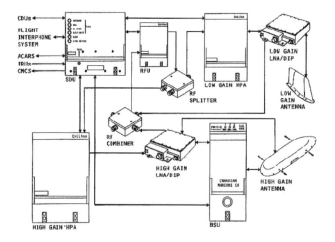

CNS / ATM (cont'd.)

Surveillance

Automatic Dependent Surveillance (ADS) – is a datalink feature of FANS-1 that permits as many as four ATS units at one time to monitor the position of the flight and other related data.

ADS is available by default as long as datalink status is Ready (FMC ATC Logon Status page). It remains available unless selected off by the flight crew. If not ATS unit is using ADS to monitor flight, ADS status is Arm. If one or more ATS units are monitoring the flight, ADS status is Active.

ADS is available by default as long as datalink status is Ready (FMC ATC Logon Status page). It remains available unless selected off by the flight crew. If not ATS unit is using ADS to monitor flight, ADS status is Arm. If one or more ATS units are monitoring the flight, ADS status is Active.

- ADS reports include the following:
- Flight Number
- Present position, time and altitude
- Next position and ETA
- Subsequent position
- Wind velocity and temperature
- Magnetic heading and indicated airspeed
- True track and groundspeed

Air Traffic Control operators, utilizing the above ADS reports, have the ability to upload flight plan changes (subject to flight crew approval) directly to the FMC. Common changes are altitude changes, route changes, and Required Time of Arrival (RTA) updates. Upon acceptance by the flight crew, the changes are automatically installed in the FMC and become part of the active flight plan.

Required Time of Arrival (RTA), is a FANS-1 function that informs the FMC what time to arrive at the next waypoint. Assuming the aircraft is operating under Autopilot (LNAV / VNAV) mode, the FMC would command the auto-throttles to maintain the speed required to reach the next waypoint at the RTA. This allows ATC to prevent aircraft congestion at any given waypoint.

Actual Navigation Performance (ANP) – An estimate of FMC position accuracy may be displayed on the FMC CDU as ANP. ANP is a statistically derived nautical mileage value that represents the radius of a circle within which the actual FMC position will be located with a 95% probability. ANP values vary as a function of the type of updating used by the FMC. For example, VOR/DME updating normally results in a higher ANP than GPS updating. With GPS updating, ANP is determined by the number of satellite signals available, satellite geometry, and signal quality. ANP values with GPS updating typically are less than one-tenth (nautical mile).

Required Navigation Performance (RNP) – RNP is a statement of the navigation performance necessary for operation within a defined airspace. RNP values are established to assure the navigation accuracy required to operate within the aircraft separation standards established by ATC within a particular airspace.

RNP relates to ANP; the units are the same. In airspace where an RNP value is specified, the ANP must remain equal to or less than that RNP. For example, on a route requiring RNP 4, the ANP value must remain at or below 4. The FMC is certified to operate with an ANP less than RNP 99.999% of the time in all phases of flight except during ILS and Localizer approaches.

CHAPTER 25 - FLIGHT MANAGEMENT COMPUTER SYSTEM (FMCS)

Flight Management Computer System Description

A principal system of the B-767 Flight Management System is the Flight Management Computer System (FMCS). The FMCS provides the crew with a "single point" area to initiate and implement a given flight plan and monitor its execution. The FMCS also provides a convenient "single point" area to initialize the three Inertial Reference Units (IRU), and supplies navigational information; i.e., NAV aid data and waypoints to the Electronic Horizontal Situation Indicator (EHSI).

These aspects allow the crew to concentrate on overall airplane management with a minimum of attention to detailed functional chores, and then execute all phases of the flight in the most economical manner.

Continued on next page ...

Figure 25-1

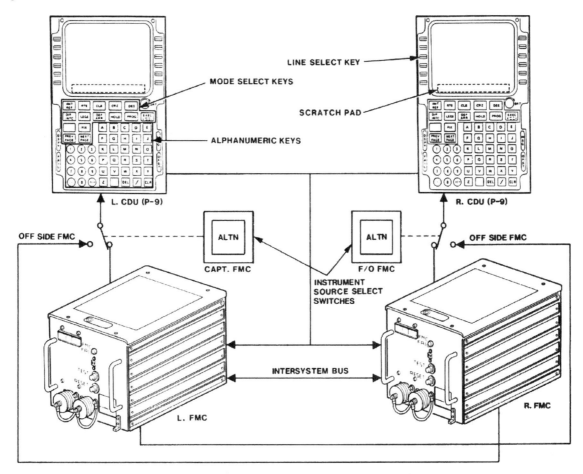

Flight Management Computer System Description (cont'd.)

The FMCS is a dual system utilizing two, four mega bit, hard disk drive Flight Management Computers (FMC); two Control Display Units (CDU); and a Data Base Loader that, when connected, allows maintenance personnel to update the FMC navigation data base.

The navigation data base is a programmable memory that stores the applicable navigational data typically found in the Jeppesen charts, and a selection of company structured routes.

Another programmable data base stored in the FMS is the performance data base. The performance data base stores engine installation data, aircraft aerodynamic limits and atmospheric models.

The Control Display Unit (CDU) houses a CRT and provides a full alpha/numeric keyboard combined with mode select and line select keys. The eleven mode select keys are used to control the type of data available or accessible on the CDU; e.g., climb, routes, and fix.

Pressing an alpha/numeric key enters the selected character into the scratch pad. The scratch pad is a holding area for data, and also is the area of the CRT where FMCS messages may appear. Once the appropriate data is entered into the scratch pad, data can then be transferred to the appropriate data field on the CRT by pressing the associated line select key, which simultaneously transfers that data to the Flight Management Computers where commands are computed and executed.

Normally, either CDU can input to both FMCs simultaneously, and each FMC will then compute and execute the command; and, if an output is appropriate, both FMCs will output to their associated "onside" CDU, giving the impression that the FMCS functions as a single system.

FMC computations are compared and kept synchronous via the intersystem bus. If a single FMS should fail, the associated "onside" FMC Instrument Source Select Switch (ISS Sw) can be actuated, allowing the "onside" CDU to receive data from the other "offside" FMC.

Flight Management Computer System Function

The Flight Management Computer is capable — when coupled to the Flight Guidance System, Thrust Management System and Electronic Flight Instrument System (EFIS) — of commanding the aircraft along a pre-selected lateral (navigation) and vertical path (performance), shortly after takeoff until the Flight Guidance System captures the localizer and glideslope. Each FMC receives twenty four digital inputs and three discrete inputs, and outputs to nine different digital customers.

The FMC performs seven major functions. The input/output function of the FMC receives and transmits digital data to and from the various systems on board the aircraft, and checks that all received data is valid. The CDU function of the FMC formats, updates and sends data to the CDU for display, and provides alerting and advisory messages to the CDU for display on the scratch pad.

Continued on next page ...

Figure 25-2

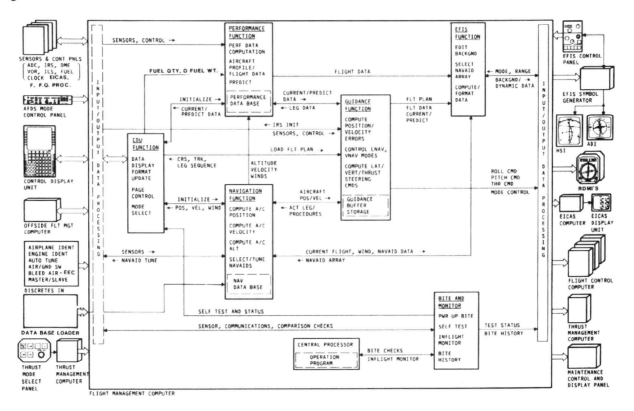

Flight Management Computer System Function (cont'd.)

The bit and monitoring function of the FMC performs a self-test of the FMC during power up and upon request, and continuously monitors the FMS during normal operation. Failures would be recorded on nonvolatile memory (disk) for retrieval at a later date.

The navigation function of the FMC houses the navigation data base, and is responsible for computing the aircraft's current position, velocity and altitude. It also selects and automatically tunes the VOR receivers and DME interrogators.

The navigation function computes the aircraft's present position by determining the distance to two auto-tuned DME stations. The intersection of the two DME, radii (rho/rho) represents the aircraft's present position.

Positional information from the three IRUs is used to solve any ambiguity that may occur, or when the aircraft is on the ground. Velocity is computed using IRU inputs, and altitude is computed using both IRU and ADC inputs. The performance function of the FMC computes performance parameters (limits) and predictions for the vertical path of the flight profile, utilizing the performance data base and the CDU input data.

The guidance function stores the active vertical and lateral flight plan input from the CDU. Using the present aircraft velocity and position information calculated by the navigation function, the guidance function compares actual and desired position, and generates steering commands which are input to the appropriate flight control computer (FCC).

Using the current computed vertical profile data from the performance function, the guidance function compares actual and desired altitude and altitude rate, and generates pitch and thrust commands which are input to the appropriate FCC and the Thrust Management Computer (TMC).

The EFIS function of the FMC provides dynamic and background data to the EFIS Symbol Generator, and provides the navigation function with a list of the closest NAV aid array for auto-tuning.

GLOSSARY

WORD/PHRASE	DEFINITION
ACARS	ARINC Communications Addressing and Reporting System
A/D	Analog to Digital
ACMP	Alternating Current Motor Pump
ADC	Air Data Computer
ADCS	Air Data Computing System
ADF	Automatic Directing Finder
ADP	Air Driven Pump
AFCS	Autoflight Control System
AFDS	Autopilot/Flight Director System
AIDS	Airborne Integrated Data System
AIL	Aileron
ALT	Altitude or Altimeter
AOA or Alpha	Angle of Attack
A/P	Autopilot
APP	Approach
APU	Auxiliary Power Unit
ARINC	Aeronautical Radio Inc.
ARR	Arrival
ASA	Autoland Status Annunciator
A/T	Autothrottle
ATC	Air Traffic Control
B/A	Bank Angle
BAT	Battery
BCD	Binary Coded Decimal
B/CRS	Back Course
BIT	Built-In-Test
BITE	Built-In-Test Equipment
BNR	Binary Numerical Representation
CAA	Civil Aviation Agency
CAS	Computed Airspeed
CDU	Control Display Unit
CLB	Climb
CMD	Command
CON	Continuous
CPU	Central Processing Unit
CRS	Course
CRT	Cathode Ray Tube
CRZ	Cruise
CSEU	Control System Electronics Unit
D/A	Digital to Analog
DAS	Directional Autopilot Servo
DADC	Digital Air Data Computer
DDM	Difference in Depth of Modulation
DEP	Departure
DES	Descent
DFDAU	Digital Flight Data Acquisition Unit
DH	Decision Height

WORD/PHRASE	DEFINITION
DITS	Digital Information Transfer System
DME	Distance Measuring Equipment
D-T/O	Derated Takeoff
EADI	Electronic Attitude Director Indicator
EAROM	Electrically Alterable ROM
EAS	Elevator Autopilot Servo
ECS	Environmental Control System
EDHP	Engine Driven Hydraulic Pump
E/E	Electrical/Electronics
EEC	Electronic Engine Control
EFCU	Elevator Feel and Centering Unit
EFIS	Electronic Flight Instrument System
EGT	Exhaust Gas Temperature
EHSI	Electronic Horizontal Situation Indicator
EHSV	Electrohydraulic Servovalve
EICAS	Engine Indication and Crew Alerting System
EMHP	Electric Motor Hydraulic Pump
EMI	Electromagnetic Interference
ENG or ENGA	Engage
EPR	Engine Pressure Ratio
FAA	Federal Aviation Administration
FCC	Flight Control Computer
F/D	Flight Director
F/E	Flight Engineer
FLCH	Flight Level Change
FMA	Flight Mode Annunicator
FMC	Flight Management Computer
FMCS	Flight Management Computer System
FMS	Flight Management System
F/O	First Officer
FPM	Feet Per Minute
FSEU	Flap/Slat Electronic Unit
FSPM	Flap/Stabilizer Position Module
G/A or GA	Go Around
G/S	Glide Slope
G.S.	Gain Schedule
HDG	Heading
HLD	Hold
HR	Height, Radio
HYD	Hydraulic
IAS	Indicated Airspeed
ILS	Instrument Landing System
INTC	Intercept
I/O	Input/Output
IRU	Inertial Reference Unit
IRS	Inertial Reference System
LCCA	Lateral Central Control Actuator
LNAV	Lateral Navigation
LOC	Localizer

WORD/PHRASE	DEFINITION
LRU	Line Replaceable Unit
LVDT	Linear Variable Differential Transducer
M	Mach
M/ASI	Mach/Airspeed Indicator
MB	Millibars
MCDP	Maintenance Control Display Panel
MCHENG	Multi-Channel Engage
MCP	Mode Control Panel
MCU	Modular Concept Unit
MMO	Maximum Operating Mach
MSL	Mean Sea Level
NAV	Navigation
NCD	No Computed Data
N1	Percentage of the Defined 100% Fan Rotational Speed
PCA	Power Control Actuator
PDU	Power Drive Unit
PROG	Program
PROM	Programmable Read Only Memory
R/A	Radio Altitude or Radio Altimeter
RAT	Ram Air Turbine
RCVR	Receiver
RDMI	Radio Distance Magnetic Indicator
REF	Reference
RF	Radio Frequency
RMI	Radio Magnetic Indicator
RNAV	Area Navigation
ROM	Read Only Memory
RTE	Route
RTG	Rating
RVDT	Rotary Variable Differential Transducer
RVR	Runway Visual Range
SAM	Stabilizer Trim and Aileron Lockout Module
SAT	Static Air Temperature
SCM	Spoiler Control Module
SG	Symbol Generator
SID	Standard Instrument Departure
SOV	Shut-Off Valve
SPD	Speed
SSM	Sign Status Matrix
STAB	Stabilizer
STAR	Standard Terminal Arrival Route
STCM	Stabilizer Trim Control Module
STLM	Stabilizer Trim Limit Switch and Position Transmitter Module
TAS	True Airspeed
TAT	Total Air Temperature
TCA	Terminal Control Area
TGT	Target
TLA	Throttle Lever Angle
TMC	Thrust Management Computer

WORD/PHRASE	DEFINITION
TMS	Thrust Management System
TMSP	Thrust Mode Select Panel
TO or T/O	Takeoff
TOC	Top of Climb
TOD	Top of Decent
TRA	Thrust Resolver Angle Transducer
TRK	Track
TRU	Transformer/Rectifier Unit
VFR	Visual Flight Rules
VHF	Very High Frequency
VMO (Vmo)	Maximum Operating Airspeed (Knots)
VNAV	Vertical Navigation
VOR	VHF Omnidirectional Range
V/S	Vertical Speed
XFMR	Transformer
XMTR	Transmitter
XPNDR	Transponder
Y/D	Yaw Damper
YDS	Yaw Damper Servo